AF443659

Razoxane and Dexrazoxane –
Two Multifunctional Agents

Kurt Hellmann · Walter Rhomberg
Editors

Razoxane and Dexrazoxane – Two Multifunctional Agents

Experimental and Clinical Results

 Springer

Editors
Prof. Dr. Kurt Hellmann
Windleshaw House
TN7 4DB Withyham, East Sussex
United Kingdom

Prof. Dr. Walter Rhomberg
Unterfeldstrasse 32
6700 Bludenz
Austria

The authors have no conflict of interest of any kind

ISBN 978-90-481-9167-3   e-ISBN 978-90-481-9168-0
DOI 10.1007/978-90-481-9168-0
Springer Dordrecht Heidelberg London New York

Library of Congress Control Number: 2010930880

© Springer Science+Business Media B.V. 2010
No part of this work may be reproduced, stored in a retrieval system, or transmitted in any form or by any means, electronic, mechanical, photocopying, microfilming, recording or otherwise, without written permission from the Publisher, with the exception of any material supplied specifically for the purpose of being entered and executed on a computer system, for exclusive use by the purchaser of the work.

Printed on acid-free paper

Springer is part of Springer Science+Business Media (www.springer.com)

Preface

There are now (in 2009) some 900 published papers on the development of razoxane (Rz) and dexrazoxane (DXRz). This book inevitably, therefore, represents an account of only a selected few of what the authors feel are the most important discoveries. Moreover, the history of the two drugs justifies, if justification were needed, that random screening is still a valid method for the discovery of novel and useful drugs. A compound is just a compound which is a long way from being a drug which in turn may be a long way from being a useful medicament.

Experience helps us to predict which substances may be toxic, but theories can not help us to decide which compounds are not toxic so that we can move on and decide post Hippocrates (and the US-FDA and EMA) to give the compound to patients. We cannot in advance tell how much of a new drug may enter a cell (normal or abnormal); how long it may stay there and how much will be rapidly excreted; where it will locate in the cell and how much of it and with what immediate or delayed consequences, and, critically, how answers to these questions will affect each organ. It was not surprising, therefore, that only experimentation could answer these questions and crucially, how will the new drugs affect different diseases and how will they interact with other drugs.

What could not be anticipated at all, were some of the extraordinary properties displayed by Rz and DXRz. Even today we have no idea why the enantiomer DXRz is at least 5 times as soluble in water as razoxane. We believe, we know, but cannot be sure why DXRz can prevent the cardiotoxicity of doxorubicin in the isolated or intact dog heart and 5 other species including humans. Also unknown is why and how the razoxanes are general cytoprotectants.

The influence of these drugs on the differentiation of a pathologic neovasculature of tumors again was not anticipated. Such differentiation or normalization of a pathological vasculature has probably a variety of implications in malignant or non-malignant diseases, but was not until recently a goal for drug development. Perhaps, the most astonishing discovery along the development route was the discovery that it is possible by a drug to 'take out' one malignant characteristic, the metastasis formation, of the three usually needed to characterise a tissue as malignant (uncontrolled proliferation, tumour dissemination, and metastasis formation) and leave the other two characteristics unimpaired.

Of considerable significance was the discovery that razoxane prevented cell division at a specific point in the cell generation cycle namely at G2/M. More detailed aspects of the chemistry and biology will be discussed in 3.2.

Surprising was also the readiness and effectiveness of the combination of radiotherapy and razoxane with or without other anticancer drugs. The combination of radiotherapy and razoxane and its clinical effectiveness will be discussed in Section 2.3.2.

Dexrazoxane was shown in the 1980s to afford protection from cardiotoxicity in women with breast cancer. The definitive papers on the cardiac protection in breast cancer patients receiving anthracyclines were those by Sandra Swain (1997). They will be commented on and discussed as to their relevance in 2009 in Section 3.6. Further trials were subsequently conducted in patients with different pediatric cancers. These trials showed cardioprotection when dexrazoxane was administered with the initial chemotherapy with no decrease in anticancer efficacy and allowing cumulative doxorubicin doses up to 600 mg/m^2.

Further developments are the cytoprotective activity of dexrazoxane in mitigating accidental anthracycline extravasation. This was first shown in animal studies by Seppo Langer in 1999. Meanwhile, this use of dexrazoxane became an FDA/EMA approved established treatment for the accidental extravasation of anthracyclines. Finally, new data emerging from Nigel Greigs Laboratory of Neuro-sciences (NIH, USA) provides a rationale for the potential clinical use of razoxane/dexrazoxane in specific neurodegenerative conditions such as Alzheimer's or Parkinson's disease.

Altogether, a wide range of interesting activities have been discovered for razoxane and dexrazoxane despite having received only limited attention from clinicians. This monograph may contribute to the knowledge of these drugs, hopefully leading to improved treatments.

Withyham, UK   Kurt Hellmann

Acknowledgments

The editors acknowledge with grateful thanks the following persons: S. Carter, Suzanne Eccles, H. Eiter, E.O. Field, C. Franks, J. Gilbert, A. Goldin, Sir A. Haddow, Jean Hartley, E. Herman, G. Mathe, Gillian Murkin, K.A. Newton, K.H. Renner, A. Salsbury, J. Speyer, R. Steiner, and Sandra Swain. Our special thanks to Dr. Rudolf Steiner, Zurich, for his review and editorial suggestions.

Contents

Contributors

Robert E. Becker Drug Design and Development Section, Laboratory of Neurosciences, National Institute on Aging, National Institutes of Health, Baltimore, MD 21124, USA, rebecker2008@comcast.net

Nigel H. Greig Drug Design and Development Section, Laboratory of Neurosciences, Intramural Research Program, National Institute on Aging, National Institutes of Health, Baltimore, MD 21124, USA, Greign@mail.nih.gov

Brian Hasinoff Faculty of Pharmacy, Apotex Centre, University of Manitoba, Winnipeg, MB R3E 0T5, Canada, B_Hasinoff@UManitoba.ca

Kurt Hellmann Windleshaw House, Withyham, East Sussex, TN7 4DB, UK

Eugene H. Herman Division of Applied Pharmacology Research, FDA, Silver Springs, MD, USA

Robin L. Jones Sarcoma and Drug Development Units, Royal Marsden Hospital, London SW3 6JJ, UK, Robin.jones@icr.ac.uk

Seppo W. Langer Internal Medicine and Medical Oncology, Thoracic and Neuroendocrine Oncology, Copenhagen University Hospital, Rigshospitalet, Denmark, swlanger@dadlnet.dk

Walter Rhomberg 6700 Bludenz, Unterfeldstrasse 32, Austria, walter.rhomberg@gmx.at

Robert Rubens Wimbledon, London SW19 7DX, UK, rdrubens@doctors.org.uk

Chapter 1
Introduction

Kurt Hellmann

Abstract In chapter 1, the preface is followed by a short overview on the historical development of two bisdioxopiperazine derivatives razoxane (ICRF-159) and dexrazoxane (ICRF-187) which are the subject of this book. Their history dates back to 1965 and gradually revealed a variety of modes of action which are as unexpected as they are unique – at present. The influence of razoxane and dexrazoxane on the differentiation of a pathologic neovasculature of tumors was totally unanticipated. Razoxane is the most effective preclinical drug for the prevention and suppression of metastases. Dexrazoxane on the other hand has been tested extensively as a cardioprotector against anthracycline-induced cardiotoxicity. Later it was also shown that dexrazoxane has an impressive cytoprotective activity in mitigating accidental anthracycline extravasation. In retrospect, the history of the two drugs justifies random screening as a valid method for the discovery of novel and effective drugs.

1.1 Overview and Historical Development of Razoxane and Dexrazoxane

1.1.1 History

Razoxane (Rz) and dexrazoxane (DXRz), the dextro enantiomer of razoxane, belong to a class of compounds, the *bis*-dioxopiperazines, first described by Geigy chemists in 1964 [1] and independently by Eastman Kodak scientists in 1965 [2]. Uses ascribed to these compounds ranged from pharmaceutical intermediates and textile levelling agents to jet fuel additives. Later it was also suggested that they might be antitumor agents [3].

K. Hellmann (✉)
Windleshaw House, Withyham, East Sussex, TN7 4DB, UK

K. Hellmann, W. Rhomberg (eds.), *Razoxane and Dexrazoxane – Two Multifunctional Agents*, DOI 10.1007/978-90-481-9168-0_1, © Springer Science+Business Media B.V. 2010

1.1.2 Influence on the Cell Cycle

It soon became clear, however, that as a class they were neither cytotoxic nor selective for tumor cells. For reasons that are still obscure, five compounds from this class, including razoxane and dexrazoxane, are cytostatic, inhibiting cell division by blocking cell cycle progression at one brief period (late G2/M) [4] with no discernible inhibitory, toxic, or destructive effect at any other phase of the cell cycle. They had no effect on noncycling or resting cells. Moreover, the cytostatic activity was nonselective, affecting normal dividing and malignant cells alike. While the exact mechanism of this cytostatic effect at the molecular level has yet to be fully clarified, it appears that this may be due at least in part to inhibition of topoisomerase II [5].

At the microscopic level, replicating cells treated with either Rz or DXRz can be seen to greatly increase in size, depending on the concentration of these two drugs and the length of time the cells are exposed to them [6]. The treated cells can double in size; if they do manage to divide, they do not appear to be able to separate. If these observations are extrapolated to cancer patients given Rz/DXRz, then there are important consequences for the assessment of the effects of these drugs on tumor response. There is the distinct possibility that if the influence of Rz/DXRz on cell size is ignored, there may be a misleading impression that the treated but growing tumor has increased in number of cells when in fact it is only the size of the tumor cells that has increased. This also would lead to erroneous estimations of the time to tumor progression. Since it is highly unlikely that any antitumor agent ever reaches 100% of all tumor cells, particularly in a solid tumor, making allowances for cells that merely increase in size among others that might be actively dividing and increasing in number would be a highly complex matter that will require technology that is not yet available. Of course, permanent complete regression is the preferred response option, but long stability of disease, which thereby increases overall survival with a tumor that may have lost or slowed its capacity to grow, may be preferred by patients over 'responses' with a shorter survival.

Although all these considerations are academic, clinical trials with Rz and DXRz have shown that even though 'responses' (i.e., tumor size shrinkages) have been meager, survival has been noticeably improved; this finding has been shown most clearly by Swain et al. [7]. These clinical results were foreshowed many years ago by Sandberg and Goldin [8], who showed that when comparing the effects of 18 of the most active agents, including doxorubicin (DOX), cyclophosphamide, and CCNU, against an experimental breast cancer, the effect on tumor volume was directly proportional to the effect on survival; i.e., the greater the effect on tumor volume, the greater the effect on survival. The only exception was razoxane, which had the best effect on survival and one of the worst on tumor volume. It had no significant effect on tumor growth and thus did not appear to 'interfere' with an increase in tumor cell numbers; yet, survival was greater than with any of the other drugs tested.

Perhaps the most important message for clinical oncologists from these findings is that no change in tumor size does not indicate that there has been no response. In the clinical context, however, stability of disease as a result of Rz/DXRz but unaccompanied by any other objective evidence of influence on tumor size is a difficult and largely impractical clinical criterion. If Rz and DXRz had been found to be just two more of the many nonselective cell division inhibitors available, neither of these compounds would have had much to recommend them.

1.1.3 Antimetastatic Activity

What has made Rz and DXRz worth further examination and what has maintained interest in them has been a series of pharmacologic activities (Table 1.1), some of which could not have been predicted from their chemical structure, particularly since most of these activities themselves were previously unknown.

Perhaps the most surprising was the antimetastatic activity. Quite by chance, razoxane was the first compound to be tested in what was also the first screening process set up specifically to find substances that might influence tumor dissemination, invasion, and metastasis [9]. This screening process used the Lewis lung carcinoma, and razoxane was found to be highly successful in preventing the spontaneous metastasis from this and many other tumors [9, 10]. A clinical trial in colorectal cancer found that in patients with resected Dukes's stage C, the incidence of subsequent hepatic secondaries was significantly reduced by Rz and in those patients who did develop them, the secondaries took twice as long to appear as in the controls [11]. This delay in tumor development (stability of disease) also has been seen with Rz treatment in phase II studies in carcinoma of the lung (non-small cell) [12], breast [13], stomach [14], and anecdotally in carcinoma of the pancreas [15].

Analysis of the mechanism of the antimetastatic action of razoxane showed it to be due to the normalization of the tumor neovasculature [16]. This led to clinical trials with razoxane in psoriasis [17], Crohn's disease [18], and Kaposi's sarcoma [19] with outstanding results. On the assumption that the changes in the neovasculature

Table 1.1 Pharmacologic spectrum of razoxane and dexrazoxane

Modes of action
Blocks cell cycle at G2/M
Normalizes tumor neovasculature
Antiinflammatory activity
Antiinvasive activity
Inhibition of topoisomerase II α
Metal chelating activity, e.g. Pb++, Cu++, Fe++, Zn++, Mg++

These modes of action are associated with
Potentiation of irradiation
Potentiation of chemotherapy (variably)
Prevention of spontaneous metastases
Cytoprotection for myocardium, pulmonary epithelium, gastrointestinal tract, kidney
Cytoprotection specifically against doxorubicin, daunorubicin, bleomycin, mitoxantrone, etoposide, cisplatin
Cytoprotection against extravasated anthracyclines

would lead to better tumor oxygenation, it also led to extensive clinical trials with razoxane in combination with radiotherapy [20]. It was shown in several randomized trials that razoxane given together with radiotherapy is able to significantly increase the response rate and local control of soft tissue sarcomas or colorectal carcinomas compared to radiotherapy alone (see Section 2.3.2).

1.1.4 Cardioprotection (Tissue Protection)

Perhaps less surprising was the highly effective protection afforded by dexrazoxane (DXRz) against the cardiotoxicity of the anthracyclines. Herman et al.'s [21] discovery of the protective effect of the closely related EDTA made DXRz a likely candidate for the same activity. More unsuspected was the protection of other tissues by DXRz and Rz against anthracycline toxicity [22]. The cytoprotection also extended not only to other tissues [23], but also to other drugs. Thus, DXRz provides cytoprotection against the lethal toxic effects of cis-Pt [24], mitoxantrone [25], bleomycin [26], and VP-16 [27]. This protection permits larger doses of these drugs to be given, thereby obtaining an antitumor activity that could not be seen at lower, but previously optimal doses.

The problem with almost all anticancer drugs is that the therapeutic dose and the maximum tolerated dose are not very different. The question for most, therefore, is: If the maximum tolerated dose could be increased, would the response rate also increase? A vast amount of research over many years concentrated on finding equally or more effective compounds than doxorubicin, but without the dose-limiting cardiotoxicity of this drug. While a number of substances were also proposed as cardioprotectors and new methods of administering doxorubicin have been developed, the problem was neatly solved by Herman et al.'s [21] discovery that DXRz could significantly reduce the doxorubicin-induced cardiotoxicity. Since much was already known about the behavior of DXRz from several phase I and five phase II studies in adults and children, Speyer et al. [28] suggested and subsequently undertook the first randomized, controlled clinical trial to determine whether DXRz would permit doses of doxorubicin greater than that limited by the conventional maximum tolerated dose to be given to women with advanced breast cancer. These investigators found that doses of doxorubicin in excess of the maximum tolerated dose of 450 mg/m^2 could indeed be given without any evidence of cardiotoxicity if DXRz (ratio 20:1) had been given 30 min beforehand. The combination produced no new side effects, nor did it exacerbate the noncardiac toxicities of doxorubicin (DOX). These results formed the basis of six further randomized clinical trials, three of which were monitored by the Food and Drug Administration.

Interim analysis of the largest of these trials [29] surprisingly appeared to show that the addition of DXRz to the 5-Fluorouracil/DOX/cyclophosphamide (FAC) regimen had reduced the response rate in women with advanced breast cancer from 96 of 152 patients (63%) to 67 of 141 patients (48%). This highly significant result ($p = 0.007$) which was, however, not in line with a reduction in time to progression in the DXRz group led to a crucial protocol amendment so that patients would only receive

DXRz in addition to the FAC regimen after they had already received 300 mg/m^2 of DOX and were thought to benefit from further doses of DOX. This decision and amendment selected doxorubicin responders and excluded the non-responders largely from the trials which at the time of analysis involved 1,008 patients. This is an important consideration in view of the survival data of subsequent analyses of the mature trial results. These were published in 1997 and represent the definitive conclusions as to the value of DXRz for patients with advanced breast cancer treated with DOX-containing chemotherapy [7, 29]. A comment on these two studies is given in Chapter 3. All subsequent trials conducted in breast cancer and other adult malignancies, however, did not confirm this reduction of antitumor efficacy of anthracyclines by dexrazoxane.

1.1.4.1 Pediatric Trials

High-dose doxorubicin (DOX) for childhood malignances is extremly important because it may be curative. Unfortunately, the curative and cardiotoxic doses are similar; therefore, the importance of an effective cardioprotector cannot be overstated. Dexrazoxane has been shown in several studies to prevent acute cardiotoxicity, providing considerable hope that the long-term chronic anthracycline cardiotoxicity may also be avoided. Wexler et al.'s [30] carefully controlled study of children with bony sarcomas found that only two of 15 children in the control group compared with eight of 18 children in the dexrazoxane (DXRz) group reached the projected dose of 410 mg/m^2 of DOX; 10 of 15 children in the control group and three of 18 children in the DXRz group developed cardiotoxicity before reaching 410 mg/m^2. Response rates, event-free and overall survival, and noncardiac toxicities were unaffected by DXRz [30].

The findings of Wexler et al. [30] were anticipated by Bu'Lock et al. [32], who compared two groups of five children each who had a variety of anthracyclines (daunorubicin, DOX, epirubicin, or a combination). These investigators showed that in the 'control' group, which received cumulative anthracycline doses varying between 600 and 1,150 mg/m^2, three of five children had severe cardiac dysfunction compared with none of the children in the DXRz group which received anthracycline doses between 550 (+marrow transplant) and 1,650 mg/m^2.

Clinical trials were subsequently conducted in patients with pediatric malignancies, including Ewing's sarcoma, osteosarcoma, Hodgkin's lymphoma, and leukemia. These trials demonstrated cardioprotection, with no decrease in anticancer efficacy, when dexrazoxane was added to chemotherapeutic protocols [31]. It has allowed treatment with cumulative doxorubicin doses up to 600 mg/m^2, without cardiac failure [33]. The recently (2009) published update on long-term results of Dana-Farber Cancer Institute ALL Consortium pediatric trials in patients with acute lymphoblastic leukaemia treated in the period of 1985–2000 unequivocally showed that prolonged infusion of doxorubicin does not give an advantage over iv bolus applications, and again that dexrazoxane does not reduce the response rate, the event-free survival and the overall survival. There was no increase of secondary malignancies with dexrazoxane [34].

1.1.5 Further Aspects

Dexrazoxane has been used in clinical trials for more than 20 years, and so far there has been no evidence of any untoward adverse effects, apart from a dose-related, rapidly reversible neutropenia. Therefore, trials of its activity in nonmalignant conditions, particularly those in which Rz has already been shown to be effective, need not be delayed.

More details on the issue of cardioprotection by dexrazoxane as well as other interesting activities of this drug that were discovered during recent years will be outlined and discussed in Chapter 3 of this monograph. Future prospects of Rz and DXRz may be found at the end of the book.

References

1. Geigy JR: UK Patent 978,724, 1964
2. Eastman Kodak Co: UK Patent 1,001,157, 1965
3. Creighton AM, Hellmann K, Whitecross S (1969) Antitumor activity in a series of bisdike-topiperazines. Nature 222:384–5
4. Sharpe HBA, Field EO, Hellmann K (1970) The mode of action of the cytostatic agent ICRF 159. Nature 226:524–6
5. Tanabe K, Ikegami Y, Ishida R, Andoh T (1991) Inhibition of topoisomerase II by antitumor agents bis (2,6-dioxopiperazine) derivatives. Cancer Res 51:4903–8
6. Hallowes RC, West DG, Hellmann K (1974) Cumulative cytostatic effect of ICRF 159. Nature 247:487–90
7. Swain SM, Whaley FS, Gerber MC, et al (1997) Delayed administration of dexrazoxane provides cardioprotection for patients with advanced breast cancer treated with doxorubicin-containing therapy. J Clin Oncol 15:1333–40
8. Sandberg J, Goldin A (1971) Use of first generation transplants of a slow growing solid tumor for the evaluation of new cancer chemotherapeutic agents. Cancer Chemother Rep 55:233–8
9. Hellmann K, Burrage K (1969) Control of malignant metastases by ICRF 159. Nature 224:273–5
10. Hellmann K, Gilbert J, Evans M, Cassell P, Taylor RH (1987) Effect of razoxane on metastases from colorectal cancer. Clin Exp Metastasis 5:3–8
11. Gilbert JM, Hellmann K, Evans M et al (1986) Randomised trial of oral adjuvant razoxane (ICRF 159) in resectable colorectal cancer. Br J Surg 73:446–50
12. Eagan BT, Carr DT, Coles DT et al (1979) ICRF 159 versus polychemotherapy in non-small cell lung cancer. Cancer Treat Rep 60:947–8
13. Ahmann DL, O'Connell MJ, Bisel HF et al (1977) Phase II study of dianhydrogalactitol and ICRF-159 in patients with advanced breast cancer previously exposed to cytotoxic chemotherapy. Cancer Treat Rep 61:81–2
14. Gilbert JM, Cassel P, Ellis H, Wastell Ch, Hermon-Taylor J, Hellmann K (1979) Adjuvant treatment with razoxane (ICRF 159) following resection of cancer of the stomach. Recent Results Cancer Res 68:217–21
15. Ward A, Sherlock D (1980) Long term survival following chemotherapy for carcinoma of the pancreas. Br J Clin Pract 34:157–9
16. Salsbury AJ, Burrage K, Hellmann K (1970) Inhibition of metastatic spread by ICRF 159: Selective deletion of a malignant characteristic. Br Med J 4:344–6
17. Horton JJ, Wells RS (1983) Razoxane – a review of 6 years' therapy in psoriasis. Br J Dermatol 109:669–73
18. Kingston RD, Hellmann K (1993) Razoxane for Crohn's colitis and non-specific proctitis. Br J Clin Pract 46:252–5

19. Olweny CL, Sikyewunda W, Otim D (1980) Further experience with razoxane (ICRF 159; NSC 129 943) in treating Kaposi's sarcoma. Oncology 37:174–6
20. Hellmann K, Rhomberg W (1991) Radiotherapeutic enhancement by razoxane. Cancer Treat Rev 18:225–40
21. Herman EH, Mhatre RM, Lee I et al (1972) Prevention of the cardiotoxic effects of Adriamycin and daunomycin in the isolated dog heart. Proc Soc Exp Biol Med 140:234–9
22. Tian Hu S, Brändle E, Zbinden G (1983) Inhibition of cardiotoxic, nephrotoxic and neurotoxic effects of doxorubicin by ICRF-159. Pharmacology 26 (4):210–20
23. Herman EH, El-Hage A, Ferrans VJ (1988) Protective effect of ICRF 187 on doxorubicin-induced cardiac and renal toxicity in spontaneously hypertensive (SHR) and normotensive (WKY) rats. Toxicol Appl Pharmacol 92:42–53
24. Woodman RJ (1974) Enhancement of antitumor effectiveness of ICRF 159 against early L1210 by combination with cis-diamminedichloroplatinum (NSC-82151). Cancer Chemother Rep 4:45–52
25. Weilbach FX, Chan A, Toyka KV, Gold R (2004) The cardioprotector dexrazoxane augments therapeutic efficacy of mitoxantrone in experimental autoimmune encephalomyelitis. Clin Exp Immunol 135(1):49–55
26. Herman EH, Hasinoff BB, Zhang J et al (1995) Morphologic and morphometric evaluation of the effect of ICRF-187 on bleomycin-induced pulmonary toxicity. Toxicology 98:163–75
27. Holm B, Jensen PB, Sehested M (1996) ICRF-187 rescue in etoposide treatment in vivo. A model targeting high-dose topoisomerase II poisons to CNS tumors. Cancer Chemother Pharmacol 38:203–9
28. Speyer JL, Green MD, Zeleniuch-Jacquotte A et al (1997) ICRF-187 permits longer treatment with doxorubicin in women with breast cancer. J Clin Oncol 10:117–27
29. Swain SM, Whaley FS, Gerber MC et al (1997) Cardioprotection with dexrazoxane for doxorubicin-containing therapy in advanced breast cancer. J Clin Oncol 15:1318–32
30. Wexler LH, Andrich MP, Venzon D et al (1996) Randomised trial of the cardioprotective agent ICRF-187 in pediatric sarcoma patients treated with doxorubicin. J Clin Oncol 14:362–72
31. Lipshultz SE, Rifai N, Dalton VM et al (2004) The effect of dexrazoxane on myocardial injury in doxorubicin-treated children with acute lymphoblastic leukemia. N Engl J Med 351:145–53
32. Bu'Lock F, Gabriel HM, Oakhill A et al (1993) Cardioprotection by ICRF 187 against high dose anthracycline toxicity in children with malignant disease. Br Heart J 70:185–8
33. Steinherz L (2008) Early breast cancer therapy and cardiovascular injury. JACC 51:1235
34. Silverman LB, Stevenson KE, O'Brien JE et al (2010) Long-term results of Dana-Farber Cancer Institute ALL Consortium protocols for children with newly diagnosed acute lymphoblastic leukaemia (1985–2000). Leukemia 24(2):320–34. Epub 17 Dec 2009.

Chapter 2
Razoxane

Kurt Hellmann and Walter Rhomberg

Abstract In Chapter 2, preclinical and clinical data are given for razoxane, the older of the two drugs. The various modes of action of razoxane are briefly summarized, e.g. the block of the cell cycle in the G2/M phase, the antimetastatic and antiinvasive activity, the ability to normalize pathological tumor blood vessels, the inhibition of topoisomerase II α, and the metal chelating activity. These activites are associated with a marked radiosensitization, suppression of remote metastases in malignant tumors, and non-specific tissue protection. The clinical part deals with studies and results in malignant tumors as well as in non-malignant diseases such as psoriasis and Crohn's disease. The clinical studies of malignant tumors range from investigations of the cytotoxic and cytostatic activity of razoxane in leukemias, lymphomas and solid tumors via its radiosensitizing efficacy in sarcomas and colorectal carcinomas to the proof of an impressive anti-metastatic activity escpecially if razoxane is combined with tubulin affinic drugs. The latter drug combination together with radiotherapy led to unrivalled clinical results in soft tissue sarcomas. A detailed analysis of the toxicity may be found at the end of the section on malignant tumors. The results of the treatment of non-malignant diseases are likewise impressive. Up to now they found limited interest and could therefore be of special interest.

2.1 Preclinical Data – In Vitro and In Vivo

Kurt Hellmann

This is no complete account of the numerous experiments which were done on razoxane. The preclinical work is, in part, summarized in different sections of this book. The experimental work performed until 1975 has been reviewed by Bakowski [1].

K. Hellmann (✉)
Windleshaw House, Withyham, East Sussex, TN7 4DB, UK

W. Rhomberg (✉)
6700 Bludenz, Unterfeldstrasse 32, Austria
e-mail: walter.rhomberg@gmx.at

K. Hellmann, W. Rhomberg (eds.), *Razoxane and Dexrazoxane – Two Multifunctional Agents*, DOI 10.1007/978-90-481-9168-0_2, © Springer Science+Business Media B.V. 2010

In vitro, razoxane exhibited cytostatic activity. Experiments with cultured human lymphoytes showed that razoxane blocks the cell cycle in the G2- and early M-phase. It arrests dividing cells in the prophase and early metaphase [2–4]. The administration of razoxane at a high dose over a short period of time is less effective than at a lower dose given over a longer period [4].

Since the G2/M phase is the most sensitive phase to ionizing irradiation, a block of the cell cycle in this phase might be one reason why the drug exhibits a strong radiosensitizing ability in animal experiments [5, 6–8] and finally also in the clinic (see Section 2.3.2). The synergism of radiotherapy and razoxane, first described in experimental tumors in 1974 [5], was confirmed also during in vitro conditions with irradiated Chinese hamster fibroblasts [9].

An outstanding feature of razoxane in vivo is its antimetastatic activity observed in a variety of animal models. The earliest experiments were performed in the Lewis lung cancer (LLC) model already some 40 years ago [10]. There it was shown that the pretreatment with razoxane almost completely suppressed the formation of distant metastases in the lungs which inevitably occur after transplantation of this tumor if the animals remain untreated. The antimetastatic effect was linked to the normalization of tumour blood vessels. This phenomenon was repeatedly discribed in LLC [11–13] and later also in a hamster lymphoma model [14].

Suppression of distant tumor spread was further seen in KHT-sarcomas in mice [15] and in murine squamous cell carcinomas [16]. In prostate cancer models such as R3327 MAT-LyLu and Pa III, an impressive degree of antimetastatic efficacy was achieved [17, 18]. That also applies to an osteosarcoma model in Sprague-Dawley rats [19]. A more detailed description of these results may be found in Section 2.3.3 and especially Section 2.3.4, where an update is given concerning the inhibition of distant tumor spread in experimental tumor systems.

Further preclinical data of razoxane, and especially dexrazoxane, are related to tissue protecting activities. They are described in Chapter 3.

References

1. Bakowski MT (1976) ICRF 159, (+/–) 1,2 bis (3,5-dioxopiperazin-1-yl) propane, NSC 129943; razoxane. Cancer Treat Rev 3:95–107 (Review)
2. Hallowes RC, West DG, Hellmann K (1974) Cumulative cytostatic effect of ICRF 159. Nature 247:487–90
3. Hellmann K, Field EO (1970) The effect of ICRF 159 on the mammalian cell cycle. Significance for its use in cancer chemotherapy. J Natl Cancer Inst 44:539
4. Sharpe HBA, Field EO, Hellmann K (1970) The mode of action of the cytostatic agent ICRF 159. Nature 226:524–6
5. Hellmann K, Murkin GE (1974) Synergism of ICRF 159 and radiotherapy in experimental tumours. Cancer 34:1033–9
6. Kovacs CJ, Evans MJ, Schenken LL, Burhalt DR (1979) ICRF 159 enhancement of radiation response in combined modality therapies. I. Time/dose relationship for tumor response. Br J Cancer 39:516–23
7. Kovacs CJ, Evans MJ, Burhalt DR, Schenken LL (1979) ICRF 159 enhancement of radiation response in combined modality therapies. II. Differential responses of tumour and normal tissues. Br J Cancer 39:524–30

8. Norpoth K, Schaphaus A, Ziegler H, Witting U (1974) Combined treatment of the Walker tumour with radiotherapy and ICRF 159. Z Krebsf 82:328–34
9. Kimler BF (1982) Interaction of razoxane and radiation on cultered Chinese hamster cells. Int J Radiat Oncol Biol Phys 8(8):1333–8
10. Hellmann K, Burrage K (1969) Control of malignant metastases by ICRF 159. Nature 224:273–75
11. Burrage K, Hellmann K, Salsbury AJ (1970) Drug induced inhibition of tumour cell dissemination. Br J Pharmacol 39:205–6
12. James SE, Salsbury AJ (1974) Effect of (+/–) 1,2-bis (3,5 dioxopiperazin-1-yl) propane on tumor blood vessels and its relationship to the antimetastatic effect in Lewis lung carcinoma. Cancer Res 34:839
13. Salsbury AJ, Burrage K, Hellmann K (1970) Inhibition of metastatic spread by ICRF 159: selective deletion of a malignant characteristic. Br Med J 4:344–6
14. Atherton Anne (1975) The effect of (+/–) 1,2-bis (3,5-dioxopiperazin-1yl) propane (ICRF 159) on liver metastases from a hamster lymphoma. Eur J Cancer 11:383–8
15. Baker D, Constable W, Elkon D, Rinehart L (1981) The influence of ICRF 159 and levamisole on the incidence of metastases following local irradiation of a solid tumor. Cancer 48:2179–83
16. Peters LJ (1975) A study of the influence of various diagnostic and therapeutic procedures applied to a murine squamous carcinoma on its metastatic behaviour. Br J Cancer 32(3): 355–65
17. Heston WDW, Kadmon D, Fair WR (1981) Effect of high dose diethylstilbestrol and ICRF 159 on the growth and metastases of the R3327 MAT-LyLu prostate-derived tumor. Cancer Lett 13:139–45
18. Pollard M, Burleson GR, Luckert PH (1981) Interference with in vivo growth and metastasis of prostate adenocarcinoma (PA-III) by ICRF 159. Prostate 2:1–9
19. Wingen F, Spring H, Schmähl D (1987) Antimetastatic effects of razoxane in a rat osteosarcoma model. Clin Exp Metastasis 5(1):9–16

2.2 Modes of Action: A Brief Summary

Walter Rhomberg

Razoxane has an intriguing and wide spectrum of activities. The exclusive focus on a possible cytotoxic activity alone in the early trials of razoxane probably contributed to an incomplete appreciation of its antitumour activity in the 1970s. Since then, several novel modes of action have been discovered facilitating our understanding of the radiosensitizing, antimetastatic and cytorallentaric abilities of the drug. In addition, some modes of action formed a link to the effective use of razoxane in several benign diseases, and led recently to the discovery of a cytoprotective biological activity which allowed the prevention of toxicity in different organs and tissues from intrinsic or external toxic influences.

2.2.1 Effects on the Cell Cycle

Razoxane blocks the cell cycle in the G2- and early M-phase. It arrests dividing cells in the prophase and early metaphase [1–3]. These experiments have been performed with cultured human lymphocytes. It was also shown that the administration of razoxane at a high dose over a short period of time is less effective than at a lower dose given over a longer period [3].

Since the G2/M phase is the most sensitive phase to ionizing irradiation, a block of the cell cycle in this phase might be one reason why the drug exhibits a strong radiosensitizing ability in animal experiments [4–7] and in the clinic (see Section 2.3.2).

2.2.2 Normalization of Tumour Blood Vessels

One of the most interesting activities of razoxane is its ability to normalize pathological blood vessels induced by experimental tumours. This was first shown in transplanted Lewis lung tumours (LLC) in mice [8–11]. The same phenomenon was observed in a hamster lymphoma model [12]. It was hypothesized and concluded that this unique 'blood vessel normalizing activity' could be the key to prevent distant metastases [9–11, 13; and Sections 2.3.3 and 2.3.4].

2.2.3 Antiinvasive Activity

Little attention has been given to an effect of razoxane that involves the invasion of tumour cells. Razoxane is not a tubulin-affinic drug and therefore, does not touch cellular motility and deformability, but Karakiulakis et al. observed that the drug is able to inhibit the collagen-degradation of basement membrane induced by a malignant tumor enzyme [14]. Some years earlier, Duncan and Reynolds already observed that the collagenase production was inhibited and TIMP (tissue inhibitor of metallo-proteinase) increased by razoxane, in a dose-dependent manner, when cells were treated daily for 3 days [15]. It has been suggested that this ability of razoxane may correlate with its effectiveness in treating psoriatic arthritis. A suppression of up-regulation of gelatinases was later described by Garbisia et al. and linked to the suppression of metastasis [16]. Welch et al. confirmed an antiinvasive potential of razoxane in the membrane invasion culture system (MICS) [17]. The antiinvasive mode of action of razoxane is rarely mentioned in the literature, yet it probably represents a further mechanism which supports our understanding of the marked antimetastatic activity of the drug.

2.2.4 Inhibition of Topoisomerase II

Tanabe et al. [18] first described an inhibition of DNA topoisomerase II by several 2,6-dioxopiperazin derivatives including razoxane (ICRF-159). The authors investigations suggested that ICRF-154, ICRF-159 and few other related compounds are specific inhibitors of topoisomerase II with different modes of action interfering with some steps before the formation of the intermediate cleavable complex in the catalytic cycle. This is a property quite distinct from previously known cleavable complex-forming type topoisomerase II-targeting antitumor agents such as acridines, anthracyclines, and epipodophyllotoxins.

The relation of *dexrazoxane* (ICRF-187) to topoisomerase II inhibition is described and discussed in detail in the Section 3.2. There seems to be a complex, dose dependent interaction between dexrazoxane and the growth inhibition mediated by doxorubicin, daunorubicin, or etoposide in experimental tumor cell systems, which is not yet well understood [19–21]. At present, the results seem to be conflicting, and they are not easy to be adopted in the planning of clinical trials.

2.2.5 Chelation of Metals

Already in 1963, authors, e.g. A. Furst et al., have argued that most drugs which retard the growth of human or experimental neoplasms are actual or potential chelating agents [22]. This idea was linked to the notion that many enzyme systems depend on or contain trace metals. Creighton, Whitecross and Hellmann published the first results of the antitumour activity of a series of *bis*-diketopiperazines in 1969 [22]. *Bis*-diketopiperazines are less polar derivatives of EDTA (ethylenediamine tetra-acetic acid), and were therefore chosen for screening of antitumour activities. EDTA, one of the most powerful chelating agents, has no significant antitumour activity because of its high polarity which does preclude its entry to critical intracellular sites [21]. From these experiments, ICRF-154 and ICRF-159 (razoxane) emerged as least toxic and most effective drugs in Leukemia 1210 and Lewis lung carinoma.

Since razoxane could be considered as a derivative of the potent chelating agent EDTA, serum calcium level determinations were frequently performed during the early clinical studies [23] but one did not find a significant decrease in the serum calcium level in any patient. Preclinical work further indicated that the iron chelating activity of razoxane seemed to be also involved in neuroprotection, and could possibly influence neurodegenerative diseases (see Section 3.7 and 3.8).

A chelating activity has likewise been confirmed for the (+)enantiomer *dexrazoxane* (ICRF-187, DXRz). In a pharmacokinetic study by Tetef et al., urinary iron and zinc excretion during a 96-h infusion of dexrazoxane increased in 12 of 18 and 19 of 19 patients by a median of 3.7- and 2.4-fold, respectively [24]. The chelating activity has mainly been associated with the well documented protective effect of ICRF-187 on the anthracycline-induced cardiotoxicity. DXRz is hydrolyzed to its active form intracellularly and binds iron to prevent the formation of superhydroxide radicals, thus preventing mitochondrial destruction [25].

Studies of the protective effects of DXRz against the pulmonary damage induced by bleomycin were done in male and female C57/BL6 mice. Pulmonary alterations, especially lung fibrosis were significantly reduced, albeit not completely, in all groups of animals treated with different doses of DXR [26]. In vitro studies indicated that both ICRF-187 and its open-ring hydrolysis product (ADR-925) remove iron slowly from the bleomycin-iron complex. According to the authors, this observation provides a basis for the concept that ICRF-187 protects by chelating iron involved in the formation of the bleomycin-Fe^{3+} complex that generates reactive oxygen radicals capable of causing pulmonary damage [26].

In contrast, the hypothesis that the protective effect of ICRF-187 on tissue damage induced by extravasation of anthracyclines is exclusively linked to the chelating activity of ICRF-187, was rejected by Langer et al. [27].

2.2.6 Cytorallentaric Mode of Action

Cytorallentaric activity means that a drug is able to slow down the growth rates of tumours without inducing spectacular objective responses in tumour size. K. Hellmann coined the term 'cytorallentaric' (from It. *rallentare*, to slow down) which was as yet not used in the oncologic terminology. The behaviour of several tumors treated by razoxane – experimentally and in the clinic – is well characterized by the term although the clinical evidence is usually gained only in retrospect or after having done statistics on a treated patient group. Preclinical data and background of the cytorallentaric activity of razoxane is discussed in detail in the Section 2.3.4.

Indeed, when razoxane was clinically tested as cytotoxic agent, it was observed in randomized studies that even when there were fewer objective responses in the group treated with razoxane compared to a control group receiving a different treatment, there still was a survival benefit in patients who received razoxane (see 2.3.1). This interesting and seemingly paradoxical phenomenon should be further elucidated. It is not clear whether it is linked to the antimetastatic activity of razoxane, its chelating activity, a hitherto unknown mode of action, or to a combination of diverse mechanisms.

References

1. Hallowes RC, West DG, Hellmann K (1974) Cumulative cytostatic effect of ICRF 159. Nature 247:487–90
2. Hellmann K, Field EO (1970) The effect of ICRF 159 on the mammalian cell cycle. Significance for its use in cancer chemotherapy. J Natl Cancer Inst 44:539
3. Sharpe HBA, Field EO, Hellmann K (1970) The mode of action of the cytostatic agent ICRF 159. Nature 226:524–6
4. Hellmann K, Murkin GE (1974) Synergism of ICRF 159 and radiotherapy in experimental tumours. Cancer 34:1033–9
5. Kovacs CJ, Evans MJ, Schenken LL, Burholt DR (1979) ICRF 159 enhancement of radiation response in combined modality therapies. I. Time/dose relationship for tumor response. Br J Cancer 39:516–23
6. Kovacs CJ, Evans MJ, Burholt DR, Schenken LL (1979) ICRF 159 enhancement of radiation response in combined modality therapies. II. Differential responses of tumour and normal tissues. Br J Cancer 39:524–30

7. Norpoth K, Schaphaus A, Ziegler H, Witting U (1974) Combined treatment of the Walker tumour with radiotherapy and ICRF 159. Z Krebsf 82:328–34

8. Burrage K, Hellmann K, Salsbury AJ (1970) Drug induced inhibition of tumour cell dissemination. Br J Pharmacol 39:205–6

9. Hellmann K, Burrage K (1969) Control of malignant metastases by ICRF 159. Nature 224:273–75

10. James SE, Salsbury AJ (1974) Effect of (+/–) 1,2-bis (3,5 dioxopiperazin-1-yl) propane on tumor blood vessels and its relationship to the antimetastatic effect in Lewis lung carcinoma. Cancer Res 34:839

11. Salsbury AJ, Burrage K, Hellmann K (1970) Inhibition of metastatic spread by ICRF 159: selective deletion of a malignant characteristic. Br Med J 4:344–6

12. Atherton Anne (1975) The effect of (+/–) 1,2-bis (3,5-dioxopiperazin-1yl) propane (ICRF 159) on liver metastases from a hamster lymphoma. Eur J Cancer 11:383–8

13. Salsbury AJ, Burrage K and Hellmann K (1974) Histological analysis of the antimetastatic effect of 1,2-bis (3,5-dioxopiperazin-yl) propane. Cancer Res 34:843–9

14. Karakiulakis G, Missirlis E, Maragoudakis ME (1989) Mode of action of razoxane: inhibition of basement membrane collagen-degradation by a malignant tumor enzyme. Methods Find Exp Clin Pharmacol 11:255–61

15. Duncan SJ, Reynolds JJ (1983) The effects of razoxane (ICRF-159) on the production of collagenase and inhibitor (TIMP) by stimulated rabbit articular chondrocytes. Biochem Pharmacol 32(24):3853–8

16. Garbisa S, Onisto M, Peron A, Perissin L, Rapozzi V, Zorzet S, Giraldi T (1997) Suppression of metastatic potential and up-regulation of gelatinases and uPA in LLC by protracted in vivo treatment with dacarbazine or razoxane. Int J Cancer 72(6):1056–61

17. Welch DR, Lobl TJ, Seftor EA, Wack PJ, Aeed PA, Yohem KH, Seftor RE, Hendrix MJ (1989) Use of the Membrane Invasion Culture System (MICS) as a screen for antiinvasive agents. Int J Cancer 43(3):449–57

18. Tanabe K, Ikegami Y, Ishida R, Andoh I (1991). Inhibition of topoisomerase II by antitumor agents bis (2, 6-dioxopiperazine) derivatives. Cancer Res 51:4903–8

19. Hasinoff BB, Yalowich JC, Ling Y, Buss JL (1996) The effect of dexrazoxane (ICRF-187) on doxorubicin- and daunorubicin-mediated growth inhibition of Chinese hamster ovary cells. Anticancer Drugs 7(5):558–67

20. Sehestedt M, Jensen PB, Soerensen BS et al (1993) Antagonistic effect of the cardioprotector (+)-1,2-bis(3,5-dioxopiperazinyl-1-yl)propane (ICRF-187) on DNA breaks and cytotoxicity induced by the topoisomerase II directed drugs daunorubicin and etoposide (VP-16). Biochem Pharmacol 46(3):389–93

21. Pearlman M, Jendiroba D, Pagliaro L, Keyhani A, Liu B, Freireich EJ (2003) Dexrazoxane in combination with anthracyclines lead to a synergistic cytotoxic response in acute myelogenous leukaemia cell lines. Leuk Res 27(7):617–26

22. Creighton AM, Hellmann K, Whitecross S (1969) Antitumor activity in a series of bisdiketopiperazines. Nature 222:384–5

23. Bellet RE, Mastrangelo MJ, Dixon LM, Yarbro JW (1973) Phase I study of ICRF 159 (NSC-129943) in human solid tumors. Cancer Chemother Rep 57:185–9

24. Tetef ML, Synold TW, Chow W et al (2001) Phase I trial of 96-hour-infusion of dexrazoxane in patients with advanced malignancies. Clin Cancer Res 7(6):1569–76

25. Seifert CF, Nesser ME, Thompson DF (1994) Dexrazoxane in the prevention of doxorubicin-induced cardiotoxicity. Ann Pharmacother 28(9):1063–72

26. Herman EH, Hasinoff BB, Zhang J et al (1995) Morphologic and morphometric evaluation of the effect of ICRF-187 on bleomycin-induced pulmonary toxicity. Toxicology 98(1–3):163–75

27. Langer SW, Sehested M, Jensen PB (2001) Dexrazoxane is a potent and specific inhibitor of anthracycline induced subcutaneous lesions in mice. Ann Oncol 12:405–10

2.3 Clinical Studies in Malignant Tumors

W. Rhomberg

2.3.1 Razoxane (ICRF-159) as Antitumor Agent

2.3.1.1 Leukaemias and Malignant Lymphomas

Shortly after the discovery of razoxane as antineoplastic agent, research focused on phase I and phase II studies with emphasis given to remission induction in leukaemias and solid tumors. The radiosensitizing efficacy and the ability of the drug to slow down the growth rate of various tumours were not analyzed or recognized at that time.

Leukaemias

Early trials in leukaemias and malignant lymphomas showed indeed some activity [1, 2]. In 1969, rapid falls in total white cell counts were observed in 7 of 9 pretreated patients with leukemias (6 patients) and lymphosarcomas (3 patients) accompanied by very little evidence of toxicity [3, 4]. Bone marrow examination of these nine cases showed that only one was in full hematological remission. Later it was observed that at daily doses of 30 mg/kg, ICRF 159 reduced the white cell count in every case, within 48 h. Usually, the blast cell component decreased more quickly than the normal white cells. Accordingly, in a study by Bakowski et al. [5] eleven patients with myeloid blast cell crisis and two patients with lymphoid blast cell crisis of chronic myeloid leukaemia (CML) were treated with razoxane. Two of the patients with myeloid blast cell crisis achieved partial bone marrow remission and survival for 8+ and 18 months. One of the two patients with lymphoid blast cell crisis reverted to a chronic phase of CML after treatment with razoxane in combination with prednisolone.

Treatment of monocytic skin infiltration in 4 patients with chronic myelomonocytic leukaemia was reported by Copplestone et al. [6]. Treatment of the rash with low dose cytarabine or etoposide was effective but razoxane produced no benefit. Superficial radiotherapy was useful to control pruritus in one patient.

A positive clinical experience came from Krepler and Pawlowsky [7]. They treated 20 children with acute leukaemia in relapse who were resistant to most of the generally used cytotoxic drugs. Despite the unfavourable selection of the cases, two complete and seven incomplete remissions lasting from 1 to 6 months were achieved.

A few years later a series of 17 consecutive cases of acute leukaemias (10 AML, 5 AMML, 2 CML with blast transformation), this time treated with razoxane and cytosine arabinoside, was reported from clinicians in Manchester [8]. The treatment consisted of 3 × 125 mg razoxane and 1 mg/kg cytarabine, each given over 3 days. The cycle was repeated every 10 days. The authors described poor remission results, 13 of 17 patients died. None of these patients had more than 4 courses of treatment. Of the 4 patients who are alive, only 2 entered remission. Every patient complained of nausea and vomiting. The authors concluded that razoxane together with cytarabine is not satisfactory treatment for AML and yield no better results than with cytarabine alone. Only the responses in the two patients with CML and blast cell crisis were regarded as 'encouraging'. This report provoked contradiction in terms of several letters to the editor of the British Medical Journal. For instance, in a series of 7 consecutive cases with AML, Shaw and Tudhope observed that 6 of the 7 patients went into partial remission after 2–4 courses of razoxane plus cytarabine [9]. Allegedly nausea and vomiting did not occur which was in sharp contrast to the observations from Manchester [8].

The controversies were considered closed since a larger series of outtreated acute leukaemias treated with razoxane was finally reported by Bakowski et al. [10]. There were 28 patients with advanced acute nonlymphocytic leukaemia, 16 with acute lymphoblastic leukaemia (ALL), and two with acute undifferentiated leukaemia. They were end stage patients, and not surprisingly, no patient achieved a complete remission and only 3 patients (6%) had a partial bone marrow remission. Five patients with acute nonlymphocytic leukaemia and one with ALL received a combination of ICRF-159 and low-dose cytosine arabinoside. There were no remissions in this group and the toxic effects were more marked than with ICRF-159 alone. This study confirmed the limited activity of ICRF-159 as a single agent (or in combination with cytosine arabinoside) in advanced adult acute leukaemia.

Malignant Lymphomas

The Western Cancer Study Group reported on 27 previously treated patients with advanced non-Hodgkin lymphomas [11]. All patients had stage III or IV disease, and 78% had pathologically documented extranodal disease, 75% had prior radiation therapy. Razoxane was given as initial dose of 1,000 mg/m^2/week in two divided doses 8 h apart. The dose was increased to 2,000 mg/m^2/week if hematologic parameters permitted. *Results.* Among 27 patients there were 3 complete (CR) and 5 partial responses (PR) corresponding to a 30% overall response rate, the majority of responses lasting >1 year and three responses continuing to the time of publication at 24, 31, and 33 months from the start of therapy. It was concluded that further trials appear warranted in lymphomas, especially together with adriamycin. Hematologic toxicity was not surprising in view of the extensive prior therapy.

After that a similar trial of ICRF-159 was performed in Hodgkin's disease and non-Hodgkin's lymphomas whose tumors had become resistant to conventional chemotherapy [12]. The patients were randomized to receive either a loading course or a weekly regimen of the drug. Among 82 evaluable patients, five of 39 (13%) treated with the loading dose schedule and six of 43 (14%) treated with the weekly schedule had objective tumor regressions. Response duration tended to be brief (median, 7 weeks). Life-threatening myelosuppression was more frequent in patients receiving the loading course regimen. Survival was somewhat longer among patients receiving the weekly schedule (median survival, 24 weeks vs. 12 weeks; $p =$ 0.04). In this study, razoxane demonstrated definite but limited therapeutic activity in patients with advanced, refractory malignant lymphomas. The results of previous studies were not reached again. Inconclusive results were achieved by Garrett et al. when razoxane was given adjunctive to combination chemotherapy [13].

If razoxane is combined with cisplatin, the results became even worse. Of 8 patients treated with razoxane and cisplatin, none achieved a response whereas 3 of 10 patients with non-Hodgkin lymphomas had achieved a partial response to cisplatin alone of 7–15 weeks duration [14]. Similar to the experience in cervical carcinoma (see below), there is suspicion of an antagonism if razoxane is being combined with cisplatin.

Only few patients with Hodgkin's disease were studied. Corder et al. administered razoxane to 11 patients with Stage IV sclerosing Hodgkin's lymphoma which were heavily pretreated [15]. One objective partial remission was observed among these patients with far advanced disease. No responses were seen among 9 patients with pretreated Hodgkin lymphomas in a report of the Western Cancer Study Group [11].

2.3.1.2 Solid Tumors, Single Agent Therapy

Colorectal Cancer

Several phase II trials in solid tumours followed. In these studies performed during the early 1970s advanced colorectal cancers were found to be sensitive towards razoxane in a proportion of 10–15% of the patients [16, 17]. Remission induction was less frequent or even not observed in other studies irrespectively whether the patients with disseminated colorectal cancer were chemonaive or previously exposed to systemic chemotherapy [18, 19]. The Eastern Cooperative Oncology Group randomized 127 patients with advanced measurable colorectal cancer who had received prior chemotherapy to receive piperazinedione (PZD), Yoshi-864, or razoxane (ICRF 159). Although no responses were seen among 38 razoxane treated patients, the median survival was longest (23 weeks) in the razoxane arm vs. 17 and 19 weeks, respectively in the PZD and Yoshi-864 arm [18].

A phase III study of ICRF-159 vs. 5-FU in the treatment of advanced metastatic colorectal carcinoma was published by Paul et al. [19]. The results of this study are summarized in the following *abstract*: 'Thirty-seven previously untreated patients with advanced metastatic colorectal carcinoma were treated in a prospective randomized fashion with either ICRF-159 or 5-FU. The ICRF-159 was administered orally at a dose of 1 g/m^2/day for 3 consecutive days every 3 weeks, and the 5-FU was given iv at a dose of 450 mg/m^2/day for 5 days every 5 weeks. All patients were evaluated for response and toxic effects after two courses of treatment. All those who failed to meet the criteria for objective response with either a complete remission or a partial response received the other drug in a crossover fashion. Three of 18 patients (16%) initially treated with 5-FU achieved a partial response while none of the 19 patients initially treated with ICRF-159 achieved a complete or partial response. Nine prior 5-FU-treated patients were crossed over to ICRF-159 and 14 prior ICRF-159-treated patients subsequently received 5-FU. No antitumor response was seen with the secondary agent in this study. The response rate for ICRF-159 (none of 19 patients) predicts that it is unlikely to produce a true response rate of $\geq 20\%$ with a rejection error of <5%, making it unsuitable as primary therapy for colon carcinoma. The toxicity of 5-FU was moderate and mainly gastrointestinal while the toxicity of ICRF-159 was severe mainly hematologic' [19].

Breast Cancer

Razoxane seems to have minimal cytotoxic effects in advanced breast cancer in terms of induction of objective tumour remissions although the drug was tested only in patients refractory to the main hormonal and cytotoxic therapies who in addition had very advanced carcinomas [20, 21].

Ahman and colleagues assessed two agents that were new in the 1970s, i.e. dianhydrogalactitol and razoxane, in patients with advanced breast cancer previously exposed to cytotoxic chemotherapy [20]. Thirty-two patients entered this randomized trial. All of the patients had previously been exposed to cytotoxic chemotherapy and 20 of the 32 patients had been exposed to two or more multiple-drug programs. Twenty-seven of the 32 patients had visceral dominant disease while the other five had soft tissue-dominant disease. All patients were postmenopausal. Razoxane was given at a dose of 750 mg/m^2/day for 3 days in divided doses every 8 h in all patients. *Results*: Of the 32 patients, 18 received dianhydrogalactitol as primary therapy and 14 received razoxane. None of these patients achieved an objective clinical response as defined. Four of 14 patients on razoxane had stability of disease lasting 5, 8, and 14 weeks respectively. Following razoxane as primary therapy, a mean survival of 11 weeks (median, 9 weeks) was achieved. Both agents produced leukopenia of a significant magnitude. The myelosuppression following dianhydrogalactitol had a mean duration considerably longer than that following razoxane: 22 days vs. 9 days. There were two drug-related deaths, one with each agent following severe myelosuppression and sepsis.

In a study of Creech, Engstrom and colleagues, razoxane was given at a dose of 300 mg/m^2 orally every 8 h for nine doses every 21 days to 40 patients with advanced breast cancer [21]. The patients were refractory to chemotherapeutic agents. Two patients with soft tissue disease had short-lived partial responses. Corresponding to this high dose drug regimen, the hematologic toxicity was severe. Four patients became septic at the nadir of leukopenia; two of these patients died while leukopenic.

Lung Cancer

The (partial) response rates in lung cancer achieved with razoxane (ICRF-159) are low (<10%). Eagan et al. treated 52 patients with non-small cell carcinoma of the lung with 1 g/m^2/day of razoxane orally for 3 consecutive days (divided into nine approximately equal doses given every 8 h), and they observed only 4 partial tumour responses [22]. In this study, razoxane was compared in a randomized form with the combination of vincristine, bleomycin and adriamycin. Seventeen patients with squamous cell carcinoma received the drug, 15 as initial treatment and two as crossover treatment. A response was noted in one of 15 patients receiving initial treatment but in neither of the patients treated with crossover treatment, for an overall response rate of 5.9%. In addition, responses were seen in 1/10 large cell undifferentiated carcinomas (crossover case), and in 2/25 adenocarcinomas when razoxane was given as initial treatment. The mean duration of response was 143 days (range, 81–258 days) and the mean survival of the responders was 212 days (range, 179–233 days). Survival times and duration of response were not significantly different between the groups.

Also ICRF-187, the more soluble (+) enantiomer of the racemic mixture ICRF-159 (razoxane), was administered in non-small cell lung cancer. In a study of Natale et al. [23], ICRF-187 at a dose of 1,250 mg/m^2 daily over 3 days, repeated every 3 weeks, was given to 29 patients with measurable non-small-cell lung cancer (NSCLC). Among 25 assessable patients including 15 who had not received prior chemotherapy, no objective responses were seen.

Cervical Carcinoma

A phase II study was performed by the Gynecologic Oncology Group in 31 patients with advanced squamous cell carcinoma of the cervix uteri [24]. Razoxane was administered orally at a dose of 2.5 g/m^2 weekly until progression, unacceptable toxicity, or death. Among 28 evaluable patients, five (18%) showed tumour responses (one CR, four PRs) ranging from 1 to 5 months. Fifteen patients with stable disease and eight with progressive disease had a median survival duration of 3.8+ and 3.5+ months, respectively. Twenty-three of the 28 patients exhibited leukopenia which in ten instances was severe (below 2,000/mm^3). Seven cases had thrombocytopenia (one case below 50,000/mm^3). Other toxicity, including fever and anorexia, was mild to moderate. A randomized controlled study performed later using *razoxane together with radiotherapy* in the primary treatment of cervical carcinomas showed no survival benefit for the combined modality treatment arm [25].

In contrast to the squamous cell variety of cervical cancer, the Gynecologic Oncology Group was not able to demonstrate useful objective responses in adenocarcinoma of the cervix uteri (1/27) or adenocarcinoma of the endometrium (0/24 patients) [26, 27].

Soft Tissue Sarcomas

In soft tissue sarcomas (STS), responses were rarely seen when razoxane was used alone. Among 37 patients with measurable STS, osteosarcomas, chondrosarcomas, and mesotheliomas 1 partial response of brief duration was reported in a patient with fibrosarcoma [28]. Contrary to this phase II evaluation, Olweny et al. reported an overall response rate of 61% (11% complete responses) among 18 mostly pretreated patients with endemic Kaposi sarcomas in Uganda [29, 30]. These data are interesting but not unexpected, since razoxane is able to normalize pathologic tumour blood vessels (see Sections 2.1, 2.2, and 2.3.4).

Malignant Melanoma

A small randomized study on razoxane (ICRF-159) vs. DTIC and cis-platinum in stage IV melanoma patients was published by Ahmann et al. in 1978 [31]. The median survival time for the razoxane group and the DTIC/cis-platinum group was 15 and 12 weeks, respectively. Although no objective regression was seen in the razoxane group (0/17) but 2/20 in the DTIC/cis-platinum group, the survival was slightly superior in the razoxane arm. At that time the conclusion was made that razoxane has no role in the treatment of disseminated melanomas in terms of remission induction, and that DTIC alone may be as effective as the combination of DTIC and cis-platinum. In this context, one has to recall the so-called cytorallentaric mode of action (see Section 2.3.4) that slows down the growth rates of tumours – hitherto predominantly seen and studied in animal systems. This process occurs probably also under clinical conditions. It could explain the data cited above by a direct but hidden antitumor effect. Such an effect had to be assumed also in our study of melanoma brain metastases that is described in the Section 2.3.2.4.

Other Tumours

Some cytotoxic activity of razoxane has seemingly been observed in *head and neck tumours* [32] but a subsequent placebo-controlled trial of razoxane together with radiotherapy revealed a detrimental effect on the response rates within a matched pair analysis [33]. In smaller series, response rates below 5% were noted when razoxane alone was used in advanced *gastric cancer* where only 1/12 patients showed stable disease [34].

Two phase II trials were reported in patients with ovarian cancer. Edmonson et al. noted 8 cases of stable disesease in 16 evaluable patients with alkylating agent-resistant epithelial *ovarian carcinomas* lasting from 2 to 6 months, but no major objective responses were seen [35]. Likewise, no major responses were observed in another series of 22 patients with ovarian cancer treated by Slayton et al. [36].

A few years ago, a phase II study of 40 patients with *renal cell cancer* (32 men, 8 women); median age: 58 years (range, 31–76) was performed with measurement of renal tumour and normal tissue perfusion [37], and exploring the antiangiogenic abilities of razoxane by the assessment of 6 potential surrogate markers of angiogenesis [38]. Twenty-nine percent of the patients had stable disease for a minimum of 4 months, and the remainder had progressive disease (PD). Median overall survival was 7.3 months. Duration of survival was significantly better in patients with stable disease compared to PD ($p = 0.003$). Serum VCAM-1 levels and urinary VEGF levels rose significantly after one cycle in patients with PD but not in those with stable disease. Serum VEGF, bFGF, VCAM-1 and v. Willebrand factor, plasma uPAsr and urinary bFGF levels were significantly higher in patients with progressive disease compared with patients showing stable disease before and/or after 1 cycle of treatment. It was concluded that razoxane is an antiangiogenic agent that has minimal toxicity and that requires further evaluation in combination with other active agents in the treatment of renal cell cancer.

The common *childhood solid tumors* were studied by the Southwest Oncology Group [39]. Fifty-three patients were evaluable for tumour response. Toxicity was primarily hematopoietic and gastrointestinal. There were no responses in any of the eight patients with osteogenic sarcoma, four with lymphoma, five with Ewing's sarcoma, ten with neuroblastoma, or six with rhabdomyosarcoma. There was a transient partial response in one of four children with Wilms' tumor.

No phase II data are available for a variety of other tumours, e.g. brain tumours, thyroid, pancreas, liver, prostate or urinary bladder.

2.3.1.3 Solid Tumors, Multiagent Therapy

Renal Cell Cancer

A phase II study in advanced/metastatic renal cell cancer using oral razoxane and cytoxan in metronomic doses was performed at the Hannover Medical University from 1976 to 1978. The median survival of 35 patients was far superior in comparison to historical controls, and although the tumors were rendered to a slower growing entity in the majority of the cases, this was not always associated with a better quality of life (Rhomberg W, unpublished). In this context, the above cited study of razoxane on surrogate markers of angiogenesis in renal cell cancer might be of interest [38].

Cervical Carcinoma

A phase I trial of cisplatin and razoxane in advanced squamous cell carcinoma of the cervix was performed by the Gynecologic Oncology Group Pilot Study [40]. Cisplatin at a dose of 50 mg/m^2 intravenously every 21 days was combined with razoxane at two dose levels – 750 mg/m^2 weekly and 1,150 mg/m^2 weekly. Three patients were treated at the first dose level and six patients were treated at the second dose level of razoxane. No objective regressions were observed, and three patients refused to continue therapy at the higher dose of razoxane because of nausea and vomiting. The outcome is intriguing since both cisplatin and razoxane [24] have shown activity in cervical cancer. There is the suspicion of an antagonism, perhaps by formation of cis-platinum-razoxane complexes [41]. At any rate, a further study of this regimen in cervical cancer was not recommended.

Gastrointestinal Malignancies

Razoxane was combined with doxorubicin in 6 patients with *hepatocellular carcinoma* (HCC) in a study from Uganda [42]. This was part of a pilot study using mainly doxorubicin either alone or together with various cytotoxic agents in 139 patients with HCC. One of the 6 patients showed a partial tumor response. The combination with razoxane did not enhance the therapeutic effectiveness of doxorubicin alone.

The combination of triazinate and razoxane was studied in advanced *colorectal cancer* after failure of 5-fluorouracil [43]. Triazinate 250 mg/m^2 iv over 1 h and razoxane 1.0 g/m^2 p.o. in divided doses were given together weekly. Of the 13 patients with measurable disease treated, 10 were ambulatory. All had prior chemotherapy with 5-FU plus at least one other drug. Only one patient achieved objective partial remission, which lasted 6 months. The median survival was 4 months. Leukopenia <2,000/μl occurred in 15%. Nausea and vomiting occurred in 85%, and moderately severe in 31%. Four patients developed diarrhea, which was severe and dose limiting in one. The activity of this combination was less than that reported for either drug alone in a comparable patient population. The authors concluded: it is unlikely that these drugs, as given together in this trial, will contribute significantly to the combination chemotherapy of colorectal cancer.

Anhydro-ara-5-fluorocytosine (AAFC) and razoxane were given together in a phase I and II clinical study of different *adenocarcinomas of digestive origin* [44]. The outcome in term of remission induction was as follows:

Gastric cancer: 14 patients, 2 partial responses
Colorectal cancer: 9 patients, no responses
Pancreatic cancer: 3 patients, no responses
Biliary tree cancer: 2 patients, no responses

In *metastatic pancreatic cancer,* a case of long term survival of a 37 year-old female patient was reported if razoxane (2 × 125 mg at 5 days of a week) and 5-FU (15 mg/kg every 3 weeks) were given concurrently. The liver metastases regressed completely and the primary tumor did not change in its size for a duration of 4.5 years [45].

Breast Cancer

The drug received an insufficient testing together with other chemotherapeutic agents in breast cancer. Cummings et al. reported a trial that used adriamycin plus vincristine alone or with dibromodulcitol or ICRF-159 in metastatic breast cancer [46]. The results were inconclusive, but they represented no signal for further investigations.

2.3.1.4 Adjuvant Use of Razoxane

Gastric Cancer

Preliminary data on adjuvant treatment with razoxane following resection of gastric cancer became available with a study of Gilbert et al. [47]. Sixteen patients have received razoxane. The tumor was resected in all of these patients except one in whom it was by-passed. The 16 patients had a mean survival of 20.6 months and a median survival of 12.5 months. Those patients in whom the operative specimen showed evidence of extra-gastric spread histologically had a mean surival of 9.8 months with a median of 9.5 months. If there was no evidence of extra-gastric spread, the survival was a mean 41 months with a median of 44 months. The single patient who had a simple by-pass procedure survived for 27 months.

The authors found the trend in the survival times encouraging when compared to the figures published for carcinoma of the stomach from other centers in England. For instance, Costello et al. [48] reported a median survival of approximately 6 months if lymph nodes were involved and about 1 year if no nodes were involved. However, there is a marked difference between results published by specialist referral centres and general hospitals which was illustrated in a table. Twelve of the 16 patients in this series had their operation at a district general hospital.

The data remain inconclusive, a further update of this at that time ongoing study which was part of a wider randomized study, did not appear in the literature.

Colorectal Cancer

Since razoxane prevented distant metastases in a variety of animal experiments and the drug was shown at that time to be one of the few agents having some activity in advanced colorectal cancer [17], a randomized controlled trial of oral adjuvant razoxane in resectable colorectal cancer was initiated by Gilbert et al. [49]. After a 5 year follow-up accrual has ceased and a total of 272 patients entered the trial (133 control, 139 treatment). Treated patients received razoxane postoperatively on a continuous, long-term basis while control patiens received identical clinical care, but no adjuvant chemotherapy. A significant prolongation of the time to recurrence for Dukes'C patients treated with razoxane has been found. This correlated with a doubling of the delay in time to development of liver metastases in this group of patients. Razoxane was well tolerated by the patients and caused minimal side effects. However, acute leukaemia occurred in three patients (2.45%) after prolonged exposure to the drug [50]. It was concluded that razoxane is suitable for further evaluation in colorectal cancer and for incorporation into combination chemotherapy regimens.

In the meantime, this conclusion received support from experiments in Sprague-Dawley rats with large bowel cancer induced by dimethylhydrazine (DMH). In this study of Gilbert et al. [51], razoxane was given intraperitoneally from week 25 until the end of the experiment (week 35) to see if this drug could inhibit the development of large bowel tumors which may be consistently induced by DMH. There were fewer malignant colorectal tumors in rats receiving razoxane than in the controls (10 vs. 24; $p = 0.04$). Some animals developed malignant tumors of the small intestine as well as the large bowel and the number of malignant tumors for the whole intestinal tract was also reduced in animals receiving razoxane (15 vs. 32; $p = 0.025$). Four animals in the DMH + control group died prematurely from progressing malignancy, whereas all the animals in the DMH + razoxane group survived to the designated end of the experiment ($p = 0.05$, Fisher's exact test). Metastases were found in four animals in the DMH+control group and in only one animal in the DMH + razoxane group but the experiment had insufficient power to show an effect on metastases because the tumors were not allowed to continue developing as control animals had started to die from the tumors. The authors previously reported a similar experiment in which 5-fluorouracil failed to show a significant reduction of malignant tumors induced by DMH [52].

Some years later, another group again evaluated the effectiveness and safety of razoxane as an adjunct to surgery in 603 patients with colorectal cancer in a controlled randomized trial [53]. All patients have been followed up for a minimum of 5 years. Statistical analysis showed that razoxane treatment had no effect either beneficial or adverse on the rates of recurrence or on 5 year survival. However, the projected duration of the treatment with razoxane was 9 months vs. an intended time of 2 years. A separate analysis of the patients with a 2 year-treatment showed a better outcome. So, it is possible that a more prolonged course of razoxane might have significantly influenced survival. The incidence of severe adverse reactions was low, and no renal, hepatic, pulmonary or cardiac toxicity was noted. But it was of concern to the authors that one patient developed leukaemia. This was also the reason that the study was shortened and the planned duration of giving razoxane was not reached. Should razoxane be considered for future use it was recommended that continuous low dose therapy be given for no longer that 12 months.

Summary and Comment

Taken together, razoxane has obviously a low efficacy in terms of remission induction in the majority of solid tumours tested so far. Exceptions might be some non-Hodgkin lymphomas and Kaposi sarcoma. The low response rates in solid tumors and the rather short responses in leukaemia and lymphomas were for the present the main reason that led to a limited interest for the drug. It remains unclear to which extent the intermittent dose schedules with higher single doses and the pretreatment of the patient populations studied may have compromised the results, but it seems that razoxane acts rather as a cytostatic than a cytotoxic agent.

In this context, the prolongation of the overall survival time which was seen in some randomized studies inspite of the lack of objective responses [18, 31], remains to be kept in mind (see also Section 2.3.4).

The activity of certain drug combinations was less than that reported for either drug alone in comparable patient populations in several pilot studies [14, 40, 43]. That applies especially to the combination of cisplatin and razoxane given concomitantly. The question arises whether razoxane is interferring with cisplatin in some form, perhaps by its chelating activity. There is reason for this assumption since the treatment of K2[PtCl 4] with razoxane in aqueous HCl solution gave the water-insoluble complex cis-PtCl-2(razoxane) [41]. On the other hand, preclinical experiments using the Lewis lung tumor showed that a preceeding fractionated treatment with razoxane greatly enhanced the activity of following cis-platinum-radiation combinations [54]. Thus, the presumable ability of razoxane to potentiate some cytotoxic drugs can not be generalized as yet. It must rather be scrutinized and has to be studied more in depth.

References

1. Hellmann K (1973) Current clinical and experimental studies with ICRF 159. In: Stacher A (ed) Leukämien und maligne Lymphome. Urban & Schwarzenberg, Wien
2. Mathe G, Amiel JL, Hayat M et al (1970) Preliminary data on acute leukaemia treatment with ICRF 159. In: Mathe G (ed) Advances in the treatment of acute leukaemia. Springer, New York, pp 54–5
3. Hellmann K, Newton KA, Whitmore DN, Hanham IWF, Bond JV (1969) Preliminary clinical assessment of ICRF 159 in acute leukaemia and lymphosarcoma. Br Med J 1:822–4
4. Hellmann K (1970) Further clinical experiences with ICRF 159. In: Mathe G (ed) Advances in the treatment of acute (blastic) leukaemias. Springer, New York, pp 52–3
5. Bakowski MT, Brearley RL, Wrigley PF (1979) Treatment of blast cell crisis of chronic myeloid leukaemia with ICRF-159 (razoxane). Cancer Treat Rep 63(11–12):2085–7
6. Copplestone JA, Oscier DG, Mufti GJ, Hamblin TJ (1986) Monocytic skin infiltration in chronic myelomonocytic leukaemia. Clin Lab Haematol 8:115–9
7. Krepler P, Pawlowsky J (1975) Clinical trials with bis-dioxopiperazine propane (ICRF 159; NSC 129943) in acute leukaemias. Oesterr Z Onkol 2(4):112–4
8. Bhavnani M, Dhir K, Delamore IW (1978) Razoxane in treatment of acute myeloid leukaemia. Br Med J 2(6140):801
9. Shaw D, Tudhope GR (1978) Razoxane in treatment of acute myeloid leukaemia. Br Med J 2(6144):1089–90
10. Bakowski MT, Prentice HG, Lister TA, Malpas JS, McElwain TJ, Powles RL (1979) Limited activity of ICRF-159 in advanced acute leukemia. Cancer Treat Rep 63(1):127–9
11. Flannery EP, Corder MP, Sheehan WW, Pajak TF, Bateman JR (1978) Phase II study of ICRF 159 in non-Hodgkin's lymphomas. Cancer Treat Rep 62:465–7
12. O'Connell MJ, Begg CB, Silverstein MN, Glick JH, Oken MM (1980) Randomized clinical trial comparing two dose regimens of ICRF-159 in refractory malignant lymphomas. Cancer Treat Rep 64(12):1355–8
13. Garrett MJ, Das S (1977) Treatment of the lymphomata with ICRF 159 (NSC 129943) adjunctive to combination chemotherapy. Ir Med J 70:460–1
14. Corder MP, Wiesenfeld M, Maguire LC, Leimert JT, Panther SK (1980) Cis-dichlorodiammine-platinum(II) with and without ICRF-159 in non-Hodgkin's lymphoma. Cancer Treat Rep 64:301–4
15. Corder MP, McFadden DB, Bell SJ (1984) A trial of razoxane (ICRF 159) in patients with prior therapy for Hodgkin's lymphoma. Cancer 54:1496–8
16. Bellet RE, Engstrom PF, Catalano RB, Creech RH, Mastrangelo MJ (1976) Phase II study of ICRF 159 in patients with metastatic colorectal carcinoma previously exposed to systemic chemotherapy. Cancer Treat Rep 60:1395–7
17. Marciniak TA, Moertel CG, Schutt AJ, Hahn RG, Reitemeier RJ (1975) Phase II study of ICRF 159 (NSC 129943) in advanced colorectal carcinoma. Cancer Chemother Rep 59: 761–3
18. Douglass HO, MacIntyre JM, Kaufman J, Von Hoff D, Engstrom PF, Klaassen D (1985) Eastern cooperative oncology group phase II studies in advanced measurable colorectal cancer. I. Razoxane, Yoshi-864, piperazinedione, and lomustine. Cancer Treat Rep 69:543–5
19. Paul AR, Catalano RB, Engstrom PF (1980) Phase-III-study of ICRF 159 versus 5-FU in the treatment of advanced mestatatic colorectal carcinoma. Cancer Treat Rep 64:1047–9
20. Ahmann DL, O'Connell MJ, Bisel HF, Edmonson JH, Hahn RG, Frytak S (1977) Phase II study of dianhydrogalactitol and ICRF-159 in patients with advanced breast cancer previously exposed to cytotoxic chemotherapy. Cancer Treatment Reports 61:81–2
21. Creech RH, Engstrom PF, Harris DT, Catalano RB, Bellet RE (1979) Phase II study of ICRF-159 in refractory metastatic breast cancer. Cancer Treat Rep 63(1):111–4
22. Eagan RT, Carr DT, Coles DT, Rubin J, Frytak S (1979) ICRF 159 versus polychemotherapy in non-small cell lung cancer. Cancer Treat Rep 60:947–8

23. Natale RB, Wheeler RH, Liepman MK, Sauder A, Bricker L (1983) Phase II trial of ICRF-187 in non-small cell lung cancer. Cancer Treat Rep 67(3):311–3
24. Conroy JF, Lewis GC, Blessing JA, Mangan Ch, Hatch K, Wilbanks G (1984) ICRF-159 (razoxane) in patients with advanced squamous cell carcinoma of the uterine cervix: for the gynecologic oncology group. Am J Clin Oncol 7(2):131–4
25. Belloni C, Mangioni C, Bortolozzi G, Carinelli S, D'Incalci M, Maggioni A, Morasca L (1983) ICRF 159 plus radiation versus radiation therapy alone in cervical carcinoma. A double blind study. Oncology 40:181–5
26. Homesley HD, Blessing JA, Berman M (1986) ICRF 159 (razoxane) in patients with advanced nonsquamous cell carcinoma of the cervix. A gynecologic oncology group study. Am J Clin Oncol 9:325–6
27. Homesley HD, Blessing JA, Conroy J, Hatch K, DiSaia PJ, Twiggs LB (1986) ICRF 159 (razoxane) in patients with advanced adenocarcinoma of the endometrium. A gynecologic oncology group study. Am J Clin Oncol 9:15–7
28. Borden EC, Ash A, Enterline HT, Rosenbaum C, Laucius JF, Paul AR, Falkson G, Lerner H (1982) Phase II evaluation of dibromodulcitol, ICRF 159 and maytansine for sarcomas. Am J Clin Oncol 5:417–20
29. Olweny CL, Masaba JP, Sikyewunda W, Toya T (1976) Treatment of Kaposis sarcoma with ICRF 159 (NSC 129943). Cancer Treat Rep 60:111–3
30. Olweny CL, Sikyewunda W, Otim D (1980) Further experience with razoxane (ICRF 159; NSC 129943) in treating Kaposi's sarcoma. Oncology 37:174–6
31. Ahmann DL, Edmonson JH, Frytak S, Kvols LK, Bisel HF, Rubin J (1978) Phase II study of ICRF-159 versus combination cis-dichlorodiammineplatinum (II) and DTIC in patients with disseminated malignant melanoma. Cancer Treat Rep 62:151–3
32. Poster DS, Penta J, Marsoni S, Bruno S, Macdonald JS (1980) Bis-diketopiperazine derivatives in clinical oncology: ICRF-159. Cancer Clin Trials 3(4):315–20
33. Bakowski MT, Macdonald E, Mould RF et al (1978) Double blind controlled clinical trial of radiation plus razoxane (ICRF 159) versus radiation plus placebo in the treatment of head and neck cancer. Int J Radiat Oncol Biol Phys 4:115–9
34. McVie JG, Soukop M, Claman KC, Stuart JFB, Trotter J (1979) Failure of razoxane (ICRF 159) in gastric carcinoma. Abstract No. 128, 5th annual meeting of the Medical Oncology Society, Nice, 1–3 Dec 1979
35. Edmonson JH, Decker DG, Malkasian GD, Webb MJ (1981) Concomitant phase II studies of pyrazofurin and razoxane in alkylating agent-resistant cases of epithelial ovarian carcinoma. Cancer Treat Rep 65(11–12):1127–9
36. Slayton RE (1982) New agents in ovarian cancer. Int J Radiat Oncol Biol Phys 8:1431–4
37. Anderson H, Yap JT, Wells P, Miller MP, Propper D, Price P, Harris AL (2003) Measurement of renal tumour and normal tissue perfusion using positron emission tomography in a phase II clinical trial of razoxane. Br J Cancer 89(2):262–7
38. Braybrooke JP, O'Byrne KJ, Propper DJ, Blann A, Saunders M, Dobbs N, Han C, Woodhull J, Mitchell K, Crew J, Smith K, Stephens R, Ganesan TS, Talbot DC, Harris AL (2000) A phase II study of razoxane, an antiangiogenic topoisomerase II inhibitor, in renal cell cancer with assessment of potential surrogate markers of angiogenesis. Clin Cancer Res 6(12): 4697–704
39. Dyment PG, Starling KA, Land VJ, Cangir A, Komp DM, Sexauer CL (1979) ICRF-159 (razoxane) in the treatment of pediatric solid tumors: a Southwest oncology group study. Cancer Treat Rep 63(8):1397–8
40. Bonomi P, Jordan E, Blessing JA (1988) Phase I trial of cisplatin and razoxane (ICRF-159) in advanced squamous cell carcinoma of the cervix. A gynecologic oncology group pilot study. Am J Clin Oncol 11(1):1–2
41. Davies HO, Brown DA, Yanovsky AI, Nolan KB (1998) The preparation and crystal and molecular structure of the complex cis-PtCl2(razoxane). Inorganica Chimica Acta 268: 313–6

42. Olweny CL, Katongole-Mbidde E, Bahendeka S, Otim D, Mugwerwa J, Kyalwazi SK (1980) Further experience in treating patients with hepatocellular carcinoma in Uganda. Cancer 46:2717–22

43. Chiuten D, Vogl SE, Kaplan BH (1979) Combination chemotherapy of advanced colorectal cancer with triazinate (T) and ICRF 159 (I) after failure of 5-fluorouracil (F). Proc Am Assoc Cancer Res. Abstract No. 415

44. Brugarolas A, Buesa JM, Lacave AJ, Gracia Marco M, Valle Pereda M (1978) Phase I and II clinical study of anhydro-ara-5-fluorocytosine (AAFC) and ICRF-159 combination in adenocarcinoma of digestive origin. Cancer Clin Trials 1(4):289–95

45. Ward A, Sherlock D (1980) Long term survival following chemotherapy for carcinoma of the pancreas. Br J Clin Pract 34:157–9

46. Cummings FJ, Gelman R, Tormey DC, DeWys W, Glick J (1981) Adriamycin plus vincristine alone or with dibromodulcitol or ICRF-159 in metastatic breast cancer. Cancer Clin Trials 4(3):253–60

47. Gilbert MJ, Cassell P, Ellis H, Wastell Ch, Hermon-Taylor J, Hellmann K (1979) Adjuvant treatment with razoxane (ICRF 159) following resection of cancer of the stomach. Recent Results Cancer Res 68:217–21, Springer, Berlin, Heidelberg

48. Costello CB, Taylor TV, Torrance B (1977) Personal experience in the management of carcinoma of the stomach. Br J Surg 64:47–51

49. Gilbert JM, Hellmann K, Evans M, Cassell PG, Stoodley B, Ellis H, Wastell C (1982) Adjuvant oral razoxane (ICRF-159) in resectable colorectal cancer. Cancer Chemother Pharmacol 8:293–9

50. Gilbert JM, Hellmann K, Evans M, Cassell PG, Taylor RH, Stoodley B, Ellis H, Wastell C (1986) Randomized trial of oral adjuvant razoxane (ICRF 159) in resectable colorectal cancer: five-year follow-up. Br J Surg 73(6):446–50

51. Gilbert JM, Thompson EM, Slavin G, Kark AE (1984) Razoxane in colorectal cancer induced by dimethylhydrazine. Br J Surg 71:600–3

52. Gilbert JM, Thompson EM, Slavin G, Kark AE (1983) Chemotherapy of chemically-induced colorectal tumors. J R Soc Med 76:467–72

53. Kingston RD, Fielding JWL, Palmer MK (1993) An evaluation of the effectiveness and safety of razoxane when used as an adjunct to surgery in colorectal cancer. Int J Colorectal Dis 8:106–10

54. Kovacs CJ, Schenken LL, Burholt DR (1979) Therapeutic potentiation of combined *cis*-dichlorodiammineplatinum (II) and irradiation by ICRF-159. Int J Radiat Oncol Biol Phys 5:1361–4

2.3.2 Razoxane as Radiosensitizer

Introduction

A clear radiosensitizing ability of razoxane was first described in animal experiments by Hellmann and Murkin [1]. Similar observations were made by serveral other groups [2–4]. According to Kovacs et al. [4], single acute doses of razoxane fail to enhance the radiation response in certain systems, e.g. Lewis lung tumors in BDF1 mice or normal intestinal epithelium, in contrast to a protracted drug administration before irradiation. The preclinical data were soon confirmed by translational reseach in England on soft tissue sarcomas and chondrosarcomas [5, 6]. The following reports of clinical studies on razoxane as radiosensitizer have some methods in common which relate to the intake and dosage of the drug and also to radiation treatment. These details of treatment remained largely the same during the past and will, therefore, be described first.

Common Methods

Razoxane. The drug was usually given 5 (sometimes 4) days before the first irradiation at a dose of 2 × 125 mg daily by mouth. The first tablet was taken 1 h before the irradiation, the rest of the dose in the evenings. The drug intake was continued on radiation days (interruption at weekends) and finished at the end of the radiotherapy if not otherwise pointed out. This scheme was used in all the protocols and studies of sarcomas and gastrointestinal tumors. In these studies, the median total doses of razoxane ranged between 6.5 and 8.5 g, they never exceeded 15 g in a single patient. The dosage of razoxane was adapted to the frequently occurring leukopenia at a nadir of 14–16 days.

Radiation therapy. The radiation treatment was performed with high energy photons and in part with electrons. As a rule, multiple fields based on CT treatment planning were used. In the early studies before 1978, CT treatment planning was not yet available, and at that time some patients also received irradiation with a telecobalt unit. Daily fractions of 170–200 cGy were applied five times a week. The median tumor doses and ranges were mentioned in the different articles.

Definition of Response. The radio-responsiveness was related to the clinical shrinkage of a tumour mass. The diagnosis of a complete response or remission (CR) required the complete disappearance of a tumour both clinically and on radiographic imaging. Partial regression (PR) was diagnosed with a reduction of the initial tumour volume by more than 50%, and disease progression was assumed if the pre-treatment tumour volume increased by 25% during or shortly after the end of the radiotherapy. Local tumour control was defined as no regrowth of the tumour at the site of irradiation as long as the patient survived.

2.3.2.1 Soft Tissue-, Osteosarcomas, Chordomas

Background and History

Sarcomas of soft tissues (STS) and bone were the first tumors studied for the radiosensitizing ability of razoxane. The urgent need to develop more effective radiotherapy of sarcomas led R.D.H. Ryall in Liverpool and K. Newton at the Westminster Hospital in London to study razoxane as an adjunct to radiation therapy [5, 6]. Previously, half of the patients were resistant or had severe late side effects that necessitated limb amputations in many patients. The responses to the combined use of radiotherapy together with razoxane that were achieved in phase I and II studies were far superior to the previous experiences with radiotherapy alone especially in fibro- and chondrosarcomas. These favourable initial experiences in England were the reason for other groups to study the radioresponsiveness of STS further and later to initiate a controlled randomized clinical trial.

Early British Trials

During the decades before 1975, soft tissue sarcomas were, in general, regarded as radioresistant. Exceptions to this view were single case reports on liposarcomas and the observations of Sir Bryan Windeyer who described a 50% rate of major responses in fibro-sarcomas with the use of high radiation doses [7] which led to shrivelled muscles and painful extremities which sometimes required amputations.

Reducing the Radioresistance of Soft Tissue Sarcomas

By using razoxane combined with radiotherapy the treatment options for soft tissue sarcomas were altered for the first time. In 1974, Ryall et al. reported on their first experience in treating bone- and soft tissue sarcomas with radiotherapy together with razoxane [5]. The following is a brief synopsis of this publication.

Original article: Ryall RDH, Hanham IWF, Newton KA, Hellmann K, Brinkley DM, Hjertaas OK (1974) Combined treatment of soft tissue and osteosarcomas by radiation and ICRF 159. Cancer 34:1040–5

(Copyright 1974 American Cancer Society. This material is, in part, reproduced with permission of Wiley-Liss, Inc., a subsidiary of John Wiley & Sons, Inc.)

Synopsis

Background. Response of soft tissue sarcomas (STS) to irradiation is unpredictible, though some histologic types respond more readily than others. As a rule, most histologic entities have to be regarded as fairly radioresistant. Even when complete responses have been obtained, recurrences are frequent and metastasis can appear at any time. In preclinical studies, ICRF-159 (razoxane) was shown to potentiate the effects of radiation treatment, among other tumors also in the S 180 sarcoma model (see Section 2.1).

Objectives. Based on this background it was intended to evaluate the radiosensitizing activity of razoxane in soft tissue and osteosarcomas. Primary outcome measure was the rate of objective tumor regressions on the combined treatment with irradiation and razoxane in patients with sarcomas.

Study design and patients. The trial was not randomized or controlled and neither were patients selected. All patients with inoperable or recurrent STS or osteosarcomas who presented to the participating clinicians were included. A total of 22 patients received the combined treatment. All patients with osteosarcomas had pulmonary metastases visible on radiographs.

Megavoltage was used. Dosage ranged from 42 Gy in 20 fractions in 4 weeks to 66 Gy in 34 fractions in 7 weeks. Lung fields received 21 Gy or less. The dose of razoxane was 125 mg twice daily on radiation days. The first tablet (125 mg) was taken 1–4 h before irradiation, the rest was given in the evenings. The treatment with razoxane commenced 4–5 days before the first irradiation.

Results. The combined treatment with radiotherapy and razoxane led to a good response (i.e. major remissions) in 18/22 patients (82%) with soft tissue and bone sarcomas. Six patients had complete responses (27%) and 12 had regressions of 50% or more. Two showed no response. Skin reactions were greater than expected, no lung reactions were seen.

Considering soft tissue sarcomas only, in 11 of 14 patients, there was a reduction of at least 75% of the original tumor.

Of particular interest was the rapid and impressive regression of the secondaries when patients are given low (but maximum possible) doses of radiation to the lung fields in addition to the drug.

Comment by the authors. That for no fewer than 11 of 14 patients to have had regressions of 75% or more is perhaps an indication that this result is unlikely to have been achieved by chance alone and that the drug potentiates the effect of radiation clinically as well as experimentally. On the basis of the preliminary clinical assessment, it seemed worthwhile to conduct a fully randomized controlled clinical trial.

A later publication of Ryall [6] reported that razoxane given during radiotherapy significantly increased the recurrence-free survival of 36 patients with a variety of soft tissue sarcomas. This increase was accompanied by a modest toxicity. A randomized controlled trial with 31 patients showed significantly improved results for those patients receiving the combination treatment, but the results were never published in detail because of the limited number of patients and the great variety of disease conditions [6]. Reviewing the experience until 1974, a total of 65 patients with STS treated by combined radiotherapy and razoxane have been described and of these 35% have had complete regressions and 74% have had complete or partial (>50%) regressions (Ryall, unpublished).

References

1. Hellmann K, Murkin GE (1974) Synergism of ICRF 159 and radiotherapy in the treatment of experimental tumors. Cancer 34:1033–9
2. Norpoth K, Schaphaus A, Ziegler H, Witting U (1974) Combined treatment of the Walker tumour with radiotherapy and ICRF 159. Z Krebsforsch 82:328–34
3. Kovacs CJ, Evans MJ, Schenken LL, Burhalt DR (1979) ICRF 159 enhancement of radiation response in combined modality therapies. I. Time/dose relationship for tumor response. Br J Cancer 39:516–23
4. Kovacs CJ, Evans MJ, Burhalt DR, Schenken LL (1979) ICRF 159 enhancement of radiation response in combined modality therapies. II. Differential responses of tumour and normal tissues. Br J Cancer 39:524–30
5. Ryall RDH, Hanham IWF, Newton KA, Hellmann K, Brinkley DM, Hjertaas OK (1974) Combined treatment of soft tissue and osteosarcomas by radiation and ICRF 159. Cancer 34:1040–5
6. Ryall RDH (1978) Radiotherapy and ICRF 159 in the treatment of soft tissue sarcomas. Int J Radiat Oncol Biol Phys 4:133–4
7. Windeyer B, Dische S, Mansfield CM (1966) The place of radiotherapy in the management of fibrosarcoma of the soft tissues. Clin Radiol 17:32–40

First Indication That Razoxane Could Alter the Radiation Response Even in Chondrosarcomas

Original article: Ryall RD, Bates T, Newton KA, Hellmann K (1979) Combination of radiotherapy and razoxane (ICRF 159) for chondrosarcoma. Cancer 44(3):891–5

(Copyright 1979 American Cancer Society. This material is, in part, reproduced with permission of Wiley-Liss, Inc., a subsidiary of John Wiley & Sons, Inc.)

Synopsis

Background. There is general agreement that treatment of chondrosarcoma is difficult. These tumors frequently recur after surgery, and they are unresponsive to radiotherapy, to chemotherapy or to a combination of these modalities. In view of the potentiation of radiotherapy with razoxane in the treatment of experimental tumors in animals and soft tissue sarcomas in man, it seemed possible that razoxane could potentiate the effects of radiation also in tumors not normally known for their sensitivity to radiation.

Study design. A similar study design and selection of patients was used as in the above article of Ryall et al. on soft tissue and osteosarcomas. The same applies to the treatment modalities.

Results. Eight patients with 12 chondrosarcomas were treated with radiotherapy and ICRF-159 (razoxane). Two tumors in one patient progressed, 3 tumors in 3 patients showed no change, and 7 tumors in 5 patients had complete or partial remissions. At least 2 complete regressions have responded for more than 2.5 years at the time the report was prepared.

Comment of the authors. The rarity of chondrosarcomas precludes any formal clinical trial, but it seems from the results obtained that the combination of razoxane (used as radiosensitizer) and irradiation have given results which are unusual enough to serve as a pointer to better initial treatment of this most refractory malignancy.

Further Trials and Developments

A Pilot Succession Study

Original article: Rhomberg W (1978) Radiotherapy combined with ICRF 159 (NSC 129943). Int J Radiat Oncol Biol Phys 4:121–6

Synopsis with Specific Details

Background

A trial at the Medical University of Hannover tried to repeat the results which were achieved in England. In a pilot study, the protocol of Ryall et al. [1] was used adopted for the dosage and timing of razoxane. The study was designed only to explore the sensitizing ability of razoxane in radiation treatment. Between 1972 and 1977, 34 patients with measurable soft tissue sarcomas (STS) including 2 patients with chondrosarcomas were treated with radiotherapy and razoxane irrespective of the stage of the disease.

Design

The study was performed as a non-randomized phase II study. Primary outcome measures were the rate of objective responses and their duration.

Results and Comments

Radiation response. Among all 34 patients, there were 7 complete responses (20%), 16 partial regressions (48%) and 10 <no change> situations corresponding to a major response rate of 68%. Only one of the 34 patients showed progressive disease on the combined treatment. It has to be kept in mind that several patients of this group were pretreated mostly with doxorubicin (13 fibrosarcoma patients, 9 were pretreated).

If only localized sarcomas or local recurrences without previous chemotherapy or radiation are considered, the results are even more favourable (Table 2.1). In 12 of such cases there were 7 complete and 3 partial tumor regressions, corresponding to 83% major responses.

Specific response phenomena. Surprisingly low radiation doses led sometimes to a partial or complete regression of lung metastases within a short time. Figure 2.1 shows a complete response in one of several metastases of a fibrosarcoma in the right lung only after 4 fractions with 200 cGy (total dose 800 cGy). Other lesions regressed after 1,400 cGy which is still a low dose.

The phenomenon of a complete liquefaction was observed even in larger tumors (Figs. 2.2 and 2.3). The mechanism of this phenomenon is unclear, it may be related to a sudden break-down of the tumor blood vessel network. This rapidly occurring liquefaction was seen not only in fibrosarcomas but also in synovial- and liposarcomas (Fig. 2.4). Sometimes, the liquefying necrosis remains incomplete.

Table 2.1 Results in 12 inoperable sarcomas without distant metastases or previous chemotherapy treated with radiation and razoxane (9 recurrences after surgery)

No.	Patient	Age/Sex	Diagnosis	Region	Dose (Gy)	Response	Duration (months)	Survival (months)
1	A.H.	43 M	Fibro-Sa	Lung	30	CR	14	14
2	W.A.	40 F	Fibro-Sa	Hypogastric	55	CR	27	31+
3	H.G.	60 F	Fibro-Sa	Jaw	60	CR	11+	11+
4	A.J.	70 F	Fibro-Sa	Shoulder	60	PR	10+	Lost to follow
5	K.H.	75 M	Undiff. Sa	Upper arm	60	CR	16+	16+
6	W.A.	21 F	Neurofibro	Upper arm	60	CR	30+	Lost
7	E.A.	28 F	Leiomyo-S	Thigh	60	CR	24	24
8	D.E.	55 F	MHC	Shoulder	55	CR	3+	3+
9	N.N.	40 F	Leiomyo-S	Uterus	40	PR	Operation	17+
10	G.T.	70 M	Chondro-S	Ribs	60	PR	8	Lost
11	K.H.	56 F	Chondro-S	Ribs	30	NC	14+	14+
12	K.I.	13 F	Undiff. Sa	Neck	30	NC	–	Too early

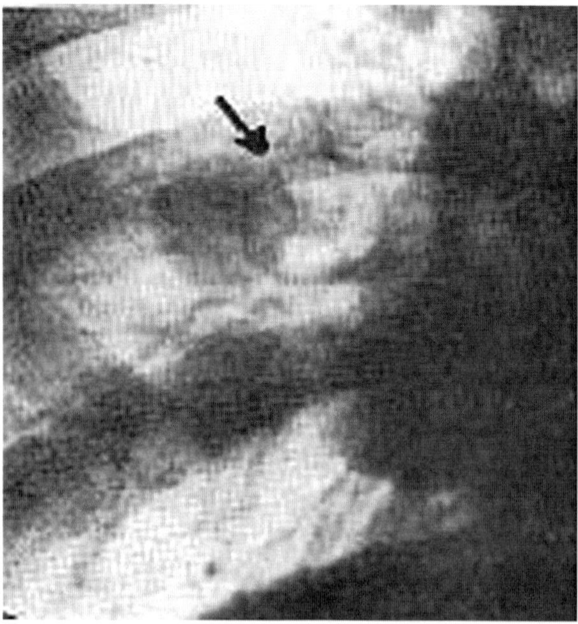

Fig. 2.1 B.A., male, 47 years, fibrosarcoma with lung metastases

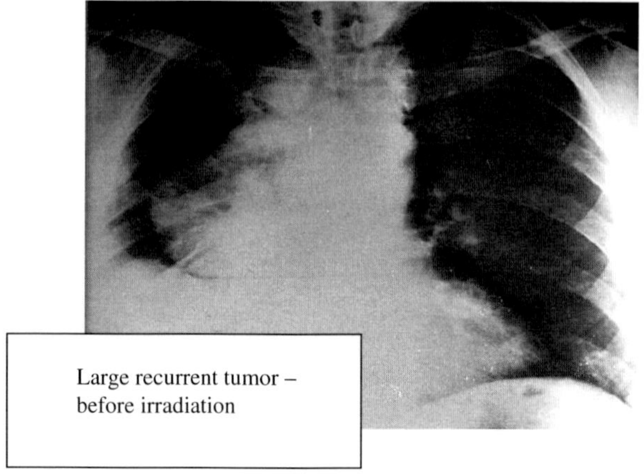

Fig. 2.2 Recurrent fibrosarcoma of the right lung and mediastinum. Male, 62 years

The situation may then be unpleasant because the tumor is under pressure and may require repeated relieving aspirations.

This kind of tumor regression under radiation and razoxane was not observed before. Apart from this type of rapid regression there remain as earlier the slowly (over months) shrinking tumor masses and the unchanged tumors. But there is no doubt: The immediate tumor progression which was common (around 10–15%) if

Fig. 2.3 Complete regression
of the fibrosarcoma after
30 Gy + razoxane

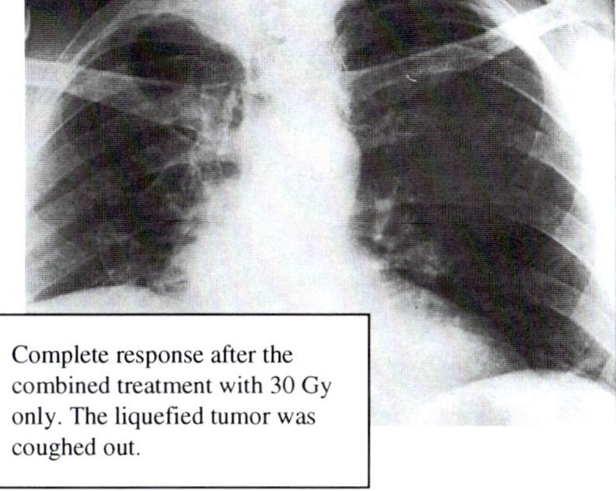

Complete response after the
combined treatment with 30 Gy
only. The liquefied tumor was
coughed out.

Fig. 2.4 Aspirate from a
liquefied liposarcoma of the
back

a STS was treated by radiotherapy alone became a rare event when the irradiation was combined with razoxane.

In brief, the results achieved in England have been confirmed in this confirmatory trial. The data clearly called for the initiation of a controlled randomized study to compare the combined treatment with radiotherapy alone.

Reference

1. Ryall RDH, Hanham IWF, Newton KA, Hellmann K, Brinkley DM, Hjertaas OK (1974) Combined treatment of soft tissue and osteosarcomas by radiation and ICRF 159. Cancer 34:1040–5

A Randomized Trial of Radiotherapy with and Without Razoxane in Soft Tissue Sarcomas

Original article: Rhomberg W, Hassenstein EOM, Gefeller D (1996) Radiotherapy vs. radiotherapy and razoxane in the treatment of soft tissue sarcomas: final results of a randomized study. Int J Radiat Oncol Biol Phys 36:1077–84

Synopsis

Background

For many soft tissue sarcomas (STS) conventional high dose radiotherapy is effective in the control of microscopic residual disease after primary surgery or complete excision of a recurrence. For inoperable advanced cases, however, the probability of local control is still considered poor. The search for improved treatment modalities might be of continued interest for this reason. Because of the favourable results seen in several uncontrolled trials using razoxane combined with radiation therapy for STS (see above), a controlled randomized trial was initiated in 1978.

Objectives

The main purpose of the study was to determine the response rates and local control of inoperable, recurrent or metastatic STS following the combined treatment compared to irradiation alone. A separate analysis of the survival time and local control was performed on patients who had surgery and were treated postoperatively with adjuvant razoxane and radiation therapy.

Design

This was a controlled randomized phase III study according to protocol H 10/78. It commenced in 1978 and was closed to entry in 1988. Treatment details, eligibility criteria, staging procedures and statistics may be found in the original article.

Patients

The median follow-up time was 7 years. The follow-up rate was 100% among the patients with radical resection and 89% among the patients with gross residual or metastatic disease.

Results

Response rates. Among 82 patients with gross disease, the treatment with radiotherapy and razoxane led to an increased response rate (CR+PR) compared to photon irradiation alone (74 vs. 49%). In Group A (n=40), the complete response rate was 26%. In Group B, the treatment with razoxane led to a complete response rate of 38%.

Progression during radiotherapy was seen in 8 of 40 patients in Group A and in only 1 of 42 patients in Group B. The median time to regrowth was 6 (0–86) months in Group A and 11 (1–49) months in Group B.

Local control. The local control rate was likewise improved by the addition of razoxane to the radiation therapy (64% vs. 30%; $p < 0.05$ in favour of Group B).

Survival. The metastatic process in STS did not seem to be influenced by the short treatment of razoxane itself. Therefore, large survival differences between Group A and B among the advanced or metastatic cases could not be expected. In fact, the overall median survival in Group A is 15 months (1.5 to 110+) and in Group B is 18 months (1.5 to 112+).

Between the patient groups treated postoperatively in an adjuvant form, there were no substantial differences in local control and survival.

Toxicity. The acute toxicity was somewhat higher in the sensitizer arm, but there was no difference in the occurrence of late complications.

Conclusions

Radiotherapy combined with razoxane led to higher objective response rates and improved the local control in inoperable, residual, or recurrent STS compared to radiotherapy alone. The combined treatment is a fairly well tolerated procedure at low costs. It can be recommended for inoperable primary STS or gross disease after incomplete resection.

Retrospective Comment

The rate of complete and partial responses (74%) achieved with radiotherapy given together with razoxane corresponds exactly with the data that were observed in several pilot and phase II studies in England and in our own previous pilot studies. In this controlled study, it was confirmed that the radioresponse of soft tissue sarcomas with razoxane given concomitantly during radiotherapy is superior to results achieved with radiotherapy alone. Thus, in terms of evidence based medicine, the level of evidence reaches the level I B.

An Angiometamorphic Drug for Angiosarcomas

A Study of Angiosarcomas of the Thyroid

Original article: Rhomberg W, Boehler F, Eiter H, Fritzsche H, Breitfellner G (2004) Treatment options for malignant hemangioendotheliomas of the thyroid. Int J Radiat Oncol Biol Phys 60(2):401–5

Synopsis

Background

Malignant hemangioendotheliomas (MHE) of the thyroid are rare tumors predominantly seen in areas with endemic goitre such as the Alpine regions. Occasionally, the disease is observed in the Andes and in the Caucasus [2] and recently also in other regions [3]. The estimated incidence of the disease is between 0.15 and 0.25 per 100,000 inhabitants per year for Western Austria. The tumour is regarded as radioresistant and its prognosis is reported to be dismal.

Objectives

To explore new treatment options for this disease, palliative or adjuvant radiotherapy was thought to be usefully combined with razoxane. The drug was chosen because it improved the radiosensitivity of sarcomas in several previous studies, and because it is able to normalize pathologic tumour blood vessels [4–6].

Characteristics of the Disease

MHE of the thyroid is a disease that has to be distinguished from other soft tissue sarcomas. In case of gross residual disease and/or distant metastases, a distinct disease entity is seen characterised by rapid weight loss, anaemia, tendency for haemorrhages and paraneoplastic syndromes like hypercalcemia, leukocytosis or an acute rise of blood sugar [1, 7–9]. As a rule, local control of the tumour is rare [9–11] and the prognosis is dismal. Only very early and radical surgery offers a chance for cure. In an earlier report on 14 evaluable patients the median survival was 2.4 months, only one patient survived 4 years [10].

Immunohistochemically, MHE of the thyroid is characterised by positive staining for factor VIII-related antigen, CD31, CD34, BMA 120, ulex europaeus agglutinin I, and Vimentin. Cytokeratin is seldom detectable in angiosarcomas, and as a rule, there is a loss of expression of thyroglobulin in the tumour cells.

Patients and Study Design

In an open prospective study, 12 patients with malignant hemangioendotheliomas of the thyroid were treated between 1982 and 1999. The treatment intention was to administer razoxane if radiation therapy was indicated. There were 8 males and 4 females with a median age of 67 years (range 55–81). All cases were immunohistochemically positive for factor VIII-related antigen and negative for thyroglobulin and cytokeratin staining. Five of the 12 patients were found to have been exposed to polymeric materials such as vinyl chloride, polyurethane and other polymeres, at their work place over a median time of 31 years.

Patients were referred from five different surgery departments of the western Austrian region Vorarlberg. Clear surgical margins were achieved in 5, microscopic residues were left in 3, and gross residual disease in 3 patients. One patient had an inoperable primary tumour. Postoperative radiotherapy was given to 8 cases, 6 of them received razoxane on radiation days, but 2 did not (they were randomised to radiation alone within an earlier soft tissue sarcoma study). Two patients with clear margins at surgery were observed only.

Radiation treatment commenced between day 12 and 23 after surgery in all but one of the patients. Daily single doses of 200 cGy at the IRCU point were given five times a week. The total tumour doses ranged between 54 and 65 Gy (median 60). One inoperable primary tumour was irradiated with 18 Gy only. The duration of the radiotherapy ranged from 1 to 13 weeks.

Two of the 6 patients mentioned above had razoxane combined with vindesine, and 2 patients received no radiation treatment after surgery but one of them took 2×125 mg razoxane for 5 days and then refused further treatment.

Results

Local tumour control was achieved in 11 of 12 patients (92%). The median survival time of all patients referred was 14 months (range, 0.5–196). If three patients with metastases at the time of referral were left out of the analysis, the median survival is 70 months (range 3–196). Five of the 12 patients survived longer than 5 years.

Postoperative irradiation of the thyroid region was given to 8 patients, 6 of them received razoxane. All 6 cases treated with razoxane remained locally controlled, among them an early local recurrence, two R-2 resections and one R-1 resection. Irradiation alone led to a local control in one patient with clear margins but not in another patient with gross residual disease.

Noteworthy is a complete regression of 2 lung metastases in a 72 year old man treated with the combination of vindesine, razoxane and radiotherapy. Interestingly, the two drugs alone resulted in a rapid cessation of haemorrhages from the lung due to the lung metastases. The patient is still in complete remission 120+ months after a 3-year maintenance therapy with vindesine and razoxane.

Some laboratory findings might be of interest: Factor VIII and factor VIII-associated antigen were increased in 3 of 4 cases. The values returned to nearly normal in 2 patients going into remission, and rose again in 1 patient after becoming resistant to the treatment. Among 5 patients with a fibrinogen analysis there were 3 increased values up to 682 μg/dl (normal values: 150–450 μg/dl). In these cases also a return to normal values was seen after successful treatment.

Toxicity

The tolerance to the combined modality treatment was good to fair. However, local chemo-radiation reactions of normal tissues have to be considered as the principal toxicity, e.g., dry desquamations of the skin and dysphagia due to radiogenic pharyngitis and esophagitis as well as focal pneumonitis. Dysphagia of grade 2 and 3 were noted in 4 of 8 cases which have been irradiated with at least 54 Gy. This led to interruptions and unplanned splits of the radiotherapy in some patients. No tube feeding was necessary but one patient received a gastric fistula because of esophageal stricture. The two patients with radiotherapy alone had practically no side effects and tolerated the treatment well. Among the patients treated with razoxane a rapidly reversible leukopenia of WHO grade 2 and 3 occurred in 2 patients.

Conclusions

This small series indicates that the course of the disease is not uniformly bleak and that the resistance to radiotherapy reported in the literature has to be questioned if the radiation treatment is combined with razoxane. It appears that adequate surgery together with rapid radiation therapy in combination with razoxane is able to improve the local control rate and thus perhaps to alter the natural history of this disease. In addition, the data offer new evidence of the occurrence of vinyl chloride induced angiosarcomas outside the liver.

Comment

The diagnoses have all been confirmed by immunohistochemistry, thereby excluding anaplastic thyroid carcinomas. Fibrinogen, factor VIII and factor VIII-related antigen in the serum proved to be potential markers to recognize phases of disease progression or remission. They could serve as surrogate markers during the follow up.

In this series the prognosis seems to be unexpectedly favourable with 5 long term survivors among 12 non-selected patients (42%). Local control was achieved in 11 of 12 patients. The only case of an uncontrolled local tumour growth was seen in 1 patient with gross residual disease which received radiotherapy alone (allocated to radiotherapy alone by randomisation within an earlier study). All 6 patients that were irradiated postoperatively together with razoxane remained locally controlled. This is by no means an obvious outcome if one considers the literature [10, 11]. Ladurner et al. underline the fact that there was no difference in the local control between their cases showing clear margins and those with microscopic or gross residual disease [10].

Razoxane is of special interest in this disease because of its proven influence on pathologic tumour blood vessels and its potential ability of a re-differentiation of those vessels – especially at a time when angiogenesis inhibitors are in the limelight and become increasingly successful.

Summarising the results of this small patient series (which is nevertheless the second largest series published), it appears that the very unfavourable natural history of this disease might be altered by photon irradiation together with the angiometamorphic substance razoxane, certainly as far as the local control at the thyroid is concerned.

References

1. Besznyak L, Besznyak H (1973) Haemangioendothelioma of the thyroid. Zentralbl Chir 98:132–4
2. Egloff B (1983) The hemangioendothelioma of the thyroid. Virchows Arch 400:119–42
3. Maiorana A, Collina G, Cesinaro AM et al (1996) Epitheloid angiosarcoma of the thyroid. Clinicopathological analysis of seven cases from non-Alpine areas. Virchows Arch 429(2–3):131–7
4. Hellmann K, Burrage K (1969) Control of malignant metastases by ICRF 159. Nature 224:273–5
5. Le Serve AW, Hellmann K (1972) Metastases and the normalization of tumor blood vessels by ICRF-159: a new type of drug action. Br Med J I:597–601
6. Salsbury AJ, Burrage K, Hellmann K (1974) Histological analysis of the antimetastatic effect of 1,2-Bis (3,5-dioxo-piperazin-1-yl)-propane. Cancer Res 34:843–9
7. Rhomberg W, Böhler FK, Eiter H et al (1998) Das maligne Hämangioendotheliom der Schilddrüse. Neue Resultate zur Pathogenese, Therapie und Prognose. Wien Klin Wochenschr 110(13–14):479–84
8. Sapinski H (1974) Das Hämangioendotheliom der Schilddrüse als Ursache eines Hämatothorax. Dtsch Med Wschr 72:93
9. Thaler W, Riccabona G, Riedler L et al (1986) Zum malignen Hämangioendotheliom der Schilddrüse. Chirurg 57:397–400
10. Ladurner D, Tötsch M, Luze T et al (1990) Das maligne Hämangioendotheliom der Schilddrüse. Pathologie, Klinik und Prognose. Wien Klin Wochenschr 102(9):256–9
11. Rösler H, Walther E (1984) Radiation therapy of the struma maligna. In: Heilmann HP (ed) Encyclopedia of medical radiology, vol XIX/part 5: radiation therapy of malignant tumours. Springer, Berlin, Heidelberg, New York, 36

The Radiation Response of Sarcomas by Histologic Subtypes: A Review with Special Emphasis Given to Results Achieved with Razoxane

Walter Rhomberg

(The original article was published in Sarcoma, Vol. 2006, Article ID 87367, Pages 1–9. DOI 10.1155/SRCM/2006/87367. All articles in Hindawi journals are open access which permits unrestricted use, distributions, and reproduction in any medium, provided the original work is properly cited.)

Abstract

Purpose: Relatively few results are available in the literature about the radiation response of unresectable sarcomas in relation to their histology. Therefore, an attempt was made to summarize the situation until the year 2006.

Materials and Methods: The report is based on a review of the literature and the author's own experience. Adult type soft tissue sarcomas, chondrosarcomas and chordomas were analysed. Radioresponse was mainly associated with the degree of tumor shrinkage i.e. objective responses. Histopathologic responses i.e. degree of necrosis, are only discussed in relation to radiation treatment reports of soft tissue sarcomas as a group.

Results: Radiation therapy alone leads to major responses in about 50% of lipo-, fibro-, leiomyo-, or chondrosarcomas. The response rate is less than 50% in malignant fibrous histiocytomas, synovial, neurogenic, and other rare soft tissue sarcomas. The response rates may increase up to 75% through the addition of radiosensitizers such as halogenated pyrimidines or razoxane, or by the use of high LET irradiation. Angiosarcomas become clearly more responsive if biologicals, angiomodulating and/or tubulin affinic substances are given together with radiation therapy. Razoxane is able to increase the duration and quality of responses even in difficult to treat tumors like chondrosarcomas or chordomas.

Conclusions: The available data demonstrate that the radioresponsiveness of sarcomas is very variable and dependent on histology, kind of radiation, and concomitantly given drugs. The rate of complete sustained remissions by radiation therapy alone or in combination with drugs is still far from satisfactory although progress has been made through the use of sensitising agents.

Introduction

Little is known about the quality and duration of objective responses in different sarcomata after definitive or palliative radiation therapy. The few reports on photon irradiation deal mainly with soft tissue sarcomas (STS) as a group [1, 2]. Radiation therapy alone leads to transient objective response rates up to 50% and local tumor control in about 30% of unresectable lesions with radiation doses in the order of 60–70 Gy [1–3]. Improvements in local control and response rates between 50 and 75% have been achieved by high LET irradiation [4–7], the use of radiosensitizers such as intravenous bromodeoxyuridine, iododeoxyuridine [8], razoxane [9], or hyperthermia. Slater et al. analyzed 72 patients with unresectable STS and found a relationship between local control rates and so called malignancy groups based on pathologic diagnosis [1]. The aim of the present review is to continue that analysis of the radioresponsiveness of the main histologic groups of sarcomas.

Materials and Methods

This review is based on the available literature and on the author's own institutional experience gained during studies of radiosensitising agents in the treatment of sarcomas from 1972 to 2004. The response to irradiation of unresectable primaries, recurrent and measurable metastatic lesions from adult type STS, chondrosarcomas and chordomas were analyzed, those from desmoids, rhabdomyosarcomas, osteogenic sarcomas and Kaposi sarcoma, were excluded.

Our radiation treatment was performed with high energy photons using shrinking field techniques. Most patients received 180–200 cGy per fraction, usually five times a week. The median total tumour dose was 58 Gy (range, 30–70) for the most frequent STS, 60 Gy (range, 54–65) for chondrosarcomas, and 63 Gy (range, 54–67 Gy) for chordomas. The sensitizer razoxane was given by mouth at a dose of 125 mg twice daily to selected patients according to study protocols. Drug treatment commenced 5 days before the first irradiation treatment and was continued on radiation days.

Definition of Radiation Response

Response to irradiation has to be seen from two aspects. Sarcomas may be radioresponsive or even radiocurable in terms of cell killing but not responsive in terms of shrinking of the mass [10, 11]. Brennan et al. wrote, 'this can be perplexing, as such masses could give the false impression that little has been accomplished, whereas in reality the mass is mainly made up of sterilized tumor cells and debris' [10]. The degree of tumor cell sterilisation (necrosis) may be of value in respect of local control and survival [12]. However, the assessment of the degree of necrosis necessitates a histopathologic investigation and is, therefore, mostly restricted to neoadjuvant preoperative studies. Moreover, considerable necrosis may preexist in many sarcomas before any treatment has been given. To be clinically useful, the degree of pathologic necrosis should reach some 95%. This value was found to be an independent prognostic factor for local control and survival [12]. Willett et al. found that moderate dose preoperative radiotherapy resulted in 80%-necrosis in 21 of 27 specimens. Grade and size (>10 vs. <10 cm) were important predictors of response to radiotherapy independent of histologic type [13].

In this review, radiation response is mainly concerned with the extent of objective tumor shrinkage defined according to standard criteria. A complete response (CR) required the complete disappearance of a tumour both clinically and by radiographic imaging. A partial regression (PR) meant a reduction of the initial tumour volume by more than 50%, and disease progression (P) was assumed if the pretreatment tumour volume increased over 25% during or shortly after the end of the radiotherapy. 'No change' (NC) described a response between PR and progression. Local tumour control was defined as freedom from symptoms and no regrowth of the tumor at the site of irradiation as long as the patient survived.

Definitions of tumour responsiveness are often a matter of debate. A proposal could be as follows: A sarcoma type is regarded as radioresponsive if the majority of tumours show an objective regression with a CR rate up to 25% and proven survival benefits in due to adjuvant irradiation. Sarcomas were designated moderately responsive if less than 50% of the tumours reduced in size in terms of a partial and occasional complete regression. Resistant sarcomas are tumors with a major objective response rate below 25%, or if the majority of them exhibit progressive disease within 3 months after the start of radiotherapy.

Radiation Response by Histologic Subtype

Fibrosarcomas

In the past, this subtype of STS was more frequently diagnosed than today since it probably included some cases of malignant fibrous histiocytoma or gastrointestinal stromal tumors. There is controversy about the radioresponses of these tumours. Fibrosarcomas were largely regarded as radioresistant, especially the subunit of dermatofibrosarcoma protuberans. This, however, was recently questioned at least for the adjuvant setting [14]. In addition, almost 40 years ago, Windeyer et al. described a satisfactory local control by radiotherapy in the adjuvant setting, and an objective response rate of 50% even in inoperable, recurrent or residual fibrosarcomas [15].

Fibrosarcomas were one of the first entities that were irradiated together with the radiosensitizer razoxane (ICRF 159) in England. Ryall et al. reported an objective response rate of 73% of STS by radiotherapy combined with razoxane. The majority of their cases were fibrosarcomas [16]. We found an overall response rate of 70% in a pilot series of 10 patients with fibrosarcomas treated with radiotherapy and razoxane between 1973 and 1977. The median response duration was 8 months.

But later, among 8 chemonaive patients with fibrosarcomas irradiated together with razoxane, there were 5 CR and 3 PR, and the median duration of response rose to 15 months emphasizing the influence of the pretreatment (e.g. in Fig. 2.5a, b). Several times a striking liquefaction of the tumors could be seen clinically. Subgroup analysis within a randomized study [9] showed that 6 of 7 patients with a fibrosarcoma irradiated together with razoxane achieved an objective response (3 CR, and 3 PR, respectively), confirming the results of our pilot study and the early results of Ryall et al. [16]. Thus, the rate of major responses in our 17 patients with measurable fibrosarcomas irradiated together with razoxane was 76%. For comparison, among 5 measurable cases randomized to be treated with radiotherapy alone we saw 2 PR and 3 unchanged tumours. Taken together, there seems to be a clear improvement of the radiation response rate in the group of fibrosarcomas treated with the radiosensitizer razoxane. Data on a doxorubicin based chemo-radiotherapy are not available.

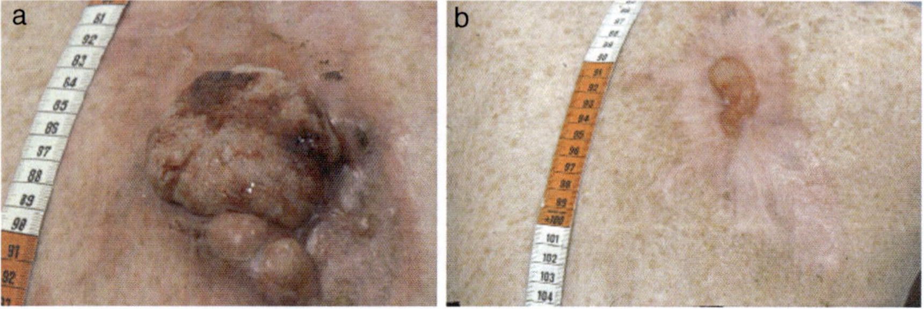

Fig. 2.5 (**a**) Fibrosarcoma of the right shoulder, before treatment. Male, 73 years. (**b**) Regression of the tumor after irradiation with 50 Gy + razoxane

Liposarcomas

Liposarcomas are known to belong to the more sensitive varieties of STS but data on quantitative measurements of remissions are sparse. One of the few studies that provided objective data of the radiation response of myxoid liposarcoma (MLS) was recently published by Pitson et al. [17]. In a series of 16 MLS tumour specimens the mean pre-treatment and post-treatment volume of the MLS tumors was 415 and 199 cm^3 corresponding to a reduction in the median tumour volume of 59%. This result achieved by radiotherapy alone was in sharp contrast to 16 control cases of malignant fibrous histiocytomas. In our own experience, among 9 patients with liposarcomas treated with radiotherapy alone there were 2 CR, 2 PR, 2 NC and 2 progressive diseases, respectively, one patient was not assessable. The objective response rate of 50% corresponds approximately to that of the MLS series of Pitson et al. [17].

These 9 patients were part of a randomized study published in 1996 [9]. In that study, another 9 patients were randomised to receive razoxane together with radiation treatment. The median radiation dose at the tumour site was 55 Gy (range, 40–60 Gy). Five of 8 assessable patients had a complete response (CR), and 3 had a partial response (PR) corresponding to a near 100% response rate. One patient was not assessable because of a gross residual unmeasurable tumour. Figure 2.6 shows a late subgroup analysis of 16 assessable liposarcomas treated within that randomized study. The median follow-up time of the 16 patients (including the patients who died) was 29 months (range, 1.5 to 110+ months).

Leiomyosarcoma

Leiomyosarcomas (LMS) occur frequently in the uterus, bowel, vascular tissues, and less commonly in somatic soft tissue or bone. LMS was proven to be the most frequent secondary tumour induced by radiation. Apart from case reports, there are no large series on the radio-responsiveness of LMS. Treatment results are usually restricted to the adjuvant setting. With the exception of LMS of the uterus [18, 19] there is agreement about the necessity and the benefit of postoperative radiation

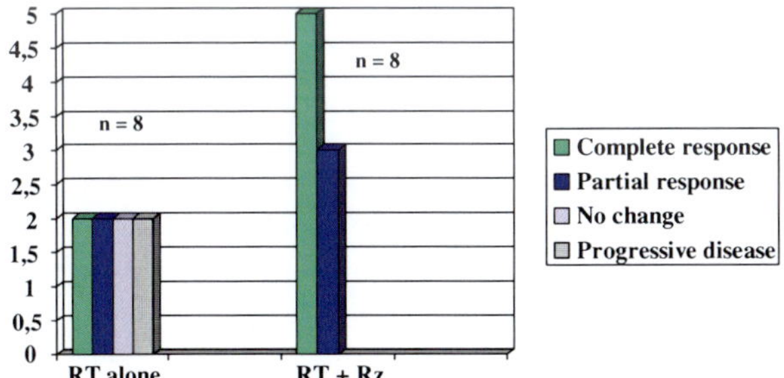

Fig. 2.6 Response of 16 assessable liposarcomas randomized between radiotherapy alone or together with razoxane

therapy in LMS. In sarcomas of the uterus, LMS were usually compared to other entities like endodermal stromal sarcomas (ESS) and mixed mesodermal sarcomas (MMS). The definitive radiotherapy of uterine leiomyosarcomas is not rewarding. Based on a review of the literature, none of 10 cases of stage II–IV LMS survived 5 years, the same was true also for the MMS and ESS varieties [19]. By combining external beam radiotherapy and brachytherapy, and total doses of 80 Gy, 2 CR of ESS by irradiation alone were observed by Weitmann et al. [20]. Perhaps, LMS of the prostate may represent a more radioresponsive entity where cures are possible in the odd case [21, 22].

We treated 9 patients with measurable LMS (primaries and metastases). Among 4 cases treated with radiotherapy alone 2 PR, 1 NC and 1 progression were seen leading to an objective response rate of 50%. The duration of the partial responses was 3 and 15 months. When razoxane was added to the radiation treatment in 5 patients, 2 CR and 3 PR with a median duration of 22 months was observed, although one female received salvage surgery. Of course, the small numbers of patients precludes firm conclusions, but the tendency to improve the radiation response by the combination with razoxane is seen also in this type of STS.

Malignant Fibrous Histiocytoma

Malignant fibrous histiocytoma (MFH) is a pleomorphic neoplastic disease with uncertain histological origin. Patients have an absolute 3-year survival rate between 43 and 72% [23, 24]. The prognosis is strongly dependent on tumour size and the depth of a muscular tumour [24, 25]. Complete surgical resection at the time of primary tumour presentation is the treatment of choice, although adjuvant radiation therapy plays an important role in achieving better local control [25–27]. Responses to primary single and combination chemotherapy occur in about 30% of the patients [24]. Most authors feel that radiotherapy alone or chemotherapy are not effective in MFH.

A literature survey in 1981 revealed that local control was obtained in only 2 of 16 patients treated by radiotherapy alone but measurement of tumour response or details of the radiation treatment were not well documented [27]. A study of 16 patients with measurement of the radiation response in MFH came from the Princess Margaret Hospital in Toronto. Fourteen tumours were Grade III–IV, the median time from the pre-RT image to the start of treatment was 25 days (range, 4–50 days). All patients received 50 Gy in 2-Gy fractions. Tumor necrosis was quantified in the pathology report after surgery in 10 MFH tumors. A median of 80% of the MFH tumors was necrotic. The mean pre-treatment and post-treatment volume of the 16 MFH's was 264 and 273 cm^3, respectively ($p = 0.804$) [17]. Thus, the radioresponsiveness of MFH tumors was clearly less than in 16 comparable cases with myxoid liposarcomas from the same institution although the observed degree of necrosis was not different [17].

Among 4 of our cases treated with radiotherapy alone, there were 1 complete response, 1 partial response and 2 unchanged tumours. Another 4 patients with measurable tumours received irradiation with razoxane, resulting in 2 PR and 2

NC with a median duration of 10.5 months. Two further cases treated with razoxane had incomplete resections with masses that were not measurable, but they remained locally controlled. Summarizing the available data, the radioresponsiveness of MFH in terms of tumor shrinkage seems to be limited. The value of the radiosensitizing agent razoxane remains unclear.

Synovial Sarcoma

This entity accounts for approximately 8% of all STS, and it typically affects the extremities. The translocation (X;18) has been noted in more than 90% of cases. Synovial sarcomas (SS) have a higher level of chemosensitivity compared with other STS [28]. They appear to be particularly sensitive to high dose ifosfamide chemotherapy [29].

In 1971, Kagan et al. observed different responses (CR, PR, and NC) in one patient with multiple metastases of a SS, dependent on the time/dose/fractionation scheme [30]. An early review from Carson et al. in 1981 indicates the existence of only a few reports on the use of radiotherapy alone or in conjunction with excision [31]. The authors reported on their own 36 patients with SS, and underlined the unfavourable outcome as soon as the patients developed infiltrative tumors of T-3 stage, or if the tumors showed poorly differentiated histologies. Among 3 macroscopic incomplete resections in primary SS which were irradiated with 40, 58, and 82 Gy, respectively, 2 died of local recurrence after 6 and 24 months, and 1 died from lung metastasis after 6 months, the primary being controlled. Measurement of the response was not performed.

Twenty years later, the question as to objective response rates following irradiation is still not solved but meanwhile there is a consensus of the need for radiotherapy following marginal resection. A large series of 271 synovial sarcomas was recently analysed by Ferrari et al. [28]. The authors found 10 cases which remained unresectable after some pre-treatment (mostly with chemotherapy), with radiotherapy being the only local treatment thereafter. Only 1 of 10 survived, response data were not given.

Our own experience comprises 7 patients. Three patients received radiotherapy alone, there were 1 histopathologically confirmed complete response (bone lesion 5 × 5 cm; 60 Gy), 1 partial response (primary at the knee; 55 Gy), and 1 progressive disease at a R-2 resected primary. The patients survived 60, 18, and 12 months, respectively. Four patients were treated with radiotherapy and razoxane by randomisation with 1 PR and 3 NC. However, 2 of the unchanged lesions were metastases that received only 30 and 36 Gy.

Sarcomas of the Blood Vessels

As a rule, *angiosarcomas* (AS) have a worse outcome in comparison to other STS. They usually present with high grade histology and have a high propensity for local recurrence and distant metastases. However, biology and prognosis are also dependent on the site of origin. Thus, tumours originating on the scalp, face or the thyroid imply a dismal prognosis compared with those in other locations [32–35]. STS in these locations are associated with 5-year survival times between 0 and 13%, especially among patients who were irradiated with clinically evident disease [32, 33, 36, 37]. Present studies including all tumor sites, however, indicate a 5-year survival of 45–60% in all angiosarcoma patients treated with curative intent [32, 37].

Most frequently, radiotherapy is given as an adjunct to surgery in the setting of minimal residual disease. Few data exist about primary or definitive radiotherapy of AS. Single case reports repeatedly describe complete durable responses with radiotherapy alone [38, 39] but the true incidence of such an outcome remains unknown, and it is probably lower than case reports suggest. In a study by Mark et al. [40], only 1 of 9 patients (11%) treated with RT $\pm$ CT was rendered free of disease. We treated 2 patients with radiotherapy alone (by randomization): One patient with a recurrent angiosarcoma of the thyroid measuring 7.5 $\times$ 4 cm had progressive disease despite of a dose of 60 Gy given in 30 fractions. A partial response was seen in a metastatic lesion on the forehead from an AS of the breast irradiated with 14 Gy only.

Angiosarcomas emerge as chemoresponsive tumors [41–43]. Radiation with innovative drug regimens, e.g. cytokines, antiangiogenic substances, and others, seems to become a promising treatment option [44–46]. Ohguri et al. used rIL-2 immunotherapy together with radiotherapy (median 70 Gy, single fractions 2–3 Gy) in 20 patients with angiosarcomas of the scalp, 15 of them had no surgery. The median overall survival time was 36 months, the median local recurrence-free survival 11 months. Local recurrences were observed in 7 patients only (35%) [44]. These results were more favorable than any previous report in the reviewed literature [44]. Unfortunately, objective response data are lacking in the article.

We used radiotherapy together with razoxane in 3 patients with unresectable AS on the scalp, the heart, and the lung. The radiation doses were 66, 30, and 40 Gy. All 3 patients responded partially, 2 with a subtotal regression of their primaries. Distant metastases were the limiting event in all 3 cases associated with survival times of 15, 9, and 8 months.

In an ongoing study of the Austrian Society of Radiooncology (ÖGRO), using radiotherapy in combination with razoxane and vindesine, a tubulin affinic antiinvasive substance [47], 3 of 7 patients with measurable angiosarcomas showed a CR, 2 had a partial response, 1 did not change (minor response) and 1 had progressive disease. The complete responses occurred at unfavourable sites like the thyroid, chest wall, and scalp. No recurrences were seen in the responding patients after a median follow up time of 17+ months for the living patients (Wink et al., preliminary observations).

The value of radiotherapy in the management of *hemangiopericytomas* (HPC) has changed. Earlier discussed as controversial, the majority of authors now favour surgery and post-operative radiotherapy as initial procedure [48, 49]. Again, little is known about the responses in case of unresectable, gross residual, or recurrent tumors. A review by Schier et al. covering a period from 1942 to 1987 identified 14 cases of HPC that were treated with radiotherapy alone. There was a fatal outcome in 91% of the cases, mostly around 3 years. Nine cases received no treatment at all:

All of them died, 85% after a mean time of 2–3 weeks only [50]. In the experience of the University of Iowa College of Medicine, only one of seven patients initially treated with surgery alone has remained continuously free of disease. In contrast, all four patients treated with surgery and postoperative radiation therapy have remained alive with NED despite the fact that three of these patients had gross residual tumour at the time of irradiation [49]. Altogether, 10 of their 15 patients received radiation therapy at some stage of the management. Local tumour control was achieved at all sites receiving greater than 5,500 cGy. Similar results were obtained at the M.D. Anderson Hospital [48].

In our institution, a recurrent HPC on the meninges (tumor volume about 4–5 ml) disappeared completely after radiotherapy alone with 60 Gy. The patient is alive without evidence of disease 20 years later. Adding razoxane to the radiation treatment, has resulted in 3 of 3 patients showing a partial response for a median time of 10 months. Adding razoxane and the vinca alkaloid vindesine to the radiotherapy, has given partial responses in 3 of 3 assessable tumors for a median time of 24+ months; 1 additional patient with non measurable residuals at the spine remained recurrence-free 40+ months.

Chondrosarcomas

Chondrosarcomas (CS) are frequently termed as radioresistant but measurements of the response are rarely presented. Location as a factor of prognosis has to be considered in the judgement of the disease. For instance, chondrosarcomas of the larynx must be seen as an own entity with a favourable prognosis [51, 52]. Likewise, chondrosarcomas of the base of the skull show favourable outcomes if protons or heavy ions were used for irradiation [53, 54]. In contrast to the majority of the literature, however, there are single reports of high local control rates of incompletely resected chondrosarcomas even at the spine and pelvic bones [55]. The authors assumed that the impression of frequent radioresistance is due to a tendency of these tumors to regress slowly and the persistence of skeletal destructions often seen in X-rays. In addition, low radiation doses were used in some earlier reports.

More exact data on the response of chondrosarcomas are found in studies which compared conventional photon irradiation with neutrons. Laramore et al. reported an objective response rate (CR+PR) of 33% with photons vs. 49% with neutrons [56]. In the experience of McNaney et al. only 1 of 7 patients treated with photons alone remained locally controlled [57].

Chondrosarcomas are regarded as not responsive to cytotoxic chemotherapy regardless whether it has been given alone or in combination with radiation treatment. The only report of an improvement of the radioresponsiveness of chondrosarcomas by chemical substances came from England: In 1979 Ryall et al. reported a complete and partial response rate of 62% when radiotherapy was combined with the sensitizer razoxane [58].

Stimulated by these observations, we undertook a long term study to explore the radioresponsiveness of these rare tumours using the protocol of Ryall et al. Between 1984 and 2003, 13 patients with chondrosarcomas were irradiated together with razoxane. Eight patients had unresectable primaries or recurrences, and 5 received postoperative irradiation (one R-0 and four R-1 resections). The median radiation dose was 60 Gy (range, 54–65 Gy), the median dose of razoxane was 8.5 g (range, 6–15 g).

Among 8 unresectable or recurrent chondrosarcomas with measurable disease, there were 1 complete and 5 partial regressions, 2 tumors did not change. The overall response rate was 75%, the median duration of response 23 months. Although 4 of these 8 patients were not locally controlled, the median survival time is not yet reached and lies presently at 45+ months (Table 2.2). Three of the 4 patients without clear surgical margins remained locally controlled. They survived up to now 12, 21+, 48+, and 160+ months, respectively. One patient with clear margins remained free of disease 87+ months, he had no further follow up. Overall, local control was achieved in 7 of 12 patients who were not radically resected.

Apart from the patients treated with razoxane, 2 other patients received radiation therapy alone by randomisation. One of them had no clear margins at surgery, and the other had a recurrence, size 8×8 cm. The recurrent tumour progressed even during the radiotherapy. The two patients survived 7 and 9 months, which was the shortest survial of all of our patients with CS.

Chordoma

Conventional megavoltage irradiation with a median dose of 50 Gy leads to a local control rate of 27% in chordomas [11], while objective tumor regressions are rarely described. For instance, in a recent series of 18 patients with skull base chordomas treated with spot scanning proton beam therapy, no complete or major (>50%) response was observed [59]. The median time to recurrence or symptoms is 3.5 years in patients with residual disease [60] but sometimes also less [61, 62]. Freedom of symptoms after 10 years was seen in less than 10% of the cases from Zurich [63], and no patient survived 10 years among 4 patients receiving radiotherapy alone in a series of 21 patients of Keisch et al. [64]. Therefore, several authors came to the conclusion that inoperable or gross residual chordomas can rarely be cured by conventional irradiation with photons [60, 61, 63]. By using protons or heavy ions the local control rates rose to around 60–70% [6, 59, 65].

We treated 5 consecutive patients with unresectable chordomas at the spine (3 cases) and the base of the skull (2 cases) with radiation and razoxane [66]. The more favourable chondroid variant of chordomas was excluded by a secondary pathology review in all but one case. The median radiation dose was 63 Gy (range, 54–67 Gy), the median total dose of razoxane was 7.6 g (range, 5.4–11.5 g).

Among 4 measurable tumours there were 2 CR, 1 PR, and 1 NC. All 5 patients survived 5 years and remained locally controlled for 5, 7+, 6.4, 13+, and 15+ years. After a potential median follow up time of 10 years, 3 of 5 patients are alive without evidence of disease proven by CT and MR imaging. The treatment had little toxicity. Mucosal reactions were predominant as the local side effects, 2 of the 5 patients showed a grade 3 leukopenia (WHO) due to razoxane. So far, no late toxicity at the CNS or the optic nerve was observed.

Table 2.2 Results of radiotherapy with razoxane in 8 patients with inoperable primary or recurrent chondrosarcomas

		Definitive irradiated tumors			Results of radiotherapy + Rz				
#	Age/gender	Primary	Size	Grade	Response	Duration (months)	Local control	Survival (months)	Remarks
1	61/m	Os ileum	20 × 15 cm	G 1	PR	20	No	38	Died from local recurrence
2	58/m	Larynx	5 × 4 cm	G 2	PR	85+	Yes	85	Intercurrent death
3	74/f	Groin	17 × 11 cm	G 3	PR	22+	Yes	22	Lung metastasis
4	80/f	Knee	15 × 14 cm	G 2	PR	90	Yes §	90	§ Secondary amputation
5	24/f	Orbita	0.8 × 0.8 cm	G 2	CR	48	No	50+	Salvage surgery after 48 months
6	65/m	Sternum	3 × 2 cm	G 1	NC	25	No	46+	Salvage surgery after 25 months
7	44/f	2nd Rib	8 × 8 cm	G 2	NC	22+	Yes	22	No follow up last 5 months
8	81/m	Humerus	6 × 5, 4 × 3 cm	G 2	PR	16	No	24+	Regrowth after 16 months

CR, complete response; PR, partial response; NC, no change of tumor size.

Neurogenic Sarcomas/Malignant Schwannomas

Because of the rarity of these tumors there is no data on larger series concerning their radioresponsiveness. Five of our patients with malignant schwannomas received definitive radiation treatment, 2 with radiation alone, and 3 with radiation and razoxane. With radiotherapy alone we observed one unchanged and one progressive tumor, with radiotherapy and razoxane there was 1 PR (25 Gy) and 2 unchanged tumors. So far, the radioresponsiveness of neurogenic sarcomas cannot be judged, for the present it seems to be low. This is in accordance with the experience of Raney et al.: None of 12 children with gross residual disease survived 3 years; and the authors stress the need for more effective treatment programs especially for those with unresectable tumors [67].

Miscellaneous Other Sarcomas

Current knowledge concerning definitive radiotherapy for various rare sarcoma entities such as clear cell, alveolar soft part, or epitheloid sarcomas is restricted to occasional case reports. Therefore, no clear picture is evolving from those data but there is the impression of a rather limited radioresponsiveness. Ferrari et al. on behalf of the Italian and German Soft Tissue Sarcoma Cooperative Group reported on 28 pediatric patients with clear cell sarcoma. Eight patients received radiotherapy postoperatively. The only two irradiated patients with gross residual disease after surgery 'did not respond significantly'. The authors concluded that radiotherapy may control microscopic residual disease after surgery, chemotherapy is ineffective and the prognosis is unfavourable for patients with unresectable and large tumours [68]. Shimm and Suit treated 2 epitheloid sarcomas with radiotherapy alone. Both tumours did not change in their size, only 1 was locally controlled, and both cases developed distant metastases [69]. A partial remission and an unchanged tumor was observed in two of our own cases with alveolar soft part sarcomas treated with radiotherapy and razoxane. Both cases remained locally controlled after radiation doses of 40 and 30 Gy but died from distant metastases after 9 months and 6 years.

Discussion

The terms radioresponsive and radiosensitive are synonymous in everyday usage. It will be difficult to associate objective responses exclusively with <radioresponsiveness> and the degree of tumor necrosis with the generic term <radiosensitivity>. In reporting on definitive radiotherapy, it should however be sufficient to describe the quality and degree of objective responses and their duration, local control and survival. Those data would speak for themselves and cover radiosensitivity (histopathologic necrosis) and radioresponsiveness (shrinking of a mass) as well. Knowledge about the radioresponsiveness of sarcomas is of value for treatment decisions, e.g., a good response may more likely permit the conservative surgical excision of tumours of borderline resectability, or for having data to be compared with results from novel treatments.

The radioresponsiveness of sarcomas as a group is moderate at best. About 40–50% of the tumours show transient objective responses to radiotherapy alone, complete responses are less frequent. Some reports deal with histopathologic response rates, i.e. the degree of tumour necrosis [12, 13, 70, 71], but the prognostic significance of a 60–90% tumor necrosis, values most frequently observed, remains unclear. As mentioned, several measures led to an increase of local control and objective responses in sarcomas. Interestingly, the use of contemporary cytotoxic chemotherapy represented by doxorubicin-based regimens together with concomitant radiotherapy seems to be of low or modest efficacy. A study with neoadjuvant doxorubicin-based chemo-radiation in STS revealed a tumour-downstaging in 18% of patients only although there was a survival gain [72]. Another work reported no radiographic partial response or disease progression among 27 patients but there were some histologic responses in the series [71]. The amount of necrosis was dependent on the doxorubicin dose given together with radiotherapy. Eilber et al. reported on neoadjuvant combined treatments for high-grade extremity STS. Major necrosis increased to 48% with the addition of ifosfamide as compared to 13% of the patients in all other protocols combined [12]. In a study using preoperative radiotherapy together with the MAID regime (MAID = mesna, adriamycin, ifosfamide, dacarbazine), DeLaney et al. saw no clinical complete responses and 10.6% partial responses among 48 patients, 36 patients (75%) did not respond [70]. Six patients (13%) had radiographic evidence of 'progression' by tumor measurements on the MRI scans. Most of the tumour in these 6 patients, however, showed pathologically confirmed necrosis of various degree. The authors suggested that the radiographic progression represented swelling of tumor secondary to the osmotic effect of necrosis [70].

This overview indicates that the radiation response of sarcomas is dependent on the histopathology and the addition of chemical modifiers of the radiation response. Radiotherapy alone leads to major but transient responses in about 50% of lipo-, fibro-, leiomyo-, or chondrosarcomas. The response rate is less than 50% in malignant fibrous histiocytomas, synovial, neurogenic, and other rare entities. No definite conclusions are allowed for the latter forms since the available data are sparse. The response rates may be increased to 75% by the use of chemical modifiers. In particular, angiosarcomas showed promising responses if biologicals, angiomodulating and/or tubulin affinic substances are given together with radiation therapy.

The radiosensitizing agent razoxane is able to increase the duration and quality of responses in fibro-, lipo- and leiomyosarcomas compared to irradiation alone. Even in difficult to treat tumors such as chondrosarcomas or chordomas, photon irradiation together with razoxane induces objective responses in the majority of patients. Compared to data from the literature, this treatment combination offers an alternative to high LET irradiation. Although the small numbers of chordomas precludes far reaching conclusions, the fact of non selected cases, the longer follow up, and the favourable responses would make it worthwhile to test razoxane together with or against a radiation therapy with protons or heavy ions.

Razoxane is a largely neglected radiosensitizer which has an interesting spectrum of modes of action. It is of value in the treatment of STS because of its proven radiosensitizing effects [9, 16, 73] and its potential to normalize pathologic tumour blood vessels [35, 74, 75]. This angiometamorphic mode of action commends the drug especially for its use in angiosarcomas. In addition, the drug induced a growth rate slowdown of transplanted tumors [76], and the development of distant metastases to the lungs were completely suppressed in preclinical trials [74, 75, 77]. Razoxane is also antiinvasive [78], and it inhibits the topoisomerase II [79]. Unfortunately, since 2004 razoxane has been discontinued by Cambridge Laboratories (UK). It is to be hoped that another pharmaceutical company will shortly take up again the unique possibilities of the drug. Meanwhile, the use of dexrazoxane could be a reasonable alternative. From our experience with radiotherapy plus razoxane in STS it would seem to be mandatory in any phase III trial of a new combination to use radiation and razoxane as the control arm.

This report represents a first inventory of radiation responses in a variety of sarcomas. It might be of value to those who intend to analyse new radiosensitizers or novel combination chemotherapies in conjunction with radiation treatment. Simple pilot trials with new drugs are needed since the rates of complete sustained regressions in different sarcomas achieved by the concomitant use of chemical modifiers and irradiation are still far from satisfactory.

References

1. Slater JD, McNeese MD, Peters LJ (1986) Radiation therapy for unresectable soft tissue sarcomas. Int J Radiat Oncol Biol Phys 12:1729–34
2. Tepper JE, Suit HD (1985) Radiation therapy alone for sarcoma of soft tissue. Cancer 56: 475–9
3. Gilbert HA, Kagan R, Winkley J (1975) Soft tissue sarcomas of the extremities: their natural history, treatment, and radiation sensitivity. J Surg Oncol 7:303–17
4. Cohen L, Hendrickson F, Mansell J et al (1984) Response of sarcomas of bone and soft tissues to neutron beam therapy. Int J Radiat Oncol Biol Phys 10:821–4
5. Kamada T, Tsujii H, Tsuji H et al (2002) Efficacy and safety of carbon ion radiotherapy in bone and soft tissue sarcomas. J Clin Oncol 20:4466–71
6. Suit HD, Goitein M, Munzenrider J et al (1990) Increased efficacy of radiation therapy by use of proton beam. Strahlenther Onkol 166:40–4
7. Wambersie A (1982) The European experience in neutron therapy at the end of 1981. Int J Radiat Oncol Biol Phys 8:2145–52
8. Kinsella TJ, Glatstein E (1987) Clinical experience with intravenous radiosensitizers in unresectable sarcomas. Cancer 59:908–15
9. Rhomberg W, Hassenstein EOM, Gefeller D (1996) Radiotherapy vs. radiotherapy and razoxane in the treatment of soft tissue sarcomas: final results of a randomized study. Int J Radiat Oncol Biol Phys 36:1077–84
10. Brennan MF, Alektiar KM, Maki RG (2001) Sarcoma of the soft tissue and bone. In: De Vita VT Jr, Hellman S, Rosenberg SA (eds) Cancer principles & practice of oncology, 6th edn. Lippincott Williams & Wilkins, Philadelphia, p 1865
11. De Vita VT Jr, Hellman S, Rosenberg SA (eds) (1993) Cancer. Principles & practice of oncology, 4th edn. JB Lippincott, Philadelphia, pp 1470, 1726–7
12. Eilber FC, Rosen G, Eckardt J, Forscher C et al (2001) Treatment induced pathologic necrosis: a predictor of local recurrence and survival in patients receiving neoadjuvant therapy for high-grade extremity soft tissue sarcomas. J Clin Oncol 9:3203–9

13. Willett CG, Schiller AL, Suit HD, Mankin HJ, Rosenberg A (1987) The histologic response of soft tissue sarcoma to radiation therapy. Cancer 60:1500–4

14. Ballo MT, Zagars GK, Pisters P, Pollack A (1998) The role of radiation therapy in the management of dermatofibrosarcoma protuberans. Int J Radiat Oncol Biol Phys 40:823–7

15. Windeyer B, Dische S, Mansfield CM (1966) The place of radiotherapy in the management of fibrosarcoma of the soft tissues. Clin. Radiol 17:32–40

16. Ryall, RDH, Hanham IWF, Newton KA et al (1974) Combined treatment of soft tissue and osteosarcomas by radiation and ICRF 159. Cancer 34:1040–5

17. Pitson G, Robinson P, Wilke D et al (2004) Radiation response: an additional unique signature of myxoid liposarcoma. Int J Radiat Oncol Biol Phys 60:522–6

18. Belgrad R, Elbadawi N, Rubin P (1975) Uterine sarcoma. Radiology 114:181–8

19. Salazar OM, Dunne ME (1980) The role of radiation therapy in the management of uterine sarcomas. Int J Radiat Oncol Biol Phys 6:899–902

20. Weitmann HD, Knocke TH, Kucera H, Pötter R (2001) Radiation therapy in the treatment of endometrial stromal sarcoma. Int J Radiat Oncol Biol Phys 49:739–48

21. Pavlov AS, Kostromina KN, Simakina EP et al (2004) Cure of a patient with prostatic leiomyosarcoma with radiotherapy. Urologiia 4:61–3

22. Sakano Y, Yonese J, Okubo Y et al (1995) Leiomyosarcoma of the prostate: a case report of remission for 9 years by radiotherapy. Hinyokika Kiyo 41:629–32

23. Jin GP, Zhao M, Qi JX, Jiang HG, Yu SG (2003) Malignant fibrous histiocytoma of head and neck: clinical analysis of 21 cases. Ai Zheng 22:523–5

24. Shinjo K (1994) Analysis of prognostic factors and chemotherapy of malignant fibrous histiocytoma of soft tissue: a preliminary report. Jpn J Clin Oncol 24:154–9

25. Peiper M, Zurakowski D, Knoefel WT, Izbicki JR (2004) Malignant fibrous histiocytoma of the extremities and trunk: an institutional review. Surgery 135:59–66

26. Belal A, Kandil A, Allam A et al (2002) Malignant fibrous histiocytoma: a retrospective study of 109 cases. Am J Clin Oncol 25:16–22

27. Reagan MT, Clowry LJ, Cox JD, Rangala N (1981) Radiation therapy in the treatment of malignant fibrous histiocytoma. Int J Radiat Oncol Biol Phys 7:311–15

28. Ferrari A, Gronchi A, Casanova M et al (2004) Synovial sarcoma: a retrospective analysis of 271 patients of all ages treated at a single institution. Cancer 101:627–34

29. Rosen G, Forscher C, Lowenbraun S et al (1994) Synovial sarcoma. Uniform response of metastases to high dose ifosfamide. Cancer 73:2506–11

30. Kagan AR, Friedman NB, Jaffe HL (1971) Synovial sarcoma: an analysis of its response to radiation. J Surg Oncol 3:379–85

31. Carson JH, Harwood AR, Cummings BJ et al (1981) The place of radiotherapy in the treatment of synovial sarcoma. Int J Radiat Oncol Biol Phys 7:49–53

32. Holden CA, Spittle MF, Jones EW (1987) Angiosarcoma of the face and scalp, prognosis and treatment. Cancer 59:1046–57

33. Ladurner D, Tötsch M, Luze T et al (1990) Das maligne Hämangioendotheliom der Schilddrüse. Pathologie, Klinik und Prognose. Wien Klin Wochenschr 102:256–9

34. Pawlik TM, Paulino AF, Mcginn CJ et al (2003) Cutaneous angiosarcoma of the scalp. A multidisciplinary approach. Cancer 98:1716–26

35. Salsbury AJ, Burrage K, Hellmann K (1974) Histological analysis of the antimetastatic effect of 1,2-bis (3,5-dioxo-piperazin-1-yl)-propane. Cancer Res 34:843–9

36. Morrison WH, Byers RM, Garden AS, Evans HL, Ang KK, Peters LJ (1995) Cutaneous angiosarcoma of the head and neck. A therapeutic dilemma. Cancer 76:319–27

37. Ward JR, Feigenberg SJ, Mendenhall NP et al (2003) Radiation therapy for angiosarcoma. Head Neck 25:873–8

38. Graham WJ, Bogardus CR (1981) Angiosarcoma treated with radiation therapy alone. Cancer 48:912–4

39. Pötter R, Baumgart P, Greve H et al (1989) Primary angiosarcoma of the heart. Thorac Cardiovasc Surg 37:374–8

40. Mark RJ, Poen JC, Tran LM, Fu YS, Juillard GF (1996) Angiosarcoma. A report of 67 patients and a review of the literature. Cancer 77:2400–6

41. Eiling S, Lischner S, Busch JO et al (2002) Complete remission of a radio-resistant cutaneous angiosarcoma of the scalp by systemic treatment with liposomal doxorubicin. Br J Dermatol 147:150–3

42. Fata F, O'Reilly E, Ilson D et al (1999) Paclitaxel in the treatment of patients with angiosarcoma of the scalp or face. Cancer 86:2034–7

43. Vogt T, Hafner C, Bross K et al (2003) Antiangiogenic therapy with pioglitazone, rofecoxib, and metronomic trofosfamide in patients with advanced malignant vascular tumors. Cancer 98:2251–6

44. Ohguri T, Imada H, Nomoto S et al (2005) Angiosarcoma of the scalp treated with curative radiotherapy plus recombinant interleukin-2 immunotherapy. Int J Radiat Oncol Biol Phys 61:1446–53

45. Sasaki R, Soejima T, Kishi K et al (2002) Angiosarcoma treated with radiotherapy: impact of tumour type and size on outcome. Int J Radiat Oncol Biol Phys 52:1032–40

46. Ulrich L, Krause M, Brachmann A, Franke I, Gollnick H (2000) Successful treatment of angiosarcoma of the scalp by intralesional cytokine therapy and surface irradiation. J Eur Acad Dermatol Venereol 14:412–5

47. Mareel MM, Storme GA, De Bruyne GK, Van Cauwenberge RM (1982) Vinblastin, vincristine and vindesine: antiinvasive effect on MO4 mouse fibrosarcoma cells in vitro. Eur J Cancer Clin Oncol 18:199–210

48. Jha N, McNeese M, Barkley HT Jr, Kong J (1987) Does radiotherapy have a role in hemangiopericytoma management? Report of 14 new cases and a review of the literature. Int J Radiat Oncol Biol Phys 13:1399–402

49. Staples JJ, Robinson RA, Wen BC, Hussey DH (1990) Hemangiopericytoma – the role of radiotherapy. Int J Radiat Oncol Biol Phys 19:445–51

50. Schier F, Langhans P (1987) Hemangiopericytoma – a review of the literature of 686 cases including 94 of children. TumorDiagnostik Ther 8:1–4

51. Gripp S, Pape H, Schmitt G (1998) Chondrosarcoma of the larynx. The role of radiotherapy revisited – a case report and review of the literature. Cancer 82:108–15

52. Thompson LD, Gannon FH (2002) Chondrosarcoma of the larynx: a clinicopathologic study of 111 cases and a review of the literature. Am J Surg Patho 26:836–51

53. Berson AM, Castro JR, Petti P, Phillips TL et al (1988) Charged particle irradiation of chordoma and chondrosarcoma of the base of skull and cervical spine: the Lawrence Berkeley Laboratory experience. Int J Radiat Oncol Biol Phys 15:559–65

54. Crockard HA, Cheeseman A, Steel T et al (2001) A multidisciplinary team approach to skull base chondrosarcomas. J Neurosurg 95:184–9

55. Harwood AR, Krajbich JI, Fornasier VL (1980) Radiotherapy of chondrosarcoma of bone. Cancer 45:2769–77

56. Laramore GE, Griffith JT, Boespflug M et al (1989) Fast neutron radiotherapy for sarcomas of soft tissue, bone, and cartilage. Am J Clin Oncol 12:320–6

57. McNaney D, Lindberg RD, Ayala AG, Barkley HT, Hussey DH (1982) Fifteen year radiotherapy experience with chondrosarcoma of bone. Int J Radiat Oncol Biol Phys 8:187–90

58. Ryall RDH, Bates T, Newton KA, Hellmann K (1979) Combination of radiotherapy and razoxane (ICRF 159) for chondrosarcoma. Cancer 44:891–5

59. Weber DC, Rutz HP, Pedroni ES et al (2005) Results of spot-scanning proton radiation therapy for chordoma and chondrosarcoma of the skull base: the Paul Scherrer Institute experience. Int J Radiat Oncol Biol Phys 63:401–9

60. Cummings BJ, Hodson I, Bush RS (1983) Chordoma: the results of megavoltage radiation therapy. Int J Radiat Oncol Biol Phys 9:633–42

61. Catton C, O'Sullivan B, Bell R et al (1996) Chordoma: long-term follow-up after radical photon irradiation. Radiother Oncol 41:67–72

62. Fuller DB, Bloom JG (1988) Radiotherapy for chordoma. Int J Radiat Oncol Biol Phys 15:331–9

63. Glanzmann C, Horst W (1978) Ergebnisse der Strahlentherapie bei Chordomen. Strahlentherapie 154:85–8
64. Keisch ME, Garcia DM, Shibuya RB (1991) Retrospective long-term follow-up analysis in 21 patients with chordomas of various sites treated at a single institution. J Neurosurg 75: 374–7
65. The Proton Therapy Working Party (2000) Proton therapy for base of skull chordoma: a report for the Royal College of Radiologists. Clin Oncol (R Coll Radiol) 12:75
66. Rhomberg W, Böhler FK, Novak H et al (2003) A small prospective study of chordomas treated with radiotherapy and razoxane. Strahlenther Onkol 179(4):249–53
67. Raney B, Schnaufer L, Ziegler M et al (1987) Treatment of children with neurogenic sarcoma. Experience at the Children's Hospital of Philadelphia, 1958–1984. Cancer 59:1–5
68. Ferrari A, Casanova M, Bisogno G et al (2002) Clear cell sarcoma of tendons and aponeuroses in pediatric patients. A report from the Italian and German soft tissue sarcoma cooperative group. Cancer 94:3269–76
69. Shimm DS, Suit HD (1983) Radiation therapy of epitheloid sarcoma. Cancer 52:1022–5
70. DeLaney TF, Spiro IJ, Suit HD et al (2003) Neoadjuvant chemotherapy and radiotherapy for large extremity soft-tissue sarcomas. Int J Radiat Oncol Biol Phys 56:1117–27
71. Pisters PWT, Patel SR, Prieto VG et al (2004) Phase I trial of preoperative doxorubicin-based concurrent chemoradiation and surgical resection for localized extremity and body wall soft tissue sarcomas. J Clin Oncol 22:3375–80
72. Sauer R, Schuchardt U, Hohenberger W et al (1999) Preoperative radiochemotherapy in soft tissue sarcomas for improving local tumor control and postoperative functional outcome. Strahlenther Onkol 175:259–66
73. Hellmann K, Murkin GE (1974) Synergism of ICRF 159 and radiotherapy in the treatment of experimental tumors. Cancer 34:1033–9
74. Hellmann K, Burrage K (1969) Control of malignant metastases by ICRF 159. Nature 224:273–5
75. Le Serve AW, Hellmann K (1972) Metastases and the normalization of tumor blood vessels by ICRF-159: a new type of drug action. Br Med J I:597–601
76. Hellmann K (2003) Dynamics of tumour angiogenesis: effect of razoxane-induced growth rate slowdown. Clin Expl Metastasis 20(2):95–102
77. Bakowski, MT (1976) ICRF 159, (+/–) 1,2-di(3,5 dioxopiperazinyl) propane; NSC 129943; razoxane. Cancer Treat Rev 3:95–107
78. Karakiulakis G, Missirlis E, Maragoudakis ME (1989) Mode of action of razoxane: inhibition of basement membrane collagen-degradation by a malignant tumor enzyme. Methods Find Exp Clin Pharmacol 11:255–61
79. Tanabe K, Ikegami Y, Ishida R, Andoh T (1991) Inhibition of topoisomerase II by antitumor agents bis(2,6-dioxopiperazine) derivatives. Cancer Res 51:4903–8

A New Dimension of Increased Efficacy: The Addition of a Tubulin-Affinic Substance. Studies Using the Combination of Radiotherapy, Razoxane and Vindesine

[Further enhancement of the radiosensitivity and the appearance of an antimetastatic effect]

Original article: Rhomberg W, Eiter H, Schmid F, Saely C (2007) Razoxane and vindesine in advanced soft tissue sarcomas: impact on metastasis, survival and radiation response. Anticancer Res 27(5B):3609–14

(The re-use of parts of this article was permitted by Internat. Institute of Anticancer Research (IIAR))

Shortened version

Abstract

Background: The treatment options in advanced soft tissue sarcomas (STS) are still limited. In this pilot study, a radiosensitizing and antimetastatic treatment concept was explored.

Materials and Methods: Twenty-one patients with unresectable and/or oligometastatic STS received the drugs razoxane and vindesine supported by radiotherapy and surgery. Long-term treatment was intended in metastatic disease. Forty-one patients with comparable stages of STS treated with contemporary chemotherapy served as non-randomised controls. The prognostic parameters of the groups were comparable.

Results: The rate of major responses after radiotherapy combined with razoxane and vindesine was 88%, and in the control group 62% ($p = 0.007$). In the study group, the median number of new metastases after 6 months was 0 (range, 0–40) and after 9 months likewise 0 (0–70). The corresponding numbers in the control group were 4.5 (range, 0–40) and 9 (0 to >100) ($p < 0.001$). The progression-free survival at 6 months was 71% in the study group and 23% in the controls, and the median survival time from the occurrence of the first metastasis was 16 months vs. 9 months. The combined treatment was associated with a low to moderate toxicity.

Conclusion: The treatment combination led to an increase of major responses, inhibited the development of remote metastases in the majority of patients with STS and prolonged the survival to some extent.

Background

Available treatment options for unresectable and disseminated soft-tissue sarcomas (STS) are still limited. Although some progress has been made in controlling inoperable or gross residual STS by neutron irradiation [1], isolated limb perfusion with biologically active agents and melphalan [2], or combinations of irradiation with radiosensitising agents such as bromodesoxyuridine and iododesoxyuridine [3] or razoxane [4, 5], distant metastases remain an obstacle to prolonged survival. From this background, an attempt to extend a merely radiosensitizing therapy with razoxane by the addition of an antiinvasive drug appeared worthwile.

Desacetyl-vinblastine-amide [Vindesine (VDS)] is a semisynthetic vinca alkaloid. It was shown to be effective in cytotoxic combination therapies in soft tissue sarcomas [6] and to have putative radiopotentiating abilities [7, 8]. In addition, VDS is a microtubule inhibitor with pronounced antiinvasive effects in vitro [9, 10] and proven antimetastatic activity in animal systems [11, 12].

The radiosensitizer [4, 5, 13] razoxane is an inhibitor of topoisomerase II [14]. The drug is of particular interest in the treatment of STS because of its potential to normalize pathological tumour blood vessels [15–17] and due to its antiinvasive effects [18]. The drug has been shown to slow down the growth rate of transplanted tumours [16] and to completely suppress the development of distant lung metastases in animals [15, 17, 19].

Patients and Methods

In a prospective study from 1996 to 2004, 21 patients with advanced adult-type STS received a combined treatment with razoxane and VDS supported by radiotherapy and, in some instances, by surgery. From these patients, 7 had unresectable primary tumours or recurrences without metastases at baseline and 14 had early metastatic disease, i.e. less than 7 distant metastases. Forty-one patients with comparable age, disease-stages and prognostic features served as non-randomized, retrospective controls.

The antimetastatic approach. Conventional cytotoxic chemotherapy is applied to induce disease regression or stable disease. In cases of progressive disease, the treatment is usually judged as not effective and will be changed or terminated. In contrast, the antimetastatic approach has the intention of preventing further metastasis – irrespective of the achievement of an objective response of existing lesions. This approach was pursued in our cohort of patients on combined razoxane/VDS treatment: If pre-existing metastases proved resistant to the combination of razoxane/VDS, this therapy was continued ('treatment beyond progression') with addition of radiation to the respective lesions and in some cases removal by surgery. In case of only a few new metastases, the razoxane/VDS treatment was also continued and local treatment measures were reinstituted. However, the combination therapy was regarded as ineffective and terminated if more than 5 new metastases appeared within 3 months.

Drug treatment. The study patients received a metronomic chemotherapy with razoxane as described earlier and small doses of VDS together with concurrent radiotherapy. The treatment was terminated on complete response of unresectable tumours, but continued if metastases were present at the time of patient referral. The median overall dose of razoxane was 14 g per patient (range, 7.25–75 g). VDS was given intravenously at weekly doses of 2 mg. The median dose of VDS per patient was 43 mg (16–302 mg).

Three of the 21 patients in the razoxane/VDS group had been pretreated with conventional chemotherapy, i.e. doxorubicin-based regimens, and 5 patients received the same treatment during the later course of their disease. Furthermore, four patients had received 2–4 doses of mitoxantrone in addition to the razoxane/VDS treatment. This initial treatment variant, however, was discontinued early because of chronic nausea.

Radiation therapy. External beam radiation therapy was used with 6 MeV and 25 MeV photons with linear accelerators and conformal planning techniques. Single tumour doses between 170 and 200 cGy were given five times a week at the ICRU (International Comission on Radiation Units) point. The median total dose to unresectable primaries or recurrences was 60 Gy (range, 50–64 Gy) and 50 Gy (range, 50–60 Gy) to solitary metastases. In case of oligotopic metastases, the average total tumour doses were below 50 Gy. Six patients received two or more radiation treatments for metastases.

Control patients. Forty-one patients of similar age and similar prognostic features, in particular with similar stages of STS, who received contemporary cytotoxic drugs (doxorubicin-based regimens) in addition to radiotherapy served as controls. The control group was selected from 121 patients with adult-type sarcomas who were referred to our department between 1993 and 2002 for adjuvant or palliative radiation therapy, and from a further set of patients who between 1978 and 1988 had served as controls in a randomised study investigating the effects of razoxane when given in addition to radiotherapy [4].

To be eligible as controls, patients had to have unresectable primaries and/or early metastatic disease with fewer than 7 distant metastases at the time of referral. This cut-off level was an arbitrary decision. Patients with multiple metastases or patients with complete tumor resections who only received adjuvant radiotherapy were excluded as controls. For all patients serving as controls, complete clinical follow-up data as well as X-rays, CT and MRT imaging had to be available.

Response evaluation and follow-up. The radio-responsiveness was related to the clinical shrinkage of a tumour mass. Response definitions were done according to standard criteria (see Section 2.3.2, 'Common Methods').

All patients were followed up until December 2005 or to their death. Abdomino-pelvic and chest CTs were performed every 3 months during the first year. Additional investigations were carried out depending on clinical needs. The number of new metastatic foci was counted every 3 months, and the cumulative incidence of new metastases after 6 and 9 months was determined. The survival time was calculated from the occurrence of the first distant metastasis, and, additionally, from the beginning of the combination therapy (razoxane/VDS/radiotherapy) in the study cases or from any systemic cytotoxic chemotherapy and/or palliative radiotherapy in the control patients.

Statistical methods. The Wilcoxon-Gehan statistic was used to compare differences in survival times between the treatment groups. Other between-group differences were tested for statistical significance with the Mann-Whitney U test for continuous variables and with the Chi-squared test for categorical variables, respectively. p-values <0.05 were considered significant. All statistical analyses were performed with the software package SPSS 11.0 for Windows (Chicago, Il., USA).

Results

The main pre-treatment characteristics of our patients including their relevant prognostic parameters are listed in Table 2.3. There were no significant differences between the patients treated with razoxane/VDS and the controls.

Table 2.3 Clinical characteristics and prognostic factors of soft tissue sarcoma patients of this study

	Razoxane + VDS ($n = 21$)	Controls ($n = 41$)
Age in years (range)	61 (31–78)	59 (23–85)
Gender		
Male	11	22
Female	10	19
Time from diagnosis to metastasis in months (range)	11.5 (0–48)	9 (0–252)
Median largest tumor diameter at diagnosis in cm (range)	12 (5.5–23)	10 (2–25)
Histological diagnoses		
Liposarcoma	3	8
Malignant fibrous histiocytoma	1	8
Leiomyosarcoma	2	7
Fibrosarcoma	1	5
Angiosarcoma	7	3
Synovial sarcoma	1	3
GIST	1	3
Other rare entities	5	4
Histological grading		
1	2	2
2	4	7
3 + 4	13 (62%)	26 (63%)
Unknown	2	6
Conventional chemotherapy, ever given	8/21 (38%)	30/39 (77%)

Radiation response. Among 17 assessable patients treated with razoxane/VDS, major clinical responses to radiation were seen in 15 patients (7 CR, 8 PR). Minor remissions were observed in 2 patients. Thus, the rate of major responses was 88%.

The response of 21 patients was assessable in the control group. Major responses were observed in 13 patients (4 CR, 9 PR; i.e. 62%); this was significantly less ($p = 0.007$) than in the razoxane/VDS group. The control tumours did not change in size in 6 patients, tumour progression was noted in 2 patients.

Development of metastases. In the razoxane/VDS group the median number of new distant metastases after 6 months was 0 (range, 0–40), and after 9 months likewise 0 (range, 0–70). The corresponding median values for the controls was 4.5 (range, 0–40) and 9 (range, 0 to >100) new metastases after 6 and 9 months, respectively. These differences in the occurrence of new metastases after 6 and 9 months were highly significant ($p = 0.001$ and $p < 0.001$, respectively).

In the subset of patients with unresectable primaries or isolated recurrences, none of the 7 patients treated with razoxane and vindesine, and 9 of 13 control patients developed distant metastases within 9 months ($p = 0.045$).

Survival. The median survival time from the occurrence of the first distant metastasis was 16 months (range 8–96+ months) in the antimetastatic treatment group and 9 months (range 2–240 months) in the control group ($p = 0.010$, Mann-Whitney U test). Survival time from the beginning of systemic drug treatment/palliative radiotherapy was 14 months (range 6–96+ months) in the study patients, and 9 months (range 2–235 months) in the controls ($p = 0.065$). The progression-free survival at 6 months was 71% in the patients treated with razoxane and vindesine and 23% in the controls, respectively ($p < 0.001$).

Among patients with unresectable primaries or recurrences without metastasis who received the antimetastatic treatment the median survival has not yet been determined. Six of these 7 patients survived longer than 1 year compared to 5 of 13 in the control group ($p = 0.043$) after the start of the systemic drug treatment/palliative radiotherapy.

Side-effects and complications (*study group only*). The main side-effect of the combined razoxane/VDS/radiotherapy was leukopenia. Leukopenia of grade 3 or 4 was noted in 36% of the patients. The nadir of leukopenia was on day 16; no case of neutropenic fever occurred. Pulmonary embolism was seen in 3 patients, one with a lethal outcome. Other systemic toxicities included mild to moderate neurotoxicity (paresthesias of the fingers which were reversible within 2–3 weeks), diarrhea grade 1–2, and nausea grade 1–2. In addition, we observed one case each of rib necrosis, Fournier necrosis of the gluteal region, severe headache, and alopecia, respectively. Normal tissue reactions were clearly enhanced by razoxane/VDS. Regional pneumonitis and esophagitis were most frequently observed when parts of the lung were irradiated. Such reactions occurred even with radiation doses of 30 Gy, but they were of limited clinical significance because most of them disappeared within days.

Discussion

From our data we conclude that the trimodal treatment with razoxane, vindesine and radiotherapy is feasible in patients with unresectable primaries and early metastatic STS. The combination leads to a high rate of objective responses at irradiated tumour sites, it inhibits the development of remote metastases in the majority of patients and seems to prolong survival.

With the exception of gastrointestinal stromal tumours (GIST) for which survival advantages have been achieved with imatinib, the median survival in advanced soft tissue sarcomas has not substantially changed for almost three decades, irrespective of whether doxorubicin alone, doxorubicin with ifosfamide or cytoxan, vincristine, doxorubicin and dacarbacine (CYVADIC regimen) has been given [6, 20–22]. The median survival time of advanced STS treated with contemporary chemotherapy ranges from 7–12 months [6, 22]; a meta-analysis which included

2,185 patients showed an overall survival time of 51 weeks [21]. In view of the unchanged prognosis of disseminated STS, a comparison of the results of this pilot trial with historical or non-randomized controls seems to be justified, especially if the prognostic features were not different among the compared groups.

Initial tumour size, histological grade, tumour site and lymph node involvement are the major prognostic factors for patients with primary STS [6]. A long disease-free interval from diagnosis to first metastasis, low histopathological grade, young age, and absence of liver metastasis are the strongest predictors for a better prognosis in patients with disseminated disease [23, 24], while a doxorubicin-based chemotherapy did not affect the survival substantially [25]. All these main prognostic factors did not differ significantly between the razoxane/VDS group and the controls.

The combination of razoxane and vindesine basically represents a combination of an angiogenesis affecting and a tubulin-inhibiting agent although the modes of action of the drugs are overlapping. Both drugs affect main steps of the metastatic cascade. By using similar combinations, e.g., long-term treatments with DC 101, a VEGF receptor-2-blocking antibody, together with the vinca-alkaloid vinblastin or the tubulin affinic drug paclitaxel, cures have been achieved in preclinical human neuroblastoma xenograft tumour models [26, 27].

The objective response rates at irradiated tumour sites on razoxane/VDS therapy were high in this study. *Complete* responses (CR) were observed in as many as 7 of 17 assessable patients (41%). For comparison, previously reported CR rates in STS treated by irradiation and intravenous radiosensitizers were 20% [3], and in STS treated with radiotherapy and razoxane 30% [4]. Limited tumour shrinkage was observed by DeLaney et al. when neoadjuvant radiation therapy was combined with modern cytotoxic chemotherapy [28]. Preoperative chemotherapy alone for extremity STS is associated with *partial* response rates between 27 and 40% [29, 30].

Even though we investigated only a small number of patients, the results indicate a statistically highly significant antimetastatic efficacy of the combination of razoxane and vindesine. The translation into a larger survival gain, however, seems to be of a modest degree as yet. Some reasons could account for this: A consistent long-term treatment with razoxane and vindesine was given to only 10 of the 21 patients of the trial. This was due to the fact that some patients were in a transient complete remission. In this situation there was no absolute need to continue the treatment, especially in a pilot study. Some patients also refused long term treatment, others asked for a second opinion and were advised against this experimental treatment. Treatment-related side-effects were never a reason not to continue the treatment. In addition, the patients of this pilot study had in general a large tumour burden.

The determination of the rate of new metastases for a given period may represent an interesting clinical trial endpoint for the assessment of antiangiogenic substances or, in general, of antimetastatic drug regimens. Presently, no basic data on the incidence and dynamics of metastases in STS are available from the literature. The retrospective evaluation of the metastatic process in the control group proved to be a cumbersome procedure. Numerous inquiries were necessary at different departments. In analyzing CT images and X-rays we had to face some imprecision, or even impossibility to count metastases exactly, especially the lesions of the peritoneum or the pleura. Most precise data on the dynamics of the metastatic process in STS can probably only be obtained by a prospective trial with repeated whole body CT's. CT scans are associated with much higher numbers of visible lesions compared to chest X-rays. Hence, there remains some imprecision and the figures given on the numbers of metastases in this study must be seen and defined as minimal numbers of detected metastases.

The trimodal combination therapy with razoxane, vindesine, and radiotherapy is easy to administer. Patient compliance and tolerability of the drugs were uncomplicated. No unsuspected toxicity was observed during long-term treatment, indicating the safety of the treatment but caution should be exercised in irradiating larger lung volumes because of the danger of pneumonitis.

In summary, the results of this study suggest that combined razoxane/vindesine treatment further increases radioresponsiveness. The treatment seems to have the potential to reduce the propensity of STS for distant metastases. Antimetastatic drug combinations supported by radiotherapy and/or surgery may become a new paradigm for the management of patients with unresectable primaries and oligometastatic STS. New rewarding areas of palliative or even curative radiation therapy of metastases may arise from this kind of treatment.

Current Note

An update of this study with slightly more patients (n = 23) and a longer follow up time revealed a statistically highly significant prolongation of survival in favour of the patients receiving VDS, razoxane and radiation therapy. The high response rate to irradiation and the antimetastatic efficacy remained unchanged [31]. A synopsis of this article is given in Section 2.3.3.

References

1. Cohen L, Hendrickson F, Mansell J, Kurup PD, Awschalom M, Rosenberg I, Tenttaken RK (1984) Response of sarcomas of bone and soft tissues to neutron beam therapy. Int J Radiat Oncolo Biol Phys 10:821–4
2. Eggermont AMM, Koops HS, Lienard D, Kroon BR, van Geel AN, Hoekstra HJ, Lejeune FJ (1996) Isolated limb perfusion with high-dose tumor necrosis factor-alpha in combination with interferon-gamma and melphalan for nonresectable extremity soft tissue sarcomas: a multicenter trial. J Clin Oncol 14:2653–65
3. Kinsella TJ and Glatstein E (1987) Clinical experience with intravenous radiosensitizers in unresectable sarcomas. Cancer 59:908–15
4. Rhomberg W, Hassenstein EOM, Gefeller D (1996) Radiotherapy vs. radiotherapy and razoxane in the treatment of soft tissue sarcomas: final results of a randomized study. Int J Radiat Oncol Biol Phys 36:1077–84
5. Ryall, RDH, Hanham IWF, Newton KA, Hellmann K, Brinkley DM, Hjertaas OK (1974) Combined treatment of soft tissue and osteosarcomas by radiation and ICRF 159. Cancer 34:1040–5

6. De Vita VT Jr, Hellman S, Rosenberg SA (eds) (2001) Cancer principles & practice of oncology, 6th edn. JB Lippincott, Philadelphia, pp 1879–83

7. Storme GA, Schallier DC, De Neve WJ, De Greve JL, Van Belle SP, De Wasch GJ, Dotremont G (1988) Vinblastine has radiosensitizing activity in limited squamous cell lung cancer. Int J Radiat Oncol Biol Phys 15(Suppl1):222

8. Rhomberg W, Eiter H, Soltesz E and Böhler F (1990) Long term application of vindesine: toxicity and tolerance. J Cancer Res Clin Oncol 116:651–3

9. Haug IJ, Siebke EM, Grimstad IA, Benestad HB (1993) Simultaneous assessment of migration and proliferation of murine fibrosarcoma cells, as affected by hydroxyurea, vinblastine, cytochalasin B, razoxane and interferon. Cell Prolif 26:251–61

10. Mareel MM, Storme GA, De Bruyne GK, Van Cauwenberge RM (1982) Vinblastin, vincristine and vindesine: antiinvasive effect on MO4 mouse fibrosarcoma cells in vitro. Eur J Cancer Clin Oncol 18:199–210

11. Atassi G, Dumont P, Vandendris M (1982) Investigation of the in vivo antiinvasive and antimetastatic effect of desacetyl vinblastine amide sulphate or vindesine. Invasion and Metastasis 2:217–31

12. Mareel MM, Bracke ME, Boghaert ER (1986) Tumor invasion and metastasis: therapeutic implications? Radiother Oncol 6:135–42

13. Hellmann K, Murkin GE (1974) Synergism of ICRF 159 and radiotherapy in the treatment of experimental tumors. Cancer 34:1033–9

14. Tanabe K, Ikegami Y, Ishida R, Andoh T (1991) Inhibition of topoisomerase II by antitumor agents *bis* (2,6-dioxopiperazine) derivatives. Cancer Res 51:4903–8

15. Hellmann K, Burrage K (1969) Control of malignant metastases by ICRF 159. Nature 224:273–5

16. Hellmann K (2003) Dynamics of tumour angiogenesis: effect of razoxane-induced growth rate slowdown. Clin Expl Metastasis 20:95–102

17. Le Serve AW, Hellmann K (1972) Metastases and the normalization of tumor blood vessels by ICRF-159: a new type of drug action. Br Med J I:597–601

18. Karakiulakis G, Missirlis E, Maragoudakis ME (1989) Mode of action of razoxane: inhibition of basement membrane collagen-degradation by a malignant tumor enzyme. Methods Find Exp Clin Pharmacol 11:255–61

19. Salsbury AJ, Burrage K, Hellmann K (1974) Histological analysis of the antimetastatic effect of 1,2-*bis* (3,5-dioxopiperazin-1-yl)-propane. Cancer Res 34: 843–9

20. Antman K, Crowley J, Balcerzak SP, Rivkin SE, Weiss GR, Elias A, Natale B, Cooper RM, Barlogie B, Trump DL et al (1993) An intergroup phase III randomized study of doxorubicin and dacarbazine with or without ifosfamide and mesna in advanced soft tissue and bone sarcomas. J Clin Oncol 11:1276–85

21. Edmonson JH, Ryan LM, Blum RH, Brooks JSJ, Shiraki M, Frytak S, Parkinson DR (1993) Randomized comparison of doxorubicin alone versus ifosfamide plus doxorubicin or mitomycin, doxorubicin, and cisplatin against advanced soft tissue sarcomas. J Clin Oncol 11:1269–75

22. Santoro A, Tursz T, Mouridsen H, Verweij J, Steward W, Somers R, Buesa J, Casali P, Spooner D, Rankin E et al (1995) Doxorubicin versus CYVADIC versus doxorubicin plus ifosfamide in first line treatment of advanced soft tissue sarcomas: a randomized study of the European Organisation for Research and Treatment of Cancer Soft Tissue and Bone Sarcoma Group. J Clin Oncol 13:1537–45

23. Rööser B, Attewell R, Berg NO, Rydholm A (1988) Prognostication in soft tissue sarcoma. A model with four risk factors. Cancer 61:817–23

24. Zagars GK, Ballo MT, Pisters PWT, Pollock RE, Patel SR, Benjamin R (2003) Prognostic factors for disease-specific survival after first relapse of soft-tissue sarcoma: analysis of 402 patients with disease relapse after initial conservative surgery and radiotherapy. Int J Radiat Oncol Biol Phys 57:739–47

25. Komdeur R, Hoekstra HJ, van den Berg E, Molenaar WM, Pras E, de Vries EGE, van der Graaf WTA (2002) Metastasis in soft tissue sarcomas: prognostic criteria and treatment perspectives. Cancer Metastasis Rev 21:167–83
26. Kerbel RS (2001) Clinical trials of antiangiogenic drugs: opportunities, problems, and assessment of initial results. J Clin Oncol 19(Suppl):45s–51s
27. Klement G, Baruchel S, Rak J, Man S, Clark K, Hicklin DJ, Bohlen P, Kerbel RS (2000) Continuous low-dose therapy with vinblastine and VEGF receptor-2 antibody induces sustained tumor regression without overt toxicity. J Clin Invest 105:R15–24
28. DeLaney TF, Spiro IJ, Suit HD et al (2003) Neoadjuvant chemotherapy and radiotherapy for large extremity soft-tissue sarcomas. Int J Radiat Oncol Biol Phys 56:1117–27
29. Pezzi CM, Pollock RE, Evans HL, Lorigan JG, Pezzi TA, Benjamin RS, Romsdahl MM (1990) Preoperative chemotherapy for soft-tissue sarcomas of the extremities. Ann Surg 211:476–81
30. Pisters PWT, Patel SR, Varma DG et al (1997) Preoperative chemotherapy for stage IIIB extremity soft tissue sarcoma: long term results from a single institution. J Clin Oncol 15:3481–7
31. Rhomberg W et al (2008) Combined vindesine and razoxane shows antimetastatic activitiy in advanced soft tissue saromas. Clin Exp Metastasis 25(1):75–80

Treatment of Vascular Soft Tissue Sarcomas with Razoxane, Vindesine and Radiation

[A phase II study of the Austrian Society of Radiooncology (ÖGRO)]

Original article: Rhomberg W, Wink A, Pokrajac B, Eiter H, Hackl A, Pakisch B, Ginestet A, Lukas P, Pötter R (2009) Treatment of vascular soft tissue sarcomas with razoxane, vindesine and radiation. Int J Radiat Oncol Biol Phys 74(1):187–91

Shortened version

Abstract

Purpose: In previous studies, razoxane and vindesine together with radiotherapy proved to be effective in soft tissue sarcomas. Since razoxane leads to a redifferentiation of pathological tumor blood vessels, it was of particular interest to study the influence of this drug combination in vascular soft tissue sarcomas.

Methods and Materials: This open multicenter phase II study was performed by the Austrian Society of Radiooncology. Among 13 evaluable patients (10 angiosarcomas and 3 hemangiopericytomas), 9 had unresectable measurable disease, 3 showed microscopic residuals, and one had a resection with clear margins. They received basic treatment with razoxane and vindesine supported by radiation therapy to all measurable lesions. Outcome measures were objective response rates, survival time, and the incidence of distant metastases.

Results: In 9 patients with measurable vascular soft tissue sarcomas (8 angiosarcomas and 1 hemangiopericytoma), 6 CR, 2 PR and 1 minor remission were achieved corresponding to a major response rate of 89%. A maintenance therapy with razoxane and vindesine of 1 year or longer led to a suppression of distant metastases. The median survival time from the start of the treatment is 23+ months (range,

3–120+) for 12 patients with macroscopic and microscopic residual disease. The progression-free survival at 6 months was 75%. The combined treatment was associated with low general toxicity but attention has to be given to increased normal tissue reactions.

Conclusions: This trimodal treatment leads to excellent response rates, and it suppresses distant metastases in malignant vascular tumors if the drugs are given as maintenance therapy.

Background

Angiosarcomas are rare aggressive malignancies with a worse prognosis compared to other soft tissue sarcomas (STS). A recently updated pilot study indicated that the addition of vindesine, a semisynthetic vinca-alkaloid, further enhances the radiosensitizing effect of razoxane, leads to a suppression of distant metastasis and prolongs survival [1]. Desacetyl-vinblastine-amide [Vindesine (VDS)] is a semisynthetic vinca alkaloid that has been shown to be effective in cytotoxic combination therapies in soft tissue sarcomas [2]. Apparently, it also has radiopotentiating abilities [3, 4] and proven antimetastatic activity in animal systems [5, 6] probably due to its microtubule inhibition which leads to pronounced antiinvasive effects in vitro [7]. VDS also has the advantage that it may be given to patients for several years without cumulative toxicity [4].

The ability of razoxane to normalize pathological tumor blood vessels makes it an ideal drug to be used in the treatment of angiosarcomas. For this reason and in view of the results of the pilot study mentioned above, the Austrian Society of Radiooncology initiated a phase II study to see whether these observations can be confirmed in angiosarcomas.

Patients and Methods

Between July 2002 and July 2005 eleven patients with sarcomas of the blood vessels entered a prospective study which consisted of a combined treatment with razoxane, vindesine and irradiation. The study was designed as non-randomized multicenter phase II study. Three patients with angiosarcomas of a preceding pilot study of soft tissue sarcomas in general were included. There were 10 male and 4 female patients with a median age of 68 years (range, 47–90 years). Eleven patients presented with angiosarcomas mostly of grade 3 and 4, three patients had malignant hemangiopericytomas. One patient with an angiosarcoma of the scalp was excluded from the treatment analysis because he received erroneously only razoxane and irradiation without vindesine. The treatment report therefore comprises 13 evaluable patients.

Trial end points in patients with unresectable measurable disease were the rate of major responses and the overall survival. In patients who received the treatment as adjunct to surgery, local control and survial were noted. Secondary endpoints were the progression-free survival at 6 months and the determination of new metastases appearing every 3 months.

Patient Selection

All patients with histologically proven angiosarcomas and hemangiopericytomas were eligible for the study. The age of the patients should have been over 18 years, there was no upper age limit. A Karnofsky performance scale over 60 and an adequate bone marrow function were required. No more than one regimen with cyto-toxic agents was allowed as pretreatment, previous radiation therapy of the lesion to be studied excluded patients from the study. The number of patients to be studied was fourteen.

Radiation Therapy

High energy photons were administered to all patients. Additionally, in three patients electrons, and in one patient with an angiosarcoma of the scalp Iridium-192 moulds were applied as part of the radiation treatment. Daily fractions of 1.8–2 Gy were given five times a week. The median total tumor dose of all patients was 56 Gy (range, 38–66 Gy).

Drug Treatment

Razoxane was given concurrently to the radiotherapy as described in previous papers. The median total dose of razoxane was 8.5 g (range, 3.7–75 g). Vindesine was given intravenously at a weekly dose of 2 mg irrespective of body weight. The median dose of all patients was 15 mg (range, 4–302 mg). The large difference in the range of the drug doses is due to the variable length of the mainentence therapy.

Results

The main pretreatment characteristics, the tumor sites, the kind of lesions irradi-ated and their response to the combined modality treatment are shown in Table 2.4. All but one patient had no pretreatment with cytotoxic agents; and only 2 patients (# 8 and #9) received any other chemotherapy during the later course of the dis-ease. Measurable disease included 4 unresectable primary tumors, 4 regrowing gross residual masses after R-2 resection, and one case with distal lung metastases. Three patients presented with microscopic residual disease (R-1 resection, lesions not measurable) and one patient had a resection of a scalp angiosarcoma with clear margins.

Response to Radiotherapy

Thirteen patients with sarcomas of the blood vessels (10 angiosarcomas and 3 hemangiopericytomas) were evaluable. The radiation response by histology was as follows: Among 8 patients with measurable angiosarcoma, 6 showed a complete regression of their tumors (75%), one had a partial response and another a minor remission (with tumor measurement by CT after 15 Gy only). One patient with microscopic residuals of an angiosarcoma of the bladder died from repeated hemor-rhages 3 months after treatment, and a patient with clear margins after resection of a scalp-tumor remained free of disease 46+ months.

Table 2.4 Clinical characteristics and treatment outcome in 13 evaluable patients

Pat. #	Age/gender	Tumor site	Histology	Grade	What has been irradiated?	Response	Time to DM or regrowth	Maintenance Tx after RT	Overall survival
1	72 M	Thyroid	Angio-Sa	G 3	Two lung metastases	CR	Alive, NED	3 years	120+
2	64 F	Thyroid	Angio-Sa	G 4	Regrowing gross residuals	CR	10 months	6 weeks	11
3	67 M	Thyroid	Angio-Sa	G 4	Regrowing gross residuals	CR	12 months	1 year	16
4	69 M	Thyroid	Angio-Sa	G 3–4	Regrowing gross residuals	PR	16 months	None[a]	19
5	61 M	Thyroid	Angio-Sa	G 3	Regrowing gross residuals	CR	Alive, NED	1 year	27+
6	56 M	Scalp/face	Angio-Sa	G x	Adjuvant radiation therapy	LC	Alive, NED	None	46+
7	71 F	Scalp/face	Angio-Sa	G 3	Primary tumor, 9×14 cm	CR	Alive, NED	1 year	41+
8	90 F	Breast	Angio-Sa	G 3	Primary tumor, >10 cm	CR	2 months	None	4
9	57 M	Left ventricle	Angio-Sa	G x	Primary tumor, 5×3 cm	MR	3 months	None[a]	$4\frac{1}{2}$
10	51 M	Bladder	Angio-Sa	G 3	Microscopic residuals	no LC	Uncontrolled	None	3
11	71 M	Omentum	HPC	G 2	Microscopic residuals	LC	30 months	2 months	36
12	47 M	Spine (L 2)	HPC	G 2	Microscopic residuals	LC	Alive, NED	None	63+
13	77 F	Pelvis	HPC	G 3	Primary tumor, 16×12 cm	PR	Intercurr death	None	46

CR, complete response; PR, partial response; MR, minor remission; LC, local control; HPC, hemangiopericytoma; DM, distant metastases; RT, radiotherapy; NED, no evidence of disease.

[a]Patient received conventional chemotherapy.

Three hemangiopericytomas were irradiated: One woman with a large tumor of the pelvis showed a subtotal, partial regression, the other two patients with microscopic residuals were locally controlled for 30 and 63+ months.

Overall, in 8 of 9 patients with gross disease of a vascular soft tissue sarcoma major regressions were observed following the trimodal therapy (Table 2.4).

Survival

Taking all patients with gross and microscopic residual vascular sarcomas (angiosarcomas + hemangiopericytomas) together, the median progression-free survival of 12 patients was 21.5+ months (range, 2–120+) and the median overall survival was 23+ months (range, 3–120+).

In the 8 patients with unresectable or metastatic angiosarcomas the median survival time was 17.5 months (range, 4–120+ months), 5 of 8 patients survived 1 year or longer (62.5%). The progression-free survival at 6 months was 75% (6/8 patients).

Metastasis

Taking all 13 assessable patients together, only 2 developed new distant metastases within the first 6 months after the start of the combined radiation treatment. Finally, 6 of 13 patients succumbed to distant metastasis, 5 are alive with no evidence of disease (NED), one died from hemorrhages without overt disease progression, and one patient died intercurrently.

Among 8 angiosarcoma patients with gross disease, 3 of 4 patients who had maintenance therapy of 1 year or longer are alive with NED whereas 4 of 4 patients without maintenance therapy finally developed distant metastasis and died.

Toxicity

Normal tissue reactions are increased with this combination therapy. The reactions had clinical significance in head and neck sarcomas. One has to deal with early mucositis, esophagitis and pneumonitis. In angiosarcomas of the thyroid, postoperative impairments add to the problems so that transient nasogastric tube feeding was necessary in 3 of 5 patients.

Rapidly reversible leukopenia was the dose limiting toxicity. Grade 3 leukopenia occurred in 40% of the patients. The subjective tolerance to the drugs, however, was good, in few instances slight nausea or paresthesias were observed.

Discussion

In this study a remarkable rate of complete responses was observed in measurable unresectable angiosarcomas, and it seems that maintenance treatment with razoxane and vindesine was able to suppress distant metastasis in these tumors.

Data on objective tumor responses to definitive or palliative radiotherapy of angiosarcomas are sparse in the literature. Single-case reports repeatedly describe complete durable responses with radiotherapy either alone [8, 9] or together with chemo/biotherapeutic agents [10–12] but the true incidence of such an outcome remains unknown, and it is probably lower than case reports suggest. In a study by Mark et al. [13], only 1 of 9 patients (11%) with gross disease treated with radiotherapy with or without chemotherapy was rendered free of disease. Garcia-Schüler et al. observed 6 partial regressions among 13 patients with macroscopic angiosarcomas (46%), and the median progression-free survival was 2.5 months in their series [14]. Thus, a complete response in 6 of 8 patients with macroscopic disease is certainly an outcome that deserves interest.

In patients with unresectable macroscopic angiosarcomas the trimodal treatment led to a median survival time of 17.5 months. If all patients with some form of residual disease are taken together ($n = 12$), the median progression-free survival was 21.5+ months. Abraham et al. described the treatment and outcome of 82 patients with angiosarcomas: Of 36 patients with advanced disease, 36% underwent a palliative operation, 78% received radiation, and 58% received chemotherapy. The median survival was 7.3 months [15]. In a larger series of 125 patients treated by Fury et al., the overall 5-year survival was 31%. For unresectable angiosarcoma, no data concerning the effectiveness of radiation was given. Doxorubicin based regimens yielded a progression-free survival of 3.7–5.4 months. Paclitaxel led to a progression-free survival of 6.8 months for scalp angiosarcoma and 2.8 months for sites below the clavicle [16]. In recent years, angiosarcomas were discovered as being sensitive to various chemo-therapeutic agents with paclitaxel and pegylated-liposomal doxorubicin being the most effective drugs but the response rate varied and does not exceed 33% [17]. The impact on long term survival remains limited. Table 2.5 shows a comparison of the few results concerning objective responses to irradiation available from the literature.

The unfavourable prognosis of angiosarcomas varies dependent on the primary site. For instance, angiosarcomas of the thyroid are associated with a dismal prognosis [18–21], and the median survival was 2.4 months in the largest series reported [20]. Other unfavourable locations are scalp and face, liver, heart and the skeletal system with 5-year survival rates of 0–15% including all stages. An extensive literature exists on the unfavourable prognosis of angiosarcomas of the scalp and face (ASF) [22–26]. No patient with endophytic ASF survived >2 years in a series

Table 2.5 Radiosensitivity and progression free survival in measurable angiosarcomas (gross disease). Review of the literature

References	n	Complete response	Partial response	Progression free survival
Garcia-Schüler et al. (radiotherapy only)	13	0%	46%	2.5 months
Mark et al. (radio/chemotherapy)	9	11%	–	–
OEGRO (radiotherapy+VDS+Rz)	8	75%	12.5%	14 months

OEGRO, Austrian society of radiotherapy; VDS, vindesine; Rz, razoxane.

of Sasaki et al. [27]. A less aggressive course was described for angiosarcomas of the nose [28, 29] or the breast where 5-year overall survival rates between 40 and 60% were reported [30]. All but one patients of our study had angiosarcomas with locations associated with a bad prognosis.

A significant suppression of the development of distant metastases was observed when razoxane and vindesine was administered in a variety of soft tissue sarcomas [1]. Both drugs affect main steps of the metastatic cascade, and both razoxane and vindesine have antimetastatic activity in animal systems [31–33]. It seems that this drug combination is able to reduce the propensity for distant metastases also in angiosarcomas provided that the drugs are continued and given as mainentance treatment. The optimum duration of such a treatment is unknown. In our study, it was observed that in case of a maintenance therapy of 1 year or longer, only 1 of 4 patients with gross disease developed distant metastasis whereas 4 of 4 patients without maintenance therapy developed distant metastasis and died.

The trimodal treatment as described has a manageable toxicity profile although the local radiation reaction to the normal tissues must not be underestimated. This, and the outstanding response rate as well as the finding of a reduction of distant metastases deserve further attention for this convenient outpatient-based treatment regimen.

References

1. Rhomberg W, Eiter H, Schmid F et al (2008) Combined vindesine and razoxane shows antimetastatic activity in advanced soft tissue sarcomas. Clin Exp Metastasis 25:75–80
2. DeVita VT Jr, Hellman S, Rosenberg SA (eds) (2001) Cancer principles & practice of oncology. JB Lippincott, Philadelphia, pp 1879–83
3. Storme GA, Schallier DC, De Neve WJ et al (1988) Vinblastine has radiosensitizing activity in limited squamous cell lung cancer. Int J Radiat Oncol Biol Phys 15(Suppl 1):222
4. Rhomberg W, Eiter H, Soltesz E et al (1990) Long term application of vindesine: toxicity and tolerance. J Cancer Res Clin Oncol 116:651–3
5. Atassi G, Dumont P, Vandendris M (1982) Investigation of the in vivo antiinvasive and antimetastatic effect of desacetyl vinblastine amide sulphate or vindesine. Invasion Metastasis 2:217–31
6. Mareel MM, Bracke ME, Boghaert ER (1986) Tumor invasion and metastasis: therapeutic implications? Radiother Oncol 6:135–42
7. Mareel MM, Storme GA, De Bruyne GK et al (1982) Vinblastin, vincristine and vindesine: antiinvasive effect on MO4 mouse fibrosarcoma cells in vitro. Eur J Cancer Clin Oncol 18:199–210
8. Graham WJ, Bogardus CR (1981) Angiosarcoma treated with radiation therapy alone. Cancer 48:912–4
9. Pötter R, Baumgart P, Greve H et al (1989) Primary angiosarcoma of the heart. Thorac Cardiovasc Surg 37:374–8
10. Holloway C, Turner AR, Dundas G (2004) Cutaneous angiosarcoma of the scalp: a case report of sustained complete response following liposomal doxorubicin and radiation therapy. Proceedings of the 18th annual meeting of the Canadian Association of Radiooncologists, Halifax, 9–12 Sept 2004

11. Ohguri T, Imada H, Nomoto S et al (2005) Angiosarcoma of the scalp treated with curative radiotherapy plus recombinant interleukin-2 immunotherapy. Int J Radiat Oncol Biol Phys 61:1446–53

12. Ulrich L, Krause M, Brachmann A et al (2000) H. Successful treatment of angiosarcoma of the scalp by intralesional cytokine therapy and surface irradiation. J Eur Acad Dermatol Venereol 14:412–5

13. Mark RJ, Poen JC, Tran LM et al (1996) Angiosarcoma. A report of 67 patients and a review of the literature. Cancer 77:2400–6

14. Garcia-Schüler H, Jensen A, Röder F et al (2005) Retrospective evaluation of treatment results after radiotherapy of angiosarcomas [Abstract]. Strahlenther Onkol 181(Suppl):65

15. Abraham JA, Hornicek FJ, Kaufmann AM et al (2007) Treatment and outcome of 82 patients with angiosarcoma. Ann Surg Oncol 14:1953–67

16. Fury MG, Antonescu CR, Van Zee KJ et al (2005) A 14-year retrospective review of angiosarcoma: clinical characteristics, prognostic factors, and treatment outcomes with surgery and chemotherapy. Cancer J 11:241–7

17. Ferrari A, Casanova M, Bisogno G et al (2002) Malignant vascular tumors in children and adolescents: a report from the Italian and German 'Soft Tissue Sarcoma Cooperative Group'. Med Pediatr Oncol 39:109–14

18. Rösler H, Walther E (1984) Die Strahlentherapie der Struma maligna. In: Heilmann HP (ed) Handbuch der medizinischen Radiologie, vol XIX/part 5: Spezielle Strahlentherapie maligner Tumoren. Springer, Berlin, Heidelberg, New York, p 36

19. Thaler W, Riccabona G, Riedler L et al (1986) Zum malignen Hämangioendotheliom der Schilddrüse. Chirurg 57:397–400

20. Ladurner D, Tötsch M, Luze T et al (1990) Das maligne Hämangioendotheliom der Schilddrüse. Pathologie, Klinik und Prognose. Wien Klin Wochenschr 102(9):256–9

21. Goh SG, Chuah KL, Goh HK et al (2003) Two cases of epitheloid angiosarcoma involving the thyroid and a brief review of non-Alpine epitheloid angiosarcoma of the thyroid. Arch Pathol Lab Med 127(2):E70–3

22. Morrison WH, Byers RM, Garden AS et al (1995) Cutaneous angiosarcoma of the head and neck. A therapeutic dilemma. Cancer 76:319–27

23. Holden CA, Spittle MF, Jones EW (1987) Angiosarcoma of the face and scalp, prognosis and treatment. Cancer 59:1046–57

24. Pawlik TM, Paulino AF, McGinn CJ et al (2003) Cutaneous angiosarcoma of the scalp. A multidisciplinary approach. Cancer 98:1716–26

25. McIntosh BC, Narayan D (2005) Head and neck angiosarcomas. J Craniofac Surg 16: 699–703

26. Ward JR, Feigenberg SJ, Mendenhall NP et al (2003) Radiation therapy for angiosarcoma. Head Neck 25:873–8

27. Sasaki R, Soejima T, Kishi K et al (2002) Angiosarcoma treated with radiotherapy: impact of tumour type and size on outcome. Int J Radiat Oncol Biol Phys 52:1032–40

28. Koontz BF, Miles EF, Rubio MA et al (2008) Preoperative radiotherapy and bevacizumab for angiosarcoma of the head and neck: two case studies. Head Neck 30:262–6

29. Hanke CW, Sterling JB (2006) Prolonged survival of angiosarcoma of the nose: a report of 3 cases. J Am Acad Dermatol 54:883–5

30. Sher T, Hennessy BT, Valero V et al (2007) Primary angiosarcoma of the breast. Cancer 110:173–8

31. Hellmann K, Burrage K (1969) Control of malignant metastases by ICRF 159. Nature 224:273–5

32. Baker D, Constable W, Elkon D et al (1981) The influence of ICRF 159 and levamisole on the incidence of metastases following local irradiation of a solid tumor. Cancer 48:2179–83

33. Peters LJ (1975) A study of the influence of various diagnostic and therapeutic procedures applied to a murine squamous carcinoma on its metastatic behaviour. Br J Cancer 32(3): 355–65

2.3.2.2 Gastro-Intestinal Malignancies

Gastric Cancer

Razoxane alone was shown to be ineffective in terms of remission induction in advanced cancer of the stomach. Between 1984 and 1993, a prospective phase II study with radiotherapy and concurrent razoxane (Rz) was performed in 28 patients with gastric carcinoma for the assessment of radiation response [1].

Original article: Rhomberg W, Boehler F, Eiter H et al (1996) Radiotherapy and razoxane in the treatment of gastric cancer. Radiat Oncol Invest 4:27–32

Synopsis

Patients/inclusion criteria. The study included patients with histologically proven inoperable primary or recurrent cancer of the stomach referred for pallation. In addition, five patients with positive surgical margins were treated postoperatively. Previous cytotoxic chemotherapy was permissible. There was no limit regarding age or performance status.

Outcome measure. Endpoints of the study were the rate of objective tumor responses and the local tumor control.

Results. Taking all patients together, there was a 89% rate of partial responses (16/18) for the measurable and evaluable tumors; 2 lesions showed no change. The local control rate for all patients was 64%, and the median time to an in-field recurrence was 7 months (9/25 patients). Rapid pain relief was achieved in all but one patients. There was no case of an immediate tumor progression at the site of the irradiation.

Among 5 unresectable patients (3 with primary tumors and 2 with gross residual disease after surgery), there were 3 partial responses and one 'no change'. One patient died intercurrently from bleeding in the region of a stent and was not evaluable.

Among 13 patients with locoregional recurrences, there were 12 measurable lesions. Eleven patients experienced a partial response (92%); one case was not evaluable. In 5 of these 13 patients the tumors showed an in-field regrowth after a median time of 7 months. After a minimum follow-up time of 14 months, the local control rate was 62%. A subgroup of 10 patients with locoregional recurrences and no distant metastases had a median survival time of 9.5 months (range, 1.5–23 months). However, 5 of these 10 patients received cytotoxic chemotherapy at further relapse, with a combination of 5-fluorouracil (5-FU), doxorubicin, and mitomycin C in most instances.

Measurable metastases were studied in 4 patients, at the abdominal wall, liver, bone, and supraclavicular nodes. All 4 patients showed partial regressions.

Finally, four of five patients with microscopic residuals after surgery remained locally controlled within the area irradiated. Three of them died from distant metastases after 6.5, 12, and 16.5 months, and two patients were alive with no evidence of disease at 108 and 14 months, respectively.

Toxicity. The treatment was fairly well tolerated by the majority of patients. None of the patients refused the treatment once started. Nausea and a reduction of the performance status may, however, cause a problem if larger fields are to be used in the postoperative setting.

The dose limiting toxicity of razoxane was leukopenia with leukocytes below 3,000/mm^3 being observed in 13 cases. The nadir of the leukopenia was reached in most instances at day 14 of therapy. Values of less than 1,000 leukocytes/mm^3 were seen 5 times. Thrombocytopenia was not observed.

Comment

Historically, radiation therapy for adenocarcinoma of the stomach was widely regarded as ineffective or at best moderately effective. Reasons for this view were probably the absence of effective diagnostic imaging as well as the limited availability of modern therapeutic equipment in radiation oncology at that time. Data from Japan and Korea, however, indicate that megavoltage therapy is able to induce objective remissions in as many as 50–70% of inoperable or recurrent lesions [2–5].

Small advantages concerning a better local control for inoperable or residual gastric cancer were reported with combined chemo-radiotherapy [6–9]. Falkson and Falkson [6] described an objective improvement in 55% of patients with advanced inoperable gastric cancer by radiotherapy with 50 Gy and 5-FU. In their study, the small gain regarding local control did not translate into an improved survival if adjuvant treatment was given. Until recently, reviews on adjuvant combined modality therapy and larger studies with either radiation or chemotherapy, denied an appreciable advantage compared to surgery alone [10–13]. As yet, only Macdonald JS et al. showed a survival advantage for radiotherapy and 5-FU in the adjuvant setting in their large US. Intergroup study (INT-0116) [14]. In an update in 2004, this survival advantages persisted at a median follow-up of 7 years [15].

Although our experience with radiotherapy and razoxane in stomach cancer is limited, the series has unselected cases, and it is remarkable that nearly all patients with measurable lesions have shown major partial responses even with relatively low radiation doses. On the other hand, there were no unbiased complete remissions. Since the tumor burden was quite high in most of our patients and the treatment was given in palliative intent (lower radiation doses), this result was not surprising. Most recent data from studies of neoadjuvant chemo-radiation therapy described complete pathological responses with radiation, infusional fluorouracil and weekly paclitaxel [16] or 5-FU and cisplatin [17] in the order of 20%. These data, however, cannot be directly compared with our palliative data since some of the patients in the neoadjuvant studies had gastric cancer of the stages IB and IIB. Nevertheless, the overall response rate of 89% achieved with razoxane seems to be clearly higher than in trials using radiotherapy alone [3–5] or in combination with cytotoxic drugs [6, 8, 9].

The combined radiation treatment with razoxane seems to be a preferable modality for the palliative treatment of gastric cancer. For instance, in a study of Tey et al. [18] only 25% of patients (2/8) with pain responded to palliative radiation therapy. To further show a potential survival gain by the combination of radiotherapy and razoxane, it would require to use the sensitizer in earlier stages of the disease, i.e., in the adjuvant setting. Studies at relapse after surgery or postmortem indicate that between 30 and 50% of patients with stomach cancer reveal only local or regional recurrences [19, 20]. In view of that, the adjuvant use of radiation together with razoxane would represent a rational approach to the treatment of high risk gastric cancer, perhaps in conjunction with other drugs, modern radiotherapy field arrangements and intensive nutritional support.

References

1. Rhomberg W, Boehler F, Eiter H, Alton R, Maier R (1996) Radiotherapy and razoxane in the treatment of gastric cancer. Radiat Oncol Invest 4:27–32
2. Abe M, Takahashi M (1981) Intraoperative radiotherapy: the Japanese experience. Int J Radiat Oncol Biol Phys 5:863–8
3. Asakawa H, Takeda T (1973) High energy X-ray therapy of gastric carcinoma. J Jpn Soc Cancer Ther 8:362
4. Kim GE, Shin HS, Seong JS, Loh JK, Suh CO, Lee JT, Roh JK et al (1994) The role of radiation treatment in management of extrahepatic biliary tract metastasis from gastric carcinoma. Int J Radiat Oncol Biol Phys 28:711–7
5. Tsukiyama I, Akine Y, Kajiura et al (1988) Radiation therapy for advanced gastric cancer. Int J Radiat Oncol Biol Phys 15:123–7
6. Falkson G, Falkson HC (1969) Fluorouracil and radiotherapy of locally unresectable gastrointestinal cancer. Lancet 2:1252–3
7. Gastrointestinal Tumor Study Group (1982) A comparison of combination chemotherapy and combined modality therapy for locally advanced gastric carcinoma. Cancer 49:1771–7
8. Moertel CG, Childs DS Jr, Reitemeier RJ, Colby MY, Holbook MA (1969) Combined 5-fluorouracil and supervoltage radiation therapy of locally unresectable gastrointestinal cancer. Lancet 2:865–7
9. O'Connell MJ, Gunderson LL, Moertel CG et al (1985) A pilot study to determine clinical tolerability of intensive combined modality therapy for locally unresectable gastric cancer. Int J Radiat Oncol Biol Phys 11:1827
10. Gunderson LL, Martin KJ, O'Connell MJ, Beart RW, Kvols LK, Nagorney DM (1985) Residual, recurrent or unresectable gastrointestinal cancer. Role of radiation in single or combined modality treatment. Cancer 55:2250–8
11. Hallissey MT, Dunn JA, Ward LC, Allum WH (1994) The British Stomach Cancer Group trial of adjuvant radiotherapy or chemotherapy in resectable gastric cancer. Lancet 343:1309–12
12. Hemans J, Bohnenkamp JJ, Boon MC (1993) Adjuvant therapy after curative resection for gastric cancer. J Clin Oncol 11:1441–7
13. Queisser W, Heim ME (1989) Combined modality of radiation and chemotherapy for the treatment of gastric carcinoma. A review. Onkologie 12:156–60
14. Macdonald JS, Smalley SR, Benedetti J et al (2001) Chemoradiotherapy after surgery compared to surgery alone for adenocarcinoma of the stomach or gastroesophageal junction. N Engl J Med 345:725–30
15. Macdonald JS (2005) Role of post-operative chemoradiation in resected gastric cancer. J Surg Oncol 90:166–70

16. Ajani JA, Winter K, Okawara GS et al (2006) Phase II trial of preoperative chemoradiation in patients with localized gastric adenocarcinoma (RTOG 9904): quality of combined modality therapy and pathologic response. J Clin Oncol 24:3953–8
17. Balandraud P, Moutardier V, Giovannini M et al (2004) Locally advanced adenocarcinomas of the gastric cardia: results of preoperative chemoradiotherapy. Gastroenterol Clin Biol 28: 651–7
18. Tey J, Back MF, Shakespeare TP et al (2007) The role of palliative radiation therapy in symptomatic locally advanced gastric cancer. Int J Radiat Oncol Biol Phys 67:385–8
19. Gunderson LL, Sosin H (1982) Adenocarcinoma of the stomach: areas of failure in a reoperation series (second or symptomatic looks). Clinicopathologic correlation and implications for adjuvant therapy. Int J Radiat Oncol Biol Phys 8:1–11
20. Timothy A (1980) Gastric cancer – prospects for radiotherapy. International congress on Diagnosis and Treatment of Upper GI Tumors, Mainz, 9–11 Sept 1980

Pancreatic Adenocarcinoma

Razoxane and Radiotherapy in Unresectable Localized Pancreatic Cancer (Unpublished)

W. Rhomberg, H. Stephan, F. Böhler, H. Eiter, and B. Schneider

Background. Unresectable pancreatic cancer has a dismal prognosis with a median survival of less than 1 year in almost all published reports. Since new treatment approaches have to be explored for this disease and razoxane had so far not yet been used as radiosensitizer in this disease, a prospective study with radiotherapy and razoxane was performed in unresectable localised pancreatic cancer.

Patients and Methods. The analysis comprises 16 patients (9 males and 7 females) with histologically confirmed localized adenocarcinomas of the pancreas which were referred for primary or additive radiation therapy and treated between 1988 and 2000. Variants with a more favourabvle prognosis such as intraductal cystic carcinomas or neuroendocrine tumors are not included in this series.

The median age was 60 years (range, 42–79 years). Ten patients were prospectively treated by radiotherapy and razoxane. Of the remaining 6 patients, 5 received complementary external beam radiation therapy (EBRT) alone after intraoperative radiotherapy (IORT) having primarily been delivered in another hospital, one patient was treated with EBRT alone. There was no randomization between the groups.

High energy photons of linear accelerators were used together with conformal treatment planning. Two patients were treated by a telecobalt unit. The median tumor dose in the razoxane treated group was 59.4 Gy (range, 54–64 Gy) at the ICRU point, in the IORT group 57 Gy (range, 50–60 Gy). Single doses between 1.7 and 2 Gy were given five times a week. The fractions of the IORT were between 20 and 30 Gy (median 20 Gy). The overall time of radiotherapy treatments ranged between 5 and 9 weeks.

Chest X-rays and CT scans of the abdomen were performed every 3 months together with serum marker measurements of CEA and Ca 19.9 at the same intervalls. Six patients received cytotoxic chemotherapy upon relapse with distant metastases.

Results. Razoxane and radiotherapy led to 5 major responses (CR + PR) among 9 assessable patients (55%), no major remissions were observed with radiotherapy alone. Local tumor control was achieved in 7 of 10 patients irradiated with razoxane, and in 3 of 6 patients who received radiotherapy alone.

The median survival of the 10 patients with localized unresectable tumors in the razoxane group was 10 months (range, 5–24 months). The 1-year survival was 40%, no patient survived beyond 5 years. Six patients with external beam radiation therapy alone, including 5 patients with IORT had a median survival time of 9 months (range, 2–39 months). The 1-year survival was 33% with no patient surviving 5 years.

Toxicity. The most frequent side effects were nausea, vomiting and loss of appetite. Only 4 of 10 patients treated with razoxane and radiotherapy had no transient symptoms in the upper GI region. A rapidly reversible leukopenia (WHO-grade 2+3) was observed twice in this group. The general tolerance to the irradiation was better without razoxane with 4 of 6 patients experiencing only a slight reduction in their general condition.

The late GI-toxicity in the razoxane group included 4 gastrointestinal bleeds (1 lethal outcome). This complication seemed to be associated with the total radiation dose. Three of the 4 cases with a GI bleed had received tumor doses of 60 Gy and above. Among the 5 cases with IORT there was 1 patient with a GI bleed. On the other hand, 4 of 6 patients without razoxane experienced radiogenic diarrheas (WHO grade 2–4).

Discussion. Regardless of treatment, the median survival in unresectable adenocarcinomas of the pancreas ranges between 5 and 12 months [1–6]. Longer survival was seen in studies with ukrain® together with gemcitabine [7] but long-term survival remains rare. Only radical resection may lead to a more favourable outcome and to occasional cures. However, in an overview of 15 surgical series up to 1993, the median survival times ranged between 6 and 23 months, and the 'median of the medians' was 11.5 months only [1]. Modifications of the radiation treatment such as interstitial brachytherapy [8], intraoperative radiotherapy [6], or novel combinations of chemotherapy and radiation [2, 5] can improve local tumor control rates to 70–80% without prolonging survival to a substantial degree. Cis-platinum and radiotherapy [9, 10], the use of taxanes [11], or other newer combinations [5] seem to be associated with rather low survival times.

The survival in our small patient series compared well with either 5-Fluorouracil or Gemcitabine [4, 12–15], or both [16–18] when combined with radiotherapy. The early outcome with razoxane is encouraging with all 10 patients surviving at least 5 months with a 1-year survival rate of 40%.

Conclusion. In our small series of patients the combination of radiotherapy and razoxane indicates no breakthrough in the treatment of unresectable pancreatic cancer, but it proved to be a costeffective readily applicable treatment modality as effective as IORT or a combination with intravenous 5-fluorouracil or even gemcitabine. To avoid gastrointestinal hemorrhages as late complication the total radiation dose has probably to be reduced to 50–55 Gy.

References

1. Evans DB, Abbruzzese JL, Willett CG (2001) Cancer of the pancreas. In: DeVita VT Jr, Hellman S, Rosenberg SA (eds) Cancer principles & practice of oncology, 6th edn. JB Lippincott, Philadelphia, 1145–9
2. Hidalgo M, Castellano D, Paz-Ares L et al (1999) Phase I–II study of gemcitabine and fluorouracil as a continuous infusion in patients with pancreatic cancer. J Clin Oncol 17:585–92
3. Hoffman JP, Lipsitz St, Pisansky T et al (1998) Phase II trial of preoperative radiation therapy and chemotherapy for patients with localized, resectable adenocarcinoma of the pancreas: an Eastern Cooperative Oncology Group Study. J Clin Oncol 16:317–23
4. Ikeda M, Okada Sh, Tokuuye K et al (2001) Prognostic factors in patients with locally advanced pancreatic carcinoma receiving chemoradiotherapy. Cancer 91:490–5
5. Kornek, Gabriela, Pötter R, Selzer E et al (2001) Combined radiochemotherapy of locally advanced unresectable pancreatic adenocarcinoma with mitomycin C plus 24-hour continuous infusional gemcitabine. Int J Radiat Oncol Biol Phys 49:665–71
6. Nishimuara Y, Hosotani R, Shibamoto Y et al (1997) External and intraoperative radiotherapy for resectable and unresectable pancreatic cancer: analysis of survival rates and complications. Int J Radiat Oncol Biol Phys 39:39–49
7. Gansauge F, Ramadani M, Pressmar J et al (2002) NSC-631570 (Ukrain) in the palliative treatment of pancreatic cancer. Results of a phase II trial. Langenbecks Arch Surg 386:570–4
8. Pfreundner L, Baier K, Schwab F et al (1998) 3D-CT-geplante interstitielle HDR-Brachytherapy und perkutane Bestrahlung und Chemotherapie bei inoperablen Pankreaskarzinomen. Strahlenther Onkol 174:133–41
9. Nguyen TD, Theobald S, Rougier Ph et al (1997) Simultaneous high-dose external irradiation and daily cisplatin in unresectable, non-metastatic adenocarcinoma of the pancreas: a phase I–II study. Radiother Oncol 45:129–32
10. Okusaka T, Okada Sh, Tokuuye K, Wakasugi H, Saisho H, Ishikawa O (2001) Lack of effectiveness of radiotherapy combined with cisplatin in patients with locally advanced pancreatic carcinoma. Cancer 91:1384–9
11. Safran H, Moore T, Iannitti D et al (2001) Paclitaxel and concurrent radiation for locally advanced pancreatic cancer. Int J Radiat Oncol Biol Phys 49:1275–9
12. Blackstock AW, Bernard StA, Richards F et al (1999) Phase I trial of twice-weekly gemcitabine and concurrent radiation in patients with advanced pancreatic cancer. J Clin Oncol 17:2208–12
13. Boz G, De Paoli A, Innocente R, Rossi C, Tosolini G, Pederzoli P et al (2001) Radiotherapy and continuous infusion 5-fluorouracil in patients with nonresectable pancreatic carcinoma. Int J Radiat Oncol Biol Phys 51:736–40
14. Budach V (1992) Kombinierte Radio-/Chemotherapie des Pankreaskarzinoms. In: Schmoll HJ, Meyer HJ, Wilke H, Pichlmayr R (eds) Aktuelle Therapie gastrointestinaler Tumoren. Springer, Berlin, Heidelberg, New York, pp 426–31
15. Murphy JD, Adusumilli S, Griffith KA et al (2007) Full-dose gemcitabine and concurrent radiotherapy for unresectable pancreatic cancer. Int J Radiat Oncol Biol Phys 68:801–8
16. Andre T, Hammel P, Selle F et al (2001) Multicentric feasibility study of bimonthly combination of leucovorin-5FU and gemcitabine followed by chemoradiotherapy with 5FU for locally advanced non metastatic pancreatic adenocarcinoma. Proc Am Soc Clin Oncol 20:155a, Abstract No 616
17. Goldstein D, Van Hazel G, Walpole E et al (2007) Gemcitabine with a specific conformal 3D 5FU radiochemotherapy technique is safe and effective in the definitive management of locally advanced pancreatic cancer. Br J Cancer 97:464–71
18. Wilkowski R, Stoffregen H, Rau H et al (2001) Chemoradiation with gemcitabine in primarily inoperable pancreatic cancer. Int J Radiat Oncol Biol Phys 51(Suppl 1):272

Cancer of the Gallbladder and Bile Ducts

An Exploration of the Radiation Sensitivity of Carcinoma of the Biliary Tree (A Phase II Study Using Radiotherapy and Razoxane)

Original article: Rhomberg W, Stephan H, Boehler F, Erhart K, Eiter H (2005) Radiotherapy and razoxane in advanced bile duct carcinomas. Anticancer Res 25:3613–8

(Reproduction of the material was permitted by IIAR)

Shortened version

Background. Radical tumor resection is presently the only way to achieve long term survival in carcinomas of the biliary tree [1–4]. The median survival in non-resected cases without distant metastases ranges between 5 and 12 months [5–9]. The majority of the literature reports confirm some survival advantages for external beam radiotherapy (EBRT) with or without intraluminal brachytherapy or chemotherapy, for advanced localized disease [5, 10–14] but loco-regional recurrence remains a frequent cause of treatment failure [10, 15, 16]. For this reason new local treatments have to be explored.

Moreover, little is known about the radiation sensitivity of bile duct carcinomas. An older review from 1984 estimates the objective response rates by irradiation in carcinomas of the gallbladder to be around 20%, and in the distal bile duct carcinomas around 50% [17]. In the few subsequent reports, the response rates in measurable disease still ranged between 30 and 54% when external beam radiotherapy and/or intraluminal brachytherapy have been used [18, 19]. Therefore, this study was also undertaken to prospectively assess the objective response rates in bile duct carcinomas treated with radiotherapy and razoxane.

Patients and Methods. Between 1986 and 2000, 23 patients (16 females and 7 males) with advanced cancer of the biliary tree were irradiated together with concurrent razoxane at a dose of 125 mg twice daily by mouth. The median total radiation dose was 48 Gy (range, 1.7–60 Gy) at the ICRU point with single fractions of 1.7–2 Gy. The median duration of the radiotherapy was 6 weeks. Intraluminal brachytherapy was not given. The patients received a median dose of 6.75 g razoxane (range, 1.5–10 g).

Twenty patients received different palliative surgical procedures prior to radiotherapy, three patients had tumour resections, 2 with no clear margins (R-1), only one had a radical (R-0) resection. Nine patients showed metastases at the referral. The patients were chemonaive, only one was previously treated with 5-fluorouracil. In 6 patients the vinca alkaloid vindesine was given with razoxane for reasons of distant metastasis. Ten patients received 5-fluorouracil or gemcitabine based chemotherapy regimes in a later phase of their disease because of progressive disease. The main clinical and pretreatment characteristics were summarized in Table 2.6. The clinicopathological correlation between stage and amount of residual tumor, and the location of the 23 carcinomas may be seen in the original article.

Table 2.6 Pretreatment data of 23 patients with biliary cancer

Number of patients	23
Gender	
Female	16
Male	7
Age, years	
Median (range)	68 (40–92)
Location	
Hepatobiliary	5
Extrahepatic bile ducts (4 Klatskin tumors)	9
Gallbladder	9
Histology	
Adenocarcinoma	13
Cholangiocellular carcinoma	4
Adeno-Ca with productive fibrosis	2
Mixed forms (mucinous, papillary)	2
Biopsy not successful	2
Grade	
G I	1
G II	5
G III	9
Unknown	8
Surgical procedures	
Radical tumor resection (R-0)	1
Tumor resection without clear margins (R-1)	2
Laparatomy, biopsy, biliodigestive anastomosis and/or gastroenterostomy	9
Stent and biopsy	5
Biopsy only	4
Others	2

The patients were followed at monthly intervals. Abdomino-pelvic CTs and chest X-rays were performed every 3 months during the first year. The patients were followed until the end of December 2002; at that time all patients have died.

All patients were analyzed, including those who did not complete therapy. The intended treatment could be completed in 17 of the 23 patients. Three of the 17 had no measurable disease after surgery, leaving 14 patients for evaluating the radiation response.

Results

Carcinomas of the gallbladder. There were nine chemonaive patients with carcinoma of the gallbladder, 7 females and 2 males. Among the 6 patients adequately irradiated there were four measurable tumours as well as one R-0 and R-1 resection, respectively. All 4 measurable tumours responded to the combined treatment (1 CR, 3 PR). Local tumour control was achieved in 4 of the 6 patients.

On an intention to treat basis, the median survival of the 9 patients (4 with distant metastases) was 12 months from diagnosis and 9.5 months from the start of the radiotherapy (range 1–31). The 1-year survival for all patients since the start of the radiotherapy was 33%.

Extrahepatic bile duct carcinomas. In this group, 9 patients without previous chemotherapy commenced treatment with razoxane and irradiation. Among 5 patients with measurable disease, 4 were shown to have objective tumour regressions (2 complete and 2 partial responses), one tumour showed no change. Local tumour control was achieved in 4 of 5 assessable patients.

Three patients did not finish the treatment after 1.7, 7, and 29 Gy, respectively (1 refusal, 2 intercurrent deaths). In the remaining 6 patients, the median survival time from the start of the irradiation was 10 months (range 3–48) with a 1-year survival rate of 33%.

Hepatobiliary carcinomas. Among 5 patients with intrahepatic bile duct carcinomas, one was pretreated with intraarterial embolisations and 5-fluorouracil. A partial tumour response was seen in one patient, three tumours showed no change, and one patient had progressive disease despite of the treatment. However, the tumours remained locally controlled in 4 of the 5 patients. The median survival was 10 months (range 2–30) from diagnosis, and 5 months (range 1.5–20) from the start of the radiotherapy.

Some overall results. If all tumors of different locations were taken together, the objective response rate is 64%. Considering only the patients with carcinomas of the gallbladder and extrahepatic bile ducts, partial and complete responses were seen in 8 out of 9 patients (89%), 3 patients (33%) showed a complete response. The overall local control rate was 75% (12 of 16 assessable patients). The objective response data are summarized in Table 2.7.

Table 2.7 Objective response rates in 14 patients with measurable carcinoma of the biliary tree

	No.	Response			
		CR	PR	NC	P
Carcinomas of the gallbladder	4	1	3	–	–
Extrahepatic bile ducts	5	2	2	1	–
Hepatobiliary cancers	5	–	1	3	1
Totals	14	3	6	4	1
			[CR + PR: 9/14 (64%)]		
Gallbladder and extrahepatic Bile ducts only			[CR + PR: 8/9 (89%)]		

CR, complete response; PR, partial response; NC, no change; P, progressive disease.

If all patients with different biliary cancers without distant metastases were selected and analysed on an intention to treat basis, the median survival was 10 months (range 1–48) from radiation treatment; the 1-year survival being 43%. No patient survived beyond 4 years.

Toxicity and complications. Most frequent side effects in this study were nausea and vomiting (61%). The reactions were of grade 1 and 2 (WHO) in all but one patients, only one refused further treatment because of intractable nausea. Reversible leukopenias of grade 3 and 4 were seen in one case, respectively, one patient developed neutropenic fever.

Complications to some extent inherent to the disease were the main reason for not having finished the combined treatment in some patients. Major complications included 2 cases of fatal pulmonary embolism during treatment, and 2 unclear cardiac deaths (one patient was in a complete remission). One patient died from a perforation of the carcinomatous gallbladder while in a subtotal remission, another patient died from a biliary cirrhosis of the liver 4 years after treatment of a Klatskin tumour with no evidence of disease at autopsy. Other events were a recurrent pulmonary embolism of minor degree, 3 cases of recurrent cholangitis, and an apoplexy with rapid recovery.

Discussion. The study showed response rates following irradiation and razoxane which are superior to those achieved with radiotherapy alone. Concomitant irradiation together with current cytotoxic agents may also enhance the local efficacy of the treatment [16] with smaller tumors (<5 cm) being more likely to respond to radiochemotherapy with gemcitabine or 5-FU [20]. What are the objective response rates? Morganti et al. reported a 33% major response rate (CR+PR) when 5-FU continuous infusion was combined with external and intraluminal radiotherapy [21]. Conformal radiotherapy combined with regional chemotherapy led to high response rates if only localised hepatobiliary cancers were considered in a study by Robertson et al. [22]. Among 11 such patients with measurable disease, there were 10 patients with a partial and one with a complete response. However, if whole liver irradiation was necessary for diffuse carcinomas, only 1 of 7 patients responded leading to an overall response in 12 of 18 patients (66%) [22]. Considering these recent literature reports, the objective response rates achieved by razoxane compare well to these data.

A variety of complications prevented 6 patients to receive the planned full radiation dose, and a third of all patients died from complications unrelated to tumour progression. Thromboembolic events were noted in five patients. The question arises whether this phenomenon is related to razoxane. We cannot exclude such an association since an increase of thromboembolic complications became known recently when antiangiogenic substances were incorporated into cytotoxic regimens [23]. On the other hand, such a connection is less likely since the spectrum of complications is wide, and comparable events were certainly not observed in other tumour entities treated in the same way (see reports on other tumor entities in Section 2.3.2). Basic data as to the kind and frequency of complications in bile duct cancer are rather sparse. Is pulmonary embolism in the order of 10% an unusual figure in bile duct carcinomas? What are the toxic consequences of high bilirubin levels in the serum? Complication reporting in cancer of the biliary tree seems to be quite variable in the literature. Some authors saw hardly any complications, others share our experience [24–27]. Cholangitis, abscess formation or GI-bleeding are observed most frequently [27, 28]. Van Gulik et al. supposed that intraluminal radiotherapy may have contributed to the rate of complications, and therefore, the authors now restrict treatment to EBRT as the survival was not different without intraluminal brachytherapy in their experience [11].

Conclusions. Combined radiotherapy and razoxane led to local response rates which are superior to data from the literature when radiotherapy alone is used. Obstacles to the treatment were complications of the disease and frequent metastasis.

References

1. Bathe OF, Pacheco JT, Ossi PB et al (2000) Management of hilar bile duct carcinoma. Hepatogastroenterology 48:1289–94
2. McMasters KM, TuttleTM, Leach SD et al (1998) Neoadjuvant chemoradiation for extrahepatic cholangiocarcinoma. Am J Surg 174:605–8
3. Schoenthaler R, Phillips TL, Castro J et al (1994) Carcinoma of the extrahepatic bile ducts. The University of California at San Francisco experience. Ann Surg 219:267–74
4. Urego M, Flickinger JC, Carr BI (1999) Radiotherapy and multimodality management of cholangiocarcinoma. Int J Radiat Oncol Biol Phys 44:121–6
5. Alden ME, Mohiuddin M (1994) The impact of radiation dose in combined external beam and intraluminal Ir-192 brachytherapy for bile duct cancer. Int J Radiat Oncol Biol Phys 28:945–51
6. Ewing H, Sali A, Kune GA (1989) Klatskin tumors: a 20 year experience. Aust N Z J Surg 59:25–30
7. Flickinger JC, Epstein AH, Iwatsuki S, Carr BI, Starzl TE (1991) Radiation therapy for primary carcinoma of the extrahepatic biliary system. Cancer 68:289–94
8. Kamada T, Saito H, Takamura A et al (1996) The role of radiotherapy in the management of extrahepatic bile duct cancer: an analysis of 145 consecutive patients treated with intraluminal and/or external beam radiotherapy. Int J Radiat Oncol Biol Phys 34:767–74
9. Veeze-Kuijpers B, Meerwaldt JH, Lameris JS et al (1990) The role of radiotherapy in the treatment of bile duct carcinoma. Int J Radiat Oncol Biol Phys 18:63–7
10. Fields JN and Emami B (1987) Carcinoma of the extrahepatic biliary system – results of primary and adjuvant radiotherapy. Int J Radiat Oncol Biol Phys 13:331–8
11. van Gulik TM, Rauws EA, Gonzales-Gonzales D et al (1997) Pre- and postoperative irradiation in the treatment of resectable Klatskins tumors. Ned Tijdschr Geneeskd 141:1331–7
12. Kresl JJ, Schild SE, Henning GT et al (2002) Adjuvant external beam radiation therapy with concurrent chemotherapy in the management of gallbladder carcinoma. Int J Radiat Oncol Biol Phys 52:167–75
13. Saito H, Takamura A (2000) Management of hilar bile duct carcinoma with high-dose radiotherapy and expandable metallic stent placement. Nippon Geka Gakkai Zasshi 101:423–8
14. Verbeek PC, Van Leeuwen DJ, Van der Heyde MN, Gonzalez-Gonzalez D (1991) Does additive radiotherapy after hilar resection improve survival of cholangiocarcinoma? An analysis in sixty-four patients. Ann Chir 45:350–4
15. Buskirk St, Gunderson LL, Adson MA et al (1984) Analysis of failure following curative irradiation of gallbladder and extrahepatic bile duct carcinoma. Int J Radiat Oncol Biol Phys 10:2013–23

16. Minsky BD, Wesson MF, Armstrong J et al (1990) Combined modality therapy of extrahepatic biliary system cancer. Int J Radiat Oncol Biol Phys 18:1157–63
17. Köster R, Scherer E (1984) Present state and possibilities of radiotherapy in the interdisciplinary treatment of malignant tumours of stomach, pancreas und biliary tract. III. Extrahepatic bile ducts and gall bladder. Strahlenther 160:403–10
18. Hendrickx P, Luska G, Junker D et al (1990) Results of percutaneous transluminal irradiation of biliary tract carcinoma with 192 Iridium. Strahlenther Onkol 166:392–6
19. Milella M, Salvetti M, Cerrotta A et al (1998) Interventional radiology and radiotherapy for inoperable cholangiocarcinoma of the extrahepatic bile ducts. Tumori 84:467–71
20. Brunner T, Grabenbauer G, Strnad V et al (2001) Nur Patienten mit kleinen irresektablen Tumoren der Gallenwege und der Gallenblase profitieren von einer simultanen Radiochemotherapie. Strahlenther Onkol Sondernr 1:88, Abstract P 4.11
21. Morganti AG, Trodella L, Valentini V et al (2000) Combined modality treatment in unresectable extrahepatic biliary carcinoma. Int J Radiat Oncol Biol Phys 46: 913–9
22. Robertson JM, Lawrence TS, Dworzanin LM et al (1993) Treatment of primary hepatobiliary cancers with conformal radiation therapy and regional chemotherapy. J Clin Oncol 11: 1286–93
23. Glade-Bender J, Kandel JJ, Yamashiro DJ (2003) VEGF blocking therapy in the treatment of cancer. Expert Opin Biol Ther 3:263–76
24. Hatano K, Cho K, Okamoto M et al (1992) Results of radiation therapy of extrahepatic bile duct carcinoma. Nippon Igaku Hoshasen Gakkai Zasshi 52:799–803
25. Hayes JK, Sapozink MD, Miller FJ (1988) Definitive radiation therapy in bile duct carcinoma. Int J Radiat Oncol Biol Phys 15:735–44
26. Kurosaki H, Karasawa K, Kaizu T et al (1999) Intraoperative radiotherapy for resectable extrahepatic bile duct cancer. Int J Radiat Oncol Biol Phys 45:635–8
27. Kuvshinoff BW, Armstrong JG, Fong Y et al (1995) Palliation of irresectable hilar cholangiocarcinoma with biliary drainage and radiotherapy. Br J Surg 82:1522–5
28. Vallis KA, Benjamin IS, Munro AJ et al (1997) External beam and intraluminal radiotherapy for locally advanced bile duct cancer: role and tolerability. Radiother Oncol 41: 61–6

Liver Metastases

A Convenient Short-Term Treatment for a Life Threatening Condition

Original article: Hellmann K, Goold M, Higgins N, Phillips RH (1992) Responses of liver metastases to radiotherapy and razoxane. J R Soc Med 85(3):136–8

Twenty-five patients with liver metastases, chiefly due to colorectal cancer, were given a loading dose of razoxane for 3 days before 5 consecutive days of radiotherapy to the whole liver. Patients also took razoxane during the radiotherapy and then for 1 months afterwards. Liver tumour volume was measured on CT scans using the ELSCINT 3D soft tissue imaging programme just before and 4 weeks after the end of the radiation treatment. Twelve of the 25 patients had tumour volume reductions of more than 50%. The overall major response rate therefore is 12/25 (48%). In two of the major responders the liver metastases were due to recurrent stomach cancer. In addition to the 12 responders, four patients had a reduction of more than 20% but less than 50%, thus giving an overall response rate of 16/25 (64%).

These results can form the basis of a formal, randomized, controlled clinical trial of radiotherapy alone (or any other treatment) compared with radiotherapy and razoxane in the difficult and life threatening condition presented by liver metastases.

Rectal Cancer

A Phase II Study with Razoxane and Irradiation in Unresectable Recurrent Rectal Cancer

Original article: Rhomberg W, Eiter H, Hergan K, Schneider B (1994) Inoperable recurrent rectal cancer: results of a prospective trial with radiation therapy and razoxane. Int J Radiat Oncol Biol Phys 30:419–25

Synopsis

Background

The treatment results of recurrent rectal cancer are not satisfactory. Unresectable disease is associated with high morbidity and a dismal prognosis. Long term survival is most likely if radical surgery with free surgical margins can be achieved. Radiotherapy alone is of limited efficacy.

Objectives

From this background, the radiosensitizer razoxane was evaluated in a phase II study in unresectable recurrent rectal cancer with or without distant metastases. *Outcome measures* were the rate of objective responses to irradiation and overall survival.

Methods and Materials

From 1984 to 1990 razoxane was given together with radiation therapy to 40 patients. Loco-regional relapses in the pelvis were isolated in 24 and associated with distant metastases in 16 patients. Special attention was given to the subset of 24 patients wih isolated local recurrences. This group was compared with historical controls who were treated with radiotherapy alone. Tumors of the sigmoid or the anorectal region were not included. Six patients had previously received cytotoxic chemotherapy including 5-fluorouracil (FU) alone or combinations of 5-FU with folinic acid or nitrosoureas. The median interval from diagnosis to relapse was 13 months (range, 4–56) which is of importance in view of the results of a subsequent randomized study of this topic.

The dosage of razoxane was 150 mg/m^2 daily orally starting 5 days before the first irradiation. The drug was then given each radiation day until the end of treatment. The median radiation dose was 60 (40–62) Gy in the evaluable patients. The minimum follow-up was 32 months and the median 72 months.

To allow for the evaluation of an objective response, serial CT scans were performed at the start of the treatment, 2–6 weeks after completion, and then every 3–6 months.

Results

Isolated locoregional recurrences. Of the 24 patients with local recurrences without distant metastases, 23 were evaluable for clinical response. There were one complete and 12 partial responses giving a major response rate of 57%. In 10 cases the tumors showed no evidence of change. No tumor progressed in any patient before 7 months from the onset of the radiation treatment. The local control rate was 43%.

The actual median survival time (MST) from the start of the radiotherapy was 24 months (12–94+) compared to 13 months in a previous study using radiotherapy only. All patients survived at least 1 year. The MST from diagnosis of the recurrence was 29 months (14–105+).

The Kaplan-Meier analysis projected a median survival of 30 months for patients treated with irradiation and razoxane and 12 months for the small historical group who received radiotherapy alone. The difference in the survival curves showed a p-value of 0.003 for the Wilkoxon test and 0.04 for the log-rank test. It has to be remembered, however, that the controls were not randomized.

Locoregional recurrence and distant metastases. The major response rate was 54% in 16 patients with recurrences and distant spread, a similar value as in isolated recurrences. Again, no patient showed a pelvic progression during or shortly after the radiation. In this patient subgroup none of the patients survived 5 years. The median survival was 10 months (range, 3.5–38).

Conclusion

It was concluded at that time that combined radiotherapy and razoxane may lead to better local control and improved survival in unresectable recurrent rectal cancer compared to historical controls and all literature reports where radiotherapy alone was given. The treatment was easy to administer and associated with a moderate toxicity.

A Small Randomized Study Confirms the Radiosensitizing Ability of Razoxane in Recurrent Rectal Cancer

Original article: Rhomberg W, Hammer J, Sedlmayer F et al (2007) Irradiation with and without razoxane in the treatment of incompletely resected or inoperable recurrent rectal cancer. Results of a small randomized multicenter study. Strahlenther Onkol 183:380–4

Synopsis

Background

An earlier phase II study using radiotherapy and razoxane in recurrent rectal cancer showed rather high local control rates and survival data which compared favourably with the literature [1]. Therefore, a randomized controlled trial was initiated by the Austrian Society of Radiooncology (OEGRO) in 1992 to compare this treatment combination vs. radiation therapy alone.

Objectives

To determine the rate of major responses induced by radiotherapy alone or in combination with razoxane in patients with unresectable recurrent rectal cancer without distant metastasis.

Outcome measures were response rates to radiotherapy, local control, overall survival, and toxicity.

Design and Intervention

This phase III study was approved by the local ethical committee in march 1992. Between 1992 and 1999, 36 patients from 4 hospitals were randomized to receive local radiotherapy without (group A) or with razoxane (group B). Exclusion criteria were previous radiation therapy of the pelvic region, age >85, Karnofsky index below 40, or radical surgery of the recurrence.

The prognostic variables of the two groups were similar except for a longer median latency period from initial surgery to local recurrence in group A compared to B (24 months vs. 12 months). In addition, more patients in group B had received a pretreatment with 5-fluorouracil based chemotherapy (44% vs. 7%).

The median total radiation dose was 60 Gy in each group. The patients of group B received a median razoxane dose of 9.6 g (range, 5–12 g) concomittantly with the radiation treatment.

Five of the 36 patients were not evaluable: two invalid randomizations because of distant metastases, one protocol violation because of concurrent 5-fluorouracil infusions during radiotherapy, one treatment refusal, and one early death before the end of the radiotherapy. All patients could be followed up until their death or the end of June 2005.

The study was closed before the planned number of patients were recruited because of slow patient accrual, and since the study was not recent enough anymore by the advent of several new chemotherapeutic agents which have a clear influence on the survival of patients with colorectal cancer.

Results

The rate of partial and complete responses was 28% in group A, and 39% in group B. No patient in the razoxane-arm showed progressive disease under or shortly after the treatment. Local tumor control was achieved in 1 of 13 cases (8%) in group A and 7 of 18 cases (39%) in group B (Chi-square $= 3.83$; $p = 0.05$).

The median survival from the start of the radiation treatment was 20 months (range, 12–52) in group A, and 20 months (range, 9–142+) in group B. The mean survival time was 26 months vs. 36 months, respectively. No patient in group A and 4 patients (22%) in group B survived 5 years.

The rate of distant metastases during the later course of the disease was 61% in patients with radiotherapy alone and 44% in patients treated with radiotherapy and razoxane.

Acute Toxicity. Main acute side effects were diarrhea, moist desquamation of small areas of the skin and dysuria. There were no differences with respect to skin reactions, diarrhea, weight loss or infections between the two groups. In general, however, the combined treatment was less well tolerated than radiation therapy alone. Leukopenia of WHO grade 2 and 3 occurred in the razoxane group exclusively. The acute side effects were transient and usually of mild to moderate degree.

Late Toxicity and Quality of Life. Radiogenic complications were of similar frequency in both treatment arms. In arm A, two cases of radiation enteritis of grade 3 and two bone fractures due to radiation treatment were noted. In arm B, there were two cases of proctitis with moderate hemorrhage, and one case of pelvic fibrosis, radiation enteritis and perineal ulcer, respectively.

In terms of frequency, surgical complications exceeded other adverse effects (21 events in both groups). In each group, three operations for adhesion ileus were necessary. Perineal fistulas, hernias, abscess formation and leaks of anastomoses were equally distributed between the groups.

The overall morbidity was high in both groups. This is reflected by a rate of 1.6 complications per patient in both groups. The median proportion of hospital days related to the remaining overall survival was 5% (range 0–23.5%) in group A and 4.5% (range 0–20%) in group B. Ultimately, many patients showed a kind of lingering illness in their terminal stage of disease that reduced the quality of life and which cannot be measured by the usual instruments of QL questionaires because then the patients are cared by family doctors at home or in different hospitals.

Discussion and Comment

In a phase II study using radiotherapy together with razoxane for recurrent non-metastatic rectal cancer [1], a median survival of 24 months was observed whereas a historical control group with radiotherapy alone showed a survival of 12 months. A median survival around 12 months was a common value in the literature until 1994, even for combination chemo-radiotherapy mostly with 5-fluorouracil based regimens [2–10]. In the present randomized study, the group with radiation therapy alone achieved a median survival which is comparable to the razoxane treated group of the previous study, the survival in the razoxane arm remained at a similar value as seen earlier. The reason for that seems not to be related to surgery for recurrence since only curative resections with clear margins are associated with a substantial survival gain [11–14], but curative (R-0) resections were not included in this study. One reason is probably the wider use of recent salvage chemotherapy regimens. In addition, there is some imbalance between the treatment groups as to prognostic variables in favour of arm A. There were more patients in the razoxane-group that were pretreated with chemotherapy for recurrent disease. As a rule, any previous treatment reduces the success rate of a following treatment. In our study, in group A only 1 patient (8%) had a previous treatment with 5 FU for 5 days whereas in group B, 8 of 18 patients (44%) were pretreated by 5-FU based regimens. This may also be the reason that the response rate achieved with radiotherapy and razoxane in this study (39%) was not as high as in the previous phase II study (57%). Last but not least, the interval from diagnosis to recurrence was shorter in the razoxane-group (12 months vs. 25 months) suggesting a more aggressive tumor behavior in the cases of group B. The feature of a longer latency period between diagnosis and local recurrence which we were not aware of at the time of randomization, seems to be – similar to breast cancer – of definite prognostic significance [12, 14–16].

In the radiotherapy-alone group there was no long term survivor, and the local control (1/13) is rather poor. Even with neutrons [17] or other radiosensitizers such as misonidazole [18], there is at present no therapeutic gain for long-lasting survival although local control and pain improvement seems to be better with neutrons than with photons. Therefore, the treatment options for localized recurrent disease are urgently to be expanded. In doing so, razoxane could well be a constituent of further experimental multimodal therapy.

Summary

Radiotherapy combined with razoxane is able to improve response rates and local control of unresectable recurrent rectal cancer compared to radiotherapy alone. That was described earlier and is confirmed in this small randomized study. Because of some imbalance of the prognostic factors in the two treatment groups, no further conclusions were possible. Although the median overall survival was not different in the two treatment arms for distinct reasons, long term survival was possible with razoxane and radiotherapy.

In future controlled studies, stratification according to prognostic parameters like the latency period to recurrence or the pretreatment with cytotoxic drugs should be considered.

References

1. Rhomberg W, Eiter H, Hergan K, Schneider B (1994) Inoperable recurrent rectal cancer: results of a prospective trial with radiation therapy and razoxane. Int J Radiat Oncol Biol Phys 30:419–25
2. Bohndorf W, Richter E, Aydin H (1984) CT diagnosis and radiotherapy of local recurrences after surgical treatment of the carcinoma of the rectum. Strahlenther 160:318–23
3. Brockmann WP, Wiegel T, Sommer K, Steiner P, Hübener KH (1993) Chemoradiotherapy of advanced colorectal cancer – results of a pilot study with 44 patients. Strahlenther Onkol 169:107–13
4. Danjoux CE, Gelber RD, Cotton GE, Klaassen DJ (1985) Combination chemoradiotherapy for residual, recurrent or inoperable carcinoma of the rectum. ECOG Study (EST 3276). Int J Radiat Oncol Biol Phys 11:765–71
5. Dobrowsky, W (1992) Mitomycin C, 5-fluorouracil and radiation in advanced, locally recurrent rectal cancer. Br J Radiol 65:143–7
6. Guiney MJ, Smith JG, Worotniuk V, Ngan S (1999) Results of external beam radiotherapy alone for incompletely resected carcinoma of rectosigmoid or rectum: Peter MacCallum Cancer Institute experience 1981–1990. Int J Radiat Oncol Biol Phys 43:531–6
7. Lybeert MLM, Martijn H, De Neve W, Crommelin MA, Ribot JG (1992) Radiotherapy for locoregional relapses of rectal carcinoma after initial radical surgery: definite but limited influence on relapse-free survival and survival. Int J Radiat Oncol Biol Phys 24:241–6

8. Overgaard M, Bertelsen K, Dalmark M, Gadeberg CC, von der Maase H, Overgaard J, Sell A (1989) A randomized trial of radiotherapy alone or combined with 5-FU in the treatment of locally advanced colorectal carcinoma. Abstract Nr. O-0624, 5th European conference on Clinical Oncology (ECCO 5), London 3–5 Sept; 1989

9. Schmidt H, Müller RP, Hildebrand D (1984) Results of radiation therapy in local recurrences of colorectal tumors. Strahlenther 160:288–92

10. Vongtama V, Douglas HO, Moore RH, Holyoke ED, Webster JH (1975) End results of radiation therapy alone and combination with 5-Fluorouracil in colorectal cancers. Cancer 36:2020–5

11. Alektiar KM, Zelefsky MJ, Paty PB, Guillem J, Saltz LB, Cohen AM, Minsky BC (2000) High-dose-rate intraoperative brachytherapy for recurrent colorectal cancer. Int J Radiat Oncol Biol Phys 48:219–26

12. James RD, Johnson RJ, Eddleston B, Zheng GL, Jones JM (1983) Prognostic factors in locally recurrent rectal carcinoma treated by radiotherapy. Br J Surg 70:469–72

13. Law WL, Chu KW (2000) Resection for local recurrence of rectal cancer: results. World J Surg 24:486–90

14. Wong CS, Cummings BJ, Brierley JD et al (1998) Treatment of locally recurrent rectal carcinoma – results and prognostic factors. Int J Radiat Oncol Biol Phys 40:427–35

15. Gunderson LL, Martin KJ, O'Conell MJ, Beart RW, Kvols LK, Nagorney DM (1985) Residual, recurrent or unresectable gastrointestinal cancer. Role of radiation in single or combined modality treatment. Cancer 55:2250–8

16. Valentini V, Morganti AG, De Franco A et al (1999) Chemoradiation with and without intraoperative radiation therapy in patients with locally recurrent rectal carcinoma. Prognostic factors and long term outcome. Cancer 86:2612–24

17. Engenhart-Cabillic R, Debus J, Prott FJ et al (1998) Use of neutron therapy in the management of locally advanced nonresectable primary or recurrent rectal cancer. Recent Results Cancer Res 150:113–24

18. Spanos WJ Jr, Wassermann T, Meoz R, Sala J, Kong J, Stetz J (1987) Palliation of advanced pelvic malignant disease with large fraction pelvic radiation and misonidazole. Final report of RTOG phase I/II study. Int J Radiat Oncol Biol Phys 13:1479–82

2.3.2.3 Lung Cancer

The response of bronchogenic carcinomas to radiotherapy and concurrent razoxane has not been assessed previously. Despite this, preliminary unpublished experiences led to the initiation of 4 randomized studies in which survival and toxicity were the primary endpoints. The rationale behind these studies, however, were the lack of effective treatment in lung cancer in general, and the hope for razoxane as a potentially cytotoxic agent, its known suppression of metastases in animal experiments and an apparently observed radiosensitizing activity.

The results of randomized studies were contradictory, and showed no significant advantages for the drug. Newman et al. found even a reduced survival with radiotherapy and razoxane compared with radiotherapy alone for inoperable lung cancer in a randomized double-blind trial [1]. The radiation dose in this study was 30–35 Gy in 10–15 fractions over 2–3 weeks. After 148 patients having been treated, the sequential design enabled the trial to be terminated after only 8 assessments. The median survival time in the razoxane group was 80 days, and in the placebo group 175 days.

However, a survival time of 80 days was not observed in other randomized studies of lung cancer treated with radiotherapy and razoxane: Corder et al. reported

a 50% Kaplan-Meier survival estimate of 9 months for patients treated with radiotherapy together with razoxane in two dose schedules [2]. In a study of Spittle et al.
[3] in which 113 patients were randomly allocated to receive either radiotherapy
or radiotherapy and razoxane, 64% of patients who had two radiation series (i.e.,
2 × 30 Gy plus razoxane) survived 12 months. This compared favourably with most
other published clinical trials of inoperable carcinoma of the bronchus. Although
the median survival was greater in those who received the combined treatment with
either one or two radiation series, neither achieved statistical significance. In the
patients with oat cell carcinoma and two radiation series, the median survival was
14 months vs. 11 months with irradiation alone. This was regarded as considerable
improvement in survival compared with other drugs or methods at that time. The
incidence of severe esophagitis reached 32% in the combined treatment arm vs. 7%
in the radiation only group.

A trend towards an improved survival in small cell lung cancer was also seen
in our own working group in Hannover [4]. Overall, there was no statistically significant difference between the treatment groups in our randomized study with 40
patients, but the median survival was around 10 months (corresponding to 300 days)
in both groups considering all histologies together. Therefore, the exceptionally low
survival time of the patients treated with razoxane in the study of Newman et al.
remains unexplained. Unfortunately, no accurate information is available on the terminal events of patients in most of the carcinoma of the bronchus studies. In the
trial of Newman et al., single radiation doses exceeded 250 cGy, and therefore,
some form of radiation toxicity (e.g. ischemic heart disease) may have contributed
to a reduced survival. Corder et al. speak of a formidable toxicity when razoxane is
combined with radiotherapy in bronchogenic carcinoma [1].

In retrospect – some 30 years later – several shortcomings of these earlier studies
may be seen. For instance, in our own study there was a considerable imbalance
(29% vs. 6%) related to the history of surgical treatment in favour of the patients
who received radiotherapy alone – which was not recognized as important stratification factor at the time of publication. But taken together, the single agent use of
razoxane together with radiotherapy in cancer of the lung revealed a limited efficacy of the drug, and there is certainly no breakthrough with the schedules used.
Some trends and details in the results, however, suggest that razoxane should still
be considered in lung cancer and incorporated into carefully selected drug combinations in multimodality concepts. Thereby, attention has to be given to the control
of the presumably increased toxicity of such procedures, perhaps by reducing the
radiation doses.

2.3.2.4 Other Solid Tumors

Malignant Gliomas

A well documented complete response of an inoperable, recurrent astrocytoma
grade II–III (WHO) was reported in a 35 year-old woman who received a radiation
dose of 51 Gy (tele-cobalt) together with razoxane by mouth [5]. Serial CT-scans

revealed a response that was not complete before 1 year after the irradiation had been finished. The woman who was pre-terminal at the start of the therapy recovered completely and was well for 3 years until a further relapse occurred.

Since complete responses to irradiation are infrequent in undifferentiated gliomas, a phase II study was performed by Eiter et al. to evaluate razoxane as a radiosensitizer in malignant gliomas between 1984 and 1990 [6]. Thirty patients (Grade IV: 19, Grade III: 11 patients) received razoxane at a dose of 125 mg/m^2 (median total dose 7.5 g) on the days of the radiotherapy. After a median follow-up of 45 months, the median survival was 10.5 months for patients with grade IV tumors and 27+ months for grade III tumors. The median overall survival for all patients was 14 months (range, 5.3 to 80+). There was no breakthrough with the single agent use of razoxane but the toxicity of the combined treatment was low, and there were, remarkably, only few early deaths (94% of the patients survived 6 months).

Melanoma

Based on the observation of two complete responses (e.g. Fig. 2.7a, b) of metastatic melanoma lesions which have been irradiated with razoxane, a randomized trial was initiated at the University Medical School in Hannover in 1978, to compare the radiation response of measurable melanoma lesions with and without razoxane.

Due to administrative reasons this trial was closed prematurely, but this kind of treatment was continued in another institution in form of an open-labeled, non-randomized comparative study. In the light of new preclinical data, e.g., the renewed confirmation of the antiangiogenetic activity in the B16F10 melanoma model where razoxane prevents tumor cell-associated cord formation [7], and the experience of some uncommon beneficial razoxane related antimelanoma effects, the results of this long lasting comparative study were analyzed retrospectively. The following data are related to the treatment of brain metastasis only [8].

Original article: Rhomberg W, Eiter H, Boehler F, Saely Ch, Strohal R (2005) Combined razoxane and radiotherapy for melanoma brain metastases. A retrospective analysis. J Neuro Oncol. doi:10.1007/s11060-004-7557-z

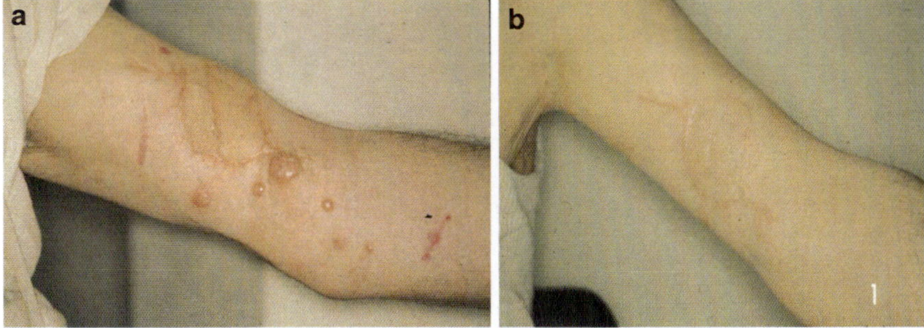

Fig. 2.7 (a) Locoregional recurrence of malignant melanoma before treatment. (b) Complete regression after radiotherapy + razoxane (30 Gy, electrons)

A Brief Synopsis

The efficacy of razoxane and radiotherapy was compared retrospectively with radiotherapy alone or in combination with a non-razoxane based medication in patients with melanoma brain metastases. From 19 assessable patients receiving whole brain irradiation with or without a boost (mean total dose 40.5 Gy) for measurable brain metastases, 8 patients underwent an additional razoxane therapy with 125 mg *per os* twice daily started 5 days before radiotherapy and given throughout the whole radiation period. The median razoxane dose was 6.25 g (range 3.2–8.0 g). Endpoints included radiation response rates, median survival time and 1-year survival rates. To generate reliable prognostic parameters for this non-randomized study population, the Score Index for Stereotactic Radiosurgery and the Radiation Therapy Oncology Group Recursive Partitioning Analysis score were applied.

The results are summarized in Table 2.8. Radiotherapy with razoxane led to higher response rates (62% vs. 27%) and a lower percentage of progressive disease (12.5% vs. 36%) if compared with radiotherapy alone or with a non-razoxane based medication. This combination was associated with a longer median survival (5 months vs. 2.2 months; $p = 0.052$) and a 1-year survival rate of 37.5% vs. 0% ($p = 0.027$). Both treatment groups belonged to similar prognosis subsets. The treatment was well tolerated. Taken together these data support the therapeutic concept of a combined razoxane radiation therapy in melanoma patients with brain metastases. The favorable treatment effects are probably due to the radiosensitizing and the cytorallentaric mode of action of razoxane. Since the patient numbers are low, confirmatory studies are certainly necessary.

Comment

The prognosis of patients with advanced melanoma still remains poor and there is a need to develop new therapeutic regimens with low toxicity. This is especially true for multiple melanoma brain metastases: median survival times around 4 months were most frequently described for this condition, better results were reported only

Table 2.8 Treatment of melanoma brain metastases: response, survival and prognostic scores for both groups

Results	With razoxane ($n = 8$)	Without razoxane ($n = 11$)
Partial responses	62%	27%
Progressive disease	12.5%	36%
Median survival time (25th and 75th percentile)* in months	5.0 (2.6–23.5)	2.2 (2.0–4.0)
1-year survival rate**	37.5%	0%
Median SIR value (range)	4.5 (2–9)	5.0 (3–7)
RPA class II and III	87.5%	82%

SIR, score index for stereotactic radiosurgery; RPA, recursive partitioning analysis; *p*-values: *$p = 0.052$; ** $p = 0.027$.

in cases of a single metastasis and/or absence of extracerebral lesions. Even by the use of DTIC or interferon alpha or recent drugs like temozolomide, the rate of objective responses after irradiation are very low. Although there were no complete responders, the rate of 62% major tumor regressions deserves attention. In accordance with the response rates, patients treated with razoxane showed a longer median survival time of 5 months (25th and 75th percentiles 2.6–23.5) vs. 2.2 months (2.0–4.0) in the non-razoxane-group of this non-randomized series.

For several reasons, we think it would be worthwhile to incorporate razoxane in further clinical research in malignant melanoma.

Bladder Cancer

A first indication for a synergistic activity of razoxane and irradiation in bladder cancer was given by Ryall in 1979 [9]. A prospective randomized controlled clinical trial of the treatment of bladder cancer by radiotherapy or radiotherapy and razoxane was performed, and data was available on the first 160 patients. There was no detectable difference between the two groups when comparing haematuria, nocturia, dysuria, pain on micturition or radiation induced diarrhea. Levels of nausea and vomiting were identical in both groups but 53% of the razoxane patients had leukopenia (WBC less than 50% of initial value). Initial actuarial survival rates show that 52% of the radiotherapy patients were alive at 18 months compared with 67.6% of the patients receiving radiotherapy and razoxane. These results were reported on congresses [9] and not published in detail.

Based on our own initial experience of 2 complete responses (duration 6 months+) out of 4 patients with recurrent bladder carcinomas previously irradiated, the influence of razoxane on the radiation response was studied further. In view of the favourable efficacy which was observed if radiotherapy and razoxane is given together with the vinca alkaloid vindesine, especially in sarcomas, this drug combination was studied in 14 unresectable, measurable bladder carcinomas from 1996 to 2004. The results of this pilot study were retrospectively compared with patients of similar stages of disease who received irradiation together with cisplatin.

The median survival time of these 14 patients with advanced disease was 17.5 months (range, 6–70 months). Local control was achieved in 11 of 12 assessable patients (91%), and a complete tumor response was reached in 8 patients as yet, which compares favourably to the controls (unpublished observations). The evaluation of these data and those of the non-randomized controls is still ongoing. The favourable response rates and local control in this small patient series clearly indicates that the triple combination of razoxane, vindesine and radiotherapy should be explored further in bladder cancer.

Ovarian Cancer

Limited and unpublished experience exists as to the application of radiation and the concomitant use of razoxane in ovarian cancer. While razoxane alone did not induce major remissions in ovarian carcinoma in two studies (see Section 2.3.1.2), its combined use together with radiotherapy led to 5 major responses among 6 patients treated at the Mersey Regional Centre for Radiotherapy and Oncology in Bebington, UK (K. Hellmann, personal communication, 1976). Two of the 5 responses were complete tumor regressions. The average survival of these 6 patients was 39 weeks, the average duration of response was 31 weeks.

We observed 2 major responses in 5 patients with advanced stages of their disease treated at the Medical School of Hannover University. Three of the 5 patients, however, were not assessable for objective remissions. Although of brief duration (7 and 16 weeks), the subtotal regressions were of a spectacular degree since there were large tumor burdens. If the experiences of the two hospitals were taken together, 7 major responses among 8 assessable patients were seen. There might be an undiscovered potential in treating ovarian cancer with radiotherapy, razoxane, and perhaps additional drugs such as tubulin affinic agents.

Miscellaneous

A double blind controlled clinical trial of irradiation plus razoxane vs. irradiation plus placebo in the treatment of *head and neck cancer* was published by Bakowski et al. in 1978 [10]. There was no advantage found for the patients treated with razoxane and radiotherapy.

The experience in the treatment of different other carcinomas with radiation and concurrent razoxane is limited. Single dramatic tumor regressions with radiation doses between 800 and 4,000 cGy make it obvious that razoxane is of interest as radiosensitizer also in different carcinomas. The following Table 2.9 shows some data of our own experience in the combined radiation of the following carcinomas:

Table 2.9 Radiation response by razoxane, different solid tumors. Anecdotal experience §

Diagnosis	Number of patients	Radiation dose (Gy)	Major response	Complete response	Previous radiotherapy	Previous chemotherapy
Adenoid cystic carcinoma	3	30–50	3	1	0	2
Breast, cancer en cuirasse	3	20–30	2	0	3	3
Melanoma	2	30, 50	2	2	0	0
Mesothelioma	1	8	1	0	0	0
Thyroid cancer (follicular)	1	50	0	0	1	0
Head and neck (squamous cell)	2	50	2	0	1	0
Teratocarcinoma	2	-	1	0	0	2

§ gained during the 1970s

References

1. Newman CE, Cox R, Ford CHJ, Johnson JR, Jones DR, Wheaton M (1985) Reduced survival with radiotherapy and razoxane compared with radiotherapy alone for inoperable lung cancer in a randomized double-blind trial. Br J Cancer 51:731–2
2. Corder MP, Tewfik HH, Clamon GH, Platz CE, Leimert JT, Herbst KD, Byfield JE (1984) Radiotherapy plus razoxane for advanced limited extent carcinoma of the lung. Cancer 53:1852–6
3. Spittle M, Bush H, James S, Hellmann K (1979) Clinical trial of razoxane and radiotherapy for inoperable carcinoma of the bronchus. Int J Radiat Oncol Biol Phys 9:1649–51
4. Hassenstein E, Rhomberg W (1977) Combined modality treatment in bronchogenic carcinoma. Report on a prospective randomized trial with radiotherapy and ICRF 159. Med Klin 72:171–5
5. Rhomberg W, Eiter H, Taxer F (1987) Complete regression of a recurrent astrocytoma grade III by combined treatment with radiation therapy and razoxane. Akt Neurol 14:168–70
6. Eiter H, Rhomberg W (1991) Undifferentiated gliomas: results with radiotherapy and razoxane. Onkologie 14:507–13
7. Rybak SM, Sanovich E, Hollingshead MG, Borgel SD, Newton DL, Melillo G, Kong D, Kaur G, Sausville EA (2003) "Vasocrine" formation of tumor cell-lined vascular spaces: implications for rational design of antiangiogenic therapies. Cancer Res 63:2812–9
8. Rhomberg W, Eiter H, Boehler F, Saely Ch, Strohal R (2005) Combined razoxane and radiotherapy for melanoma brain metastases. A retrospective analysis. J Neuro Oncol DOI 10.1007/s11060-004-7557-z
9. Ryall RDH (1979) Razoxane (ICRF-159) as a radiosensitizer for carcinoma of the bladder. Proceedings of the 11th international congress of chemotherapy, Boston, 1–5 Oct 1979, Abstract 389
10. Bakowski MT, McDonald E, Mould RF, Cawte P, Sloggens J, Barrett A, Dalley V, Newton KA, Westbury G, James SE, Hellmann K (1978) Double blind controlled clinical trial on radiation plus razoxane vs. radiation plus placebo in the treatment of head and neck cancer. Int J Radiat Oncol Biol Phys 4:115–9

2.3.3 Antimetastatic Efficacy of Razoxane

2.3.3.1 Preclinical Evidence

W. Rhomberg

The discovery of the antimetastatic potential of razoxane was connected with the earliest experiments done with the drug in the Lewis lung cancer (LLC) model in 1968 [1]. In this model, the suppression of distant metastases to the lung was indeed impressive: The almost inevitable spread to the lung was completely inhibited, and even 40 years later, Chinese authors were able to repeat these results (>90% of the animals had no metastases in the lung) when they tested a variety of bis-diketopiperazines in LLC [2]. The antimetastatic effect was linked to the normalization of tumour blood vessels which was itself another interesting feature of the drug action. This phenomenon was repeatedly discribed in LLC [3–5] and also in a hamster lymphoma model [6].

Baker et al. reported on studies of razoxane (ICRF-159) in KHT sarcomas in mice. Courses of irradiation consisting of 60 Gy in ten equal fractions over 12 days delivered to KHT sarcomas controlled 55% of the local tumors but 83% of the mice died from metastases. ICRF-159 was used with the intention of partially synchronizing the tumor growth fraction in a radiosensitive state of the growth cycle and of promoting normalization of the tumor vasculature. Levamisol was used to stimulate the immune system. The combination of ICRF-159 with an eight-fraction radiation course proved to be effective for both increasing local control and decreasing the incidence of metastases. The addition of lev-amisol did not improve the results obtained with a combination of ICRF-159 and irradiation [7].

Early investigations on experimental prostate cancer models (R3327 MAT-LyLu, PA III) revealed most impressive results concerning the reduction of distant metastases by razoxane [8, 9] (see Section 2.3.4).

The antimetastatic activities of razoxane were also studied in a transplantable, slowly growing osteosarcoma in Sprague-Dawley rats [10]. This tumour model is characterized by osteoid formation and spontaneous metastasis to lungs, kidneys and lymph nodes. Razoxane given intraperitoneally (i.p.) from 2 days before to 14 days after tumour transplantation (30 mg/kg or 10 mg/kg/day) resulted in a dose-dependent prolongation of the median survival time (83 or 48 days, respectively, vs. 38 days for the control group), but showed no influence on the growth of the primary tumour. Early treatment with 30 mg/kg i.p. showed a greater inhibition of pulmonary metastases than later treatment from day 14–28 after transplantation. Whereas 59.9% of the total sectional area of the lungs in the control animal was covered by osteosarcoma metastases, only 3.4 and 26.1%, respectively, was affected in the early and late razoxane treatment groups [10].

An inhibition of the development of distant metastases by razoxane was further described by Peters et al. in a murine squamous cell carcinoma model [11]. One of the few negative results as to the inhibition of a distant tumour spread by razoxane was reported by Pimm et al. who used a rat epithelioma model for their investigations [12].

Until 1989, razoxane has been found to be an effective inhibitor of metastases in 8/10 experimental tumours which metastasize spontaneously [13]. The experience concerning the outstanding ability of razoxane to inhibit metastasis in experimental tumor systems has since been expanded. An update of these investigations with interesting details and tables can be found in the Section 2.3.4.

Razoxane exhibits a variety of modes of action, but where is the true link to the antimetastatic activity of the drug? While, in theory, its antiinvasive efficacy could also contribute to the phenomenon [14], the more obvious relation seems to lie in its capability of normalizing pathological tumour blood vessels. The following section and then the Section 2.3.4, therefore deal with and discuss the basics of tumor angiogenesis and its possible relation to metastasis.

References

1. Hellmann K, Burrage K (1969) Control of malignant metastases by ICRF 159. Nature 224:273–75
2. Lu DY, Huang M, Xu CH, Zhu H, Xu B, Ding J (2006) Medicinal chemistry of probimane and MST-16: comparison of anticancer effects between bisoxopiperazines. Med Chem 2(4): 369–75, 2006
3. Burrage K, Hellmann K, Salsbury AJ (1970) Drug induced inhibition of tumour cell dissemination. Br J Pharmacol 39:205–6
4. James SE, Salsbury AJ (1974) Effect of (+/–) 1,2-bis (3,5 dioxopiperazin-1-yl) propane on tumor blood vessels and its relationship to the antimetastatic effect in Lewis lung carcinoma. Cancer Res 34:839
5. Salsbury AJ, Burrage K, Hellmann K (1970) Inhibition of metastatic spread by ICRF 159: selective deletion of a malignant characteristic. Br Med J 4:344–6
6. Atherton Anne (1975) The effect of (+/–) 1,2-bis (3,5-dioxopiperazin-1yl) propane (ICRF 159) on liver metastases from a hamster lymphoma. Eur J Cancer 11:383–8
7. Baker D, Constable W, Elkon D, Rinehart L (1981) The influence of ICRF 159 and levamisole on the incidence of metastases following local irradiation of a solid tumor. Cancer 48:2179–83
8. Heston WDW, Kadmon D, Fair WR (1981) Effect of high dose diethylstilbestrol and ICRF 159 on the growth and metastases of the R3327 MAT-LyLu prostate-derived tumor. Cancer Lett 13:139–45
9. Pollard M, Burleson GR, Luckert PH (1981) Interference with in vivo growth and metastasis of prostate adenocarcinoma (PA-III) by ICRF 159. Prostate 2:1–9
10. Wingen F, Spring H, Schmähl D (1987) Antimetastatic effects of razoxane in a rat osteosarcoma model. Clin Exp Metastasis 5(1):9–16
11. Peters LJ (1975) A study of the influence of various diagnostic and therapeutic procedures applied to a murine squamous carcinoma on its metastatic behaviour. Br J Cancer 32(3): 355–65
12. Pimm MV, Baldwin RW (1975) Influence of ICRF 159 and Triton WR 1339 on metastases of a rat epithelioma. Br J Cancer 31:62–7
13. Editorial (1987) Razoxane, metastasis and adjuvant chemotherapy. Clin Exp Metastasis 5: 1–2
14. Garbisa S, Onisto M, Peron A, Perissin L, Rapozzi V, Zorzet S, Giraldi T (1997) Suppression of metastatic potential and up-regulation of gelatinases and uPA in LLC by protracted in vivo treatment with dacarbazine or razoxane. Int J Cancer 72(6):1056–61

2.3.3.2 Metastasis and the Entry of Cancer Cells into the Vasculature – Prevention by Razoxane

Kurt Hellmann

Introduction

Access by tumour cells to the circulation whether through lymphatics, blood vessels or by invasion of adjacent tissues has critical consequences for the patient, the most important being tumour dissemination and the possibility of metastasis formation – prevention of access is therefore an important therapeutic goal. Entry of tumour cells into the circulation can be achieved in two fundamentally different ways. The commonest being the active invasion by tumour cells of small veins, venules or of the tumour microvasculature. Less common is the detachment of tumour cells or tumour emboli from the walls of clefts within tumours through which blood flows to join the venous return. In this latter case, access of tumour cells depends on the strength of the adhesion between tumour cells lining the clefts and the force of the blood stream through the blood channels formed by the clefts. No invasive activity on the part of the tumour cells is required. Although this method is common for sarcomas it is not so common for carcinomas.

There is a widespread belief that the entry of tumour cells into the circulation is governed in the first place by the development of a tumour neovasculature which is thought has to be formed if tumour growth is to continue beyond an early stage [1–4].

The Development of a Neovasculature

Several features characterize the tumour neovasculature. The vessels are generally poorly endothelialized, dilated sinusoidal channels [5]. The incomplete endothelial lining of these channels leads to frequent areas of haemorrhage which can be seen by the naked eye and more readily on histology. The vessels when viewed by contrast angiography have a characteristic corkscrew appearance, and they are also as might be expected much more permeable than intact normal vessels [6]. The recognition of these characteristics is important so that the results obtained on what are claimed to be models for tumour angiogenesis can be seen in perspective. Despite the controls which most experiments using such models employ, it is difficult to avoid the conclusion on seeing the sharp, straight, intact and well endothelialized vessels which tumours or tumour cells in such models of tumour angiogenesis induce, that the new vessels in these systems are anything more than inflammatory reactive vessels. A large variety of angiogenic stimulators has been discovered so that it has become difficult to believe that all of them are involved normally in tumour angiogenesis and it is doubtful whether any are specific *tumour* angiogenesis factors [7–9].

The questions which have to be answered in defining tumour angiogenesis is not only what (preferably in chemical terms) is the specific stimulus to the development of this process, but what determines its time course, its direction, its sensitivity and what, if anything, diminishes or stops it.

Antiangiogenesis as a means of inhibiting tumour growth, and preventing tumour dissemination is an attractive idea but since it has become clear that angiogenesis induced in many highly artefactual experiments are probably little more than a generalized response to foreign body implantation, this avenue to tumour and especially metastasis inhibition has become much less attractive. It will be dealt with a greater length later on.

It is difficult to know why it is necessary to invoke tumour angiogenesis factors or autocrine growth factors when rapidly growing tumours are surrounded by one of the most potent stimulators to angiogenesis in the form of hypoxia. Hypoxia is known to be a powerful stimulant to the development of a collateral circulation in any hypoxic ischaemic area, for example following myocardial infarction or diabetic retinopathy [10].

The faster a tumour grows the more likely is it to develop a rapid angiogenic response, and the more rapid this response, the more imperfect will be the vasculature that has been induced – a conclusion also reached by that astute and meticulous observer R.A. Willis during the course of some 500 cancer postmortems [11].

Tumour Size and the Neovasculature

It has frequently been asserted that the development of a neovasculalture is an absolute requirement if tumours are to grow beyond 2 mm. If they fail to develop a blood supply, it has been claimed growth will stop and no tumour dissemination can take place [12].

Primary of Unknown Origin

Such generalizations take no account of the realities of tumour pathology particularly of the well known, but rarely found primary of unknown origin (PUO) which is generally less than 2 mm and can still rapidly disseminate cells thoughout the body. Metastases arising from the PUO's show that this is not the pecularity of one type of tumour, but that the primary can be in any of the number of organs such as the breast, lung, colon or stomach whose characteristic secondaries may be found without tracing the primary tumour even after the most dilligent searches and investigations both ante and postmortem. Because the primary cannot be found, nothing can be stated about the way in which the cells from such a tumour enter the circulation except that it is unlikely that they required a tumour neovasculature [13].

A related problem with PUO's is the question of age of the tumour. It is a widely held belief that the longer a tumour has been growing, the greater the size and the more it is likely to have disseminated. While this may be true, sometimes it is not an absolute rule since a PUO may give rise to metastasis with long intervals between the appearance of the secundaries. It seems unfortunate that the characteristics of this particular type of tumour has not attracted greater attention and that it has not been studied in greater detail, particularly for its experimental, invasive and disseminating ability. One might predict that it would overcome the difficulty of demonstrating consistent ability to metastasize in human tumours transplanted to nude mice. However, no report of such an experiment seems to have appeared. Furthermore, the invasive properties of this tumour can confidently be predicted to be extremly high and this too would make it a useful experimental model for testing for invasive capacity.

The limitation of using secondaries from a PUO is that the age of the primary tumour cannot be determined and the natural history of the tumour is unknown. It could be that the tumour cells from such a PUO have an inherently high invasive potential ab initio and establish themselves effectively and quickly when they are trapped in the distant organs where they form metastases. On the other hand, it could be that they are selected from a large population which grows rapidly and also dies rapidly. Access to the circulation therefore is not necessarily dependent on the size or age of the tumour.

Krukenberg Tumours

Another important example which clearly demonstrates that neither size nor the neovasculature of a primary tumour have predictive value for judging the invasive or disseminating character of a tumour is the Krukenberg tumour [14]. This well known tumour has its primary in the wall of the stomach, but may make its first clinical appearance as a tumor of the ovary. The tumours are often bilateral with haemorrhages and small cysts. The primary growth responsible for these Krukenberg metastases, although frequently a gastric carcinoma of a particularly diffusely infiltrating type may also be from other sites. The primary is so small and inconspicuous that patients have no symptoms or signs referable to the stomach. Even a barium meal may not show any gastric abnormality and it is of course beyond the resolution of computed tomography. Indeed, the primary tumours may be so small as to escape detection altogether. Presumably, if it escapes detection at microscopic level it is certainly less than 2 mm, and no special vasculature seems to be required for this tumour to enter the blood stream and disseminate. Clearly, this is an extreme example of an unusual behaviour but it demonstrates again that no absolutes are possible with malignant tumours.

In this connection it is also noteworthy that although most gastric neoplasms disseminate to the liver, in the case of the Krukenberg metastases the liver appears to be free of any involvement. It should be emphasized moreover that similar Krukenberg tumours can arise from intestinal, biliary or quite frequently mammary carcinomas, demonstrating again that neither size of primary nor a tumour neovasculature are a bar to tumour entry into the circulation.

Direct Visualization of the Tumour Neovasculature

There are major problems for any attempt to demonstrate the tumour neovasculature by direct means. This has not prevented elegant attempts being made to achieve this by making casts [15] and examining the casts by electron scanning micrography. The difficulty in this technique is that artefactual impressions can be produced as a result of applying pressures to the permeating fluid for developing the casts which are in excess of those which the tumour vasculature would normally experience. This pressure would also under in vivo conditions be expected to change as the tumour develops; as can be clearly seen from the fact that the central part of most tumours becomes necrotic as a result of the growth of the tumour, and consequent narrowing of the lumen of the vessels supplying it to the point of total occlusion. This inevitably leads to concomitant opening of collateral channels which may be either through the substance of the tumour or through the opening or growth of smaller vessels on the periphery of the tumour. This in turn has the effect of producing neovascular pathways ultimately forming peripheral arteriovenous shunts. These changes in the morphology of the neovascular architecture are a dynamic process with constant remodelling which have important consequences for the entry of tumour cells into the vascular system. Casts, on the other hand, are made at one time point in the life of the tumour. It ought to be pointed out at this juncture that no lymph channels with their extremely delicate walls can survive these changes and none can be demonstrated. Access from tumours into the circulatory system is therefore via the blood stream and may only reach the lymphatics at a later stage.

It is also appropriate to point out at this juncture that new vessel formation by tumours in response to the demands of an increasing cell mass are more likely to come from the arterial system than from the venous and if it comes from the arterioles direct connections with the venous side through arteriovenous shunts could develop from the vascular system on the tumour periphery. The existence of the shunts are frequently overlooked by clinicians but have important therapeutic implications not only for the access of tumour cells into the circulation but the access of drugs into the tumour. Labelled anticancer drugs injected into arteries supplying tumours frequently circulate on the periphery of the tumour but do not penetrate into the necrotic central areas of the tumours. These necrotic central areas are however made up not only of dead, dying and doomed cancer cells but also of dormant cells which when transplanted to fresh sites can grow with great vigour. Thus, once lytic enzymes have reduced the pressure on the central necrotic areas and permit the collapsed blood vessels in these areas to partially re-open, dormant cells can enter the circulation.

Adhesion and the Access of Tumour Cells into the Circulation

An important factor which influences the access of tumour cells into the neovasculature is the degree of difficulty with which tumour cells detach themselves from the main body of the developing tumour [16]. Much depends on the adhesive forces between cells of the tumour, and these have been studied in great detail [17]. Most recently these studies have concerned themselves with the adhesion of tumour cells to preparation of cells from a number of organs and to endothelial cells. Earlier, Coman [18] measured the adhesion between cells of a tumour mass and the forces required to separate individual cells or clumps of cells from the main body of the tumour.

Tumours implanted in a muscle are subject to constantly changing forces. Under these conditions, although the growth rate of the tumour may not be very different from that when it is implanted subcutaneously, the development of lung metastases occurs much more quickly. This observation has been made repeatedly when comparing the time required for lung metastases to appear after implanting the Lewis lung carcinoma either in leg muscle or subcutaneously. Although metastases will appear approximately 23 days following subcutaneous implantation, none can be observed if the tumour is excised earlier than 5–6 days after implantation [19].

Adhesiveness can be altered by many substances, and amongst these EDTA and other chelating agents which remove essential ions required for adhesion can be very effective.

Tumour Emboli and Single Cell Detachment

It is usually thought that invasion into the microcirculation appears to be confined to single cancer cells and that the disparity between cancer cell and capillary diameters requires that either the cells and/or the capillaries be deformable [20]. From direct observations, however, it is apparent that although single cells may enter the microcirculation and can be traced in the circulation, few of them survive. A study of circulating Lewis lung tumour cells by Salsbury et al. [21] has shown that clumps consisting of as many as 50 tumour cells may be circulating at any one time and that this is more likely as the tumour grows progressively. The fate of these clumps is not certain. Neither is it certain whether they broke off from the tumour as a clump or whether they aggregated in the blood stream. It would seem possible, however, that in view of the gross appearance of the tumour sinusoids, where tumour cells line clefts in the primary tumour that clumps of several tumour cells could escape into the circulation on the post capillary venous side. It would be quite easily possible for such a tumour embolus to circulate and reach the lungs which would be the organ of first encounter.

It is known that the metastatic efficiency for most tumours is extremely low and that most tumour cells injected intravenously will die. What is not clear is why they die in such very large numbers and whether it requires an aggregate or clump of them to survive. It may be that emboli which go on to form metastases are aggregates of tumour cells with circulating white blood cells and/or platelets that have special characteristics that allow them to survive. There is of course a trapping advantage with cell aggregates as compared with single cells, but this alone cannot account for the greater degree of viability of tumor emboli.

It is possible that one of the reasons why the Lewis lung tumour so regularly, consistently and in all animals inoculated with a primary implant produces lung metastases at the same time and in the same organ, is due to the fact that its growth patterns in the primary implant permit cell clumps to be detached from the primary and to reach the lungs intact.

Antiangiogenesis

If one accepts the claims of Folkman [22] and his collaborators that tumour growth is not possible beyond 2 mm without the development of a neovasculature, then prevention or destruction of the angiogenesis process should theoretically result in inhibition of tumour growth and, since there would be no channels for dissemination, no tumour metastases. However, the neovasculature which they induce and on which the claims are based, seems to have neither morphological nor biochemical similarity with characteristic tumour vessels. They have neither the poorly endothelialized, tortuous sinusoidal appearance nor the leaky properties of tumour blood vessels. It also does not resemble in any shape or form the tumour clefts lined by tumour cells and no endothelium but through which blood can flow carrying with it detached tumour cells or emboli. Nevertheless, in view of the problems experienced with most chemotherapeutic agents, with hormones and with biological response modifyers, the idea of influencing tumour growth, invasion and dissemination by influencing tumour angiogenesis is certainly worthy of serious examination.

The first approach to control of tumour angiogenesis seemed to be very clear. It was the isolation of what was hoped to be a *tumour* angiogenesis factor (TAF), though from what has been said above, it would have been perhaps more precise to restrict the concept to the isolation of an angiogenesis factor. The main problem since then, however, has been that too many substances [23] have been found with very effective angiogenesis activity to allow one to believe that they are all specifically involved in *tumour* angiogenesis. Following the original concept, it would have been attractive to have isolated a tumour angiogenesis factor and then use it to produce a specific antibody to control tumour angiogenesis. None of this has materialised and if TAF had been discovered, it seems doubtful whether an antibody to it would have been effective in controlling the entry of tumour cells into the circulation since as was pointed out earlier, in tumours such as the Krukenberg, massive and destructive dissemination can take place without apparently the necessity of organizing around themselves a tumour neovasculature.

Heparin and Cortisone

More recently, Folkman and his group have claimed that tumour proliferation and metastases can be completely prevented by a combined treatment with heparin and cortison [24]. It is very doubtful, however, whether the doses of heparin and cortisone required for treatment of tumour dissemination in man could possibly have been reached or maintained. It has also not been possible for other groups to repeat these results [25], and it has since been stated by Folkman that only the special type of heparin used at the time of his experiments could obtain the results which he obtained. Unfortunately, the batch from which his heparin was derived is no longer available.

Madarnas and coworkers used cortisone and maltose tetrapalmitate with which they obtained prevention of tumour growth [26]. They believed that their results were due to lack of tumour vascularization. However, it is not clear from their experiments which came first, vascularization inhibition or tumour inhibition. If the combination prevented tumour proliferation per se, clearly angiogenesis would also have ceased.

Flavone Acetic Acid

A more recent approach derived from experiments with flavone acetic acid (FAA) which was shown to be highly effective as a tumour inhibitor in a variety of mouse tumours [27]. Phase 2 studies in man revealed no activity in any tumour tested, and subsequent examination of the reasons for the discrepancy uncovered the fact that FAA rapidly and selectively destroyed the tumour vasculature in the implanted tumours causing necrosis of the whole of the tumor [28]. Many questions remain to be answered about this unusual situation. In particular, the role of tumour necrosis factor has yet to be clarified. If destruction of the vasculature were the main mode of action then the drug should be ineffective in the development of ascitic tumours, and if it were given after tumours had disseminated into the blood stream, no effect should be seen on the metastases. The results are in line with these expectations [29].

Thus, FAA seems to be a member of a new class of drugs which specifically destroy the tumour vasculature leaving normal blood vessels and other tissues unaffected, but it is not clear whether it is species specific, since the drug has been most disappointing in man.

Razoxane

In the course of a series of experiments to discover the mechanism by which razoxane inhibited the development of spontaneous pulmonary metastases from a subcutaneous implant of the Lewis lung carcinoma, it was found that razoxane treatment prevented entry of tumour cells into the circulation and that it apparently normalized the developing tumour neovasculature [30–32]. At the same time it appeared to have little influence on the growth of the primary implant. Of course, many substances will prevent secondaries by their cytotoxic action in an undiscriminating inhibition or destruction of dividing cells. Razoxane, however, showed for the first time that it was possible to delete specifically tumour dissemination and metastases formation without interfering with tumour proliferation and possibly even invasion, by preventing tumour cells escaping into the circulation. Whether, as seems likely, the normalization of the developing neovasculature is directly responsible for the failure of the Lewis lung tumour cells to escape, remains to be established. No other substance has yet been shown to have a similar inhibitory action, though it is difficult to believe that razoxane is unique in this respect. Failure of tumour cells to enter the circulation could clearly account for the subsequent inhibition of lung metastases. It also eliminated the necessity of considering the possibility of some quite complex interaction between razoxane and circulating malignant cells such as inhibition of aggregation with other tumour cells or with platelets or with other blood formed elements. Moreover, it eliminated the necessity of considering the possibility that razoxane might have influenced circulating cells in such a way that it prevented their implantation in the lung. Histological examination of the primary tumour demonstrated however that razoxane treatment resulted in a normalization of the developing neovasculature in the primary tumour which was compatible with the results of direct cytological examination of blood from the ipsilateral draining vein of the primary tumour and which had shown that razoxane treatment prevented escape of tumour cells into the circulation [32].

Salsbury [30] also examined other tumours before and after razoxane treatment and showed that the sarcoma 180 which disseminates widely but does not form metastases (probably because of immunological destruction of the disseminating S180 cells), could be influenced by razoxane treatment in an identical fashion to that of the Lewis lung tumour. That is, there appeared to be a 'normalization' of the *developing* neovasculature, a feature already shown for other tumours by LeServe and Hellmann [6] using angiographic and other techniques. These authors realised that the normalization of blood vessel development in a tumour could have profound consequences not only for the escape of tumour cells into the circulation but also for the oxygenation of the tumour. However, the more immediate and important question was whether a normalized tumour neovasculature could prevent the escape of malignant cells from primary and indeed from secondary tumours and prevent recurrences in patients with tumours which could either not be resected or who had been treated by without achieving complete regression. Razoxane has been found to be an effective inhibitor of metastases in 8/10 experimental tumours which metastasize spontaneously [33]. Since it had also shown some effect in advanced colorectal cancer [34], a fully randomized prospective clinical trial was set up to see if it prevented recurrences when given as adjuvant treatment following surgical resection. The trial was terminated after 7 years. Throughout, the recurrence free interval was statistically significantly greater in the razoxane treated Dukes'C group (44 patients all of whom had abdominal lymphatic spread, but not yet liver secondaries) compared with the controls (44 patients). Survival was also increased though the statistical significance of this was somewhat less than for recurrence [35]. Two major conclusions have been able to be drawn from this trial. The first one being that only Dukes'C patients benefitted. This patient population is at greatest risk of recurrence because tumour is almost certainly left behind in some of the draining lymph nodes and can then spread to the liver, and also micrometastases might already be present at the time of resection of the primary. Most patients who recurred with metastases developed them in the liver, but only half the number of Dukes' C patients treated with razoxane developed liver metastases compared with controls, and those that did develop them took twice as long to do so [36]. Dukes' B patients showed no benefit, and this is not altogether surprising since an unknown number of patients in this category is cured by surgery alone and no adjuvant treatment can improve on this. These patients, however, will dilute the figures and make comparisons difficult.

Conclusions

Malignant tumour growth, invasion and dissemination do not seem to be subject to a universal law with regard to the vascular system which they organise into themselves. Some tumours cannot grow without this neovasculature, but most can and find ways of doing so such as invasion of adjacent tissues, lymph or blood channels and may never grow beyond a microscopic collection of cells which proliferate at an unknown rate but which have all the power and malignancy to disseminate widely and kill the host. On the other hand, other tumours need a vasculature in order to grow and survive, and for these it appears that there are tumour growth factors which can, if interfered with, prevent the development of the tumour. Initially it was thought that there might be a single specific tumour angiogenesis factor (TAF), but such a factor has not been isolated, although a variety of angiogenic substances have been isolated and some of them might play a part in the development of tumour angiogenesis. Several disparate substances have been identified which will interfere with or destroy tumour blood vessels, and other substances are able to normalise the development of tumour blood vessel growth. Of these latter group, razoxane has been shown to normalise the neovasculature with the result that tumour cells are unable to gain entry into the circulation and thus prevent the primary tumour from disseminating, although it continues to grow.

A much larger effort ought to be made to screen for substances which prevent tumour cells from entering the circulation. This could be done without major technical problems and it might lead to a much more rational and less sterile approach than that which is currently being used to detect new anticancer drugs. The systems which have been employed in the past have acted on the uniform assumption that the cancer problem could be solved by finding drugs which could selecetively prevent cell division. Theoretically such a programme is almost bound to fail and this has been borne out by the vast number – possibly more than 1 million new and old chemical entities which have been examined. It is time that new approaches to cancer research and treatment were adopted, and the discovery of substances which prevent tumour cells from disseminating could be of immense assistance in this respect.

References

1. D'Amore PA (1986) Growth factors, angiogenesis and metastasis. In Cancer metastasis: experimental & clinical strategies. A.R. Liss, New York, pp 269–83
2. Folkman J, Long DM, Becker FF (1963) Growth and metastasis of tumor in organ culture. Cancer 16:453–67
3. Gimbrone MA, Aster RH, Cotran RS, Corkery J, Folkman J (1969) Preservation of vascular integrity in organs perfused in vitro with a platelet-rich medium. Nature 222:33
4. Folkman J (1986) How is blood vessel growth regulated in normal and neoplastic tissue? Cancer Res 46:467–73
5. Warren BA (1979) The vascular morphology of tumors. In: Peterson HI (ed) Tumor blood circulation. CRC Press, Boca Raton, Chapter 1
6. Le Serve AW, Hellmann K (1972) Metastases and the normalization of tumour blood vessels by ICRF-159: a new type of drug action. Br Med J 1:597–601
7. Atherton A (1977) Growth stimulation of endothelial cells by simultaneous culture with Sarcoma 180 cells in diffusion chambers. Cancer Res 37:3619–22
8. Rifkin DB, Gross JL, Moscatelli D, Jaffe E (1982) Proteases and angiogenesis: production of plasminogen activator and collagenase by endothelial cells. In: Nossel HL, Vogel HJ (eds) Pathobiology of the endothelial cell. Academic Press, New York, p 191ff.
9. Weiss JB, Brown RA, Kumar S, Phillips P (1979) An angiogenic factor isolated from tumours. A potent low-molecular weight compound. Br J Cancer 40:493–6
10. Garner A (1986) Ocular angiogenesis. Int Rev Exp Pathol 28:249–306
11. Willis RA (1972) The spread of tumours in the human body. Butterworth, London, p 113
12. Gimbrone MA, Leapman SB, Cotran RS, Folkman J (1972) Tumor dormancy in vivo by prevention of neovascularization. J Exp Med 136:261–76
13. Willis RA (1972) The spread of tumours in the human body. Butterworth, London, pp 127–44
14. Krukenberg F (1895) Über das Fibrosarcoma ovarii mucocellulare (carcinomatodes). Arch Gynaekol 50:287

15. Grunt TW, Larnetschwandtner A, Karrer K (1986) The characteristic structural features of the blood vessels of the Lewis lung carcinoma. Scanning Electron Microscopy, SEM Inc., Chicago, pp 575–89

16. Weiss L, Orr FW, Honn KV (1989) Interactions between cancer cells and the microvasculature: a rate regulator for metastasis. Clin Exp Metastasis 7:127–67

17. Weiss L, Orr FW, Honn KV (1988) Interactions of cancer cells with the microvasculature during metastasis. FASEB J, 2:12–21

18. Coman DR (1944) Decreased mutual adhesiveness: a property of cells from squamous cell carcinomas. Cancer Res 4:625–29

19. James SE, Salsbury AJ (1974) Effect of (+/–)-1,2-bis(3,5-dioxopiperazin-1-yl) propane on tumor blood vessels and its relationship to the antimetastatic effect in the Lewis lung carcinoma. Cancer Res 34:839–42

20. Weiss L, Clement DS (1989) Studies on cell deformability: some rheological considerations. Exp Cell Res 58:379–87

21. Salsbury AJ, Burrage K, Hellmann K (1974) Histological analysis of the antimetastatic effect of (+/–)-1,2-bis(3,5-dioxopiperazin-1-yl) propane. Cancer Res 34:843–49

22. Folkman J, Cotran R (1976) Regulation of vascular proliferation to tumor growth. In: Richter GW, Epstein NA (eds) International review of experimental pathology. Academic Press, New York, pp 207–48

23. Kull FC, Brent DA, Farikh I, Cautrecasas P (1987) Chemical identification of a tumor-derived angiogenic factor. Science 236:843–45

24. Folkman J, Langer R, Linhardt R, Haudenschild C, Taylor S (1983) Angiogenesis inhibition and tumor regression caused by heparin or a heparin fragment in the presence of cortisone. Science 221:719–25

25. Penhaligon M, Camplejohn RS (1985) Combination heparin plus cortisone treatment of two transplanted tumors in C3H/He mice. J Natl Cancer Inst 74:869–73

26. Madarnas P, Benrezzak O, Nigam VN (1989) Prophylactic antiangiogenic tumor treatment. Anticancer Res 9:897–902

27. Corbett TH, Bissery YC, Woziniak A (1986) Activity of flavone acetic acide (NSC 347 512) against solid tumours of mice. Invest New Drugs 4:207–22

28. Bibby MC, Double JA, Loadman PM, Duke CV (1989) Reduction of tumor blood flow by flavone acetic acid: a possible component of therapy. J Natl Cancer Inst 81: 216–20

29. Zwi LJ, Baguley BC, Gavin JB, Wilson WR (1989) Blood flow failure as a major determinant in the antitumour action of flavone acetic acid. J Natl Cancer Inst 81(13):1005–13

30. Salsbury AJ, Burrage K, Hellmann K (1970) Inhibition of metastatic spread by ICRF-159: selective deletion of a malignant characteristic. Br Med J 4:344–6

31. Hellmann K, Burrage K (1969) Control of malignant metastases by ICRF-159. Nature 224:273–5

32. Hellmann K, Salsbury AJ, Burrage K, LeServe AW, James SE (1973) Drug induced inhibition of haematogenously spread metastasis. In: Garattini S, Franchi E (eds) Chemotherapy of cancer dissemination and metastases. Raven Press, New York, pp 355–59

33. Editorial (1987) Razoxane, metastasis and adjuvant chemotherapy. Clin Exp Metastasis 5:1–2

34. Marciniak DA, Moertel CG, Schutt AJ (1975) Phase II study of ICRF-159 (NSC 129 943) in advanced colorectal carcinoma. Cancer Chemother Rep 59:761–63

35. Gilbert JM, Hellmann K, Evans M et al (1986) Randomised trial of oral adjuvant razoxane (ICRF-159) in resectable colorectal cancer: five-year follow up. Br J Surg 73: 446–50

36. Hellmann K, Gilbert JM, Evans M, Cassell P, Taylor R (1987) Effect of razoxane on metastases from colorectal cancer. Clin Exp Metastasis 5:3–8

2.3.3.3 Clinical Evidence

Walter Rhomberg

A considerable amount of preclinical research has explored the prevention of metastasis by razoxane. It was shown that razoxane is one of the, if not the most active antimetastatic agent. In contrast to this, only few data exist that relate to antimetastatic treatment approaches in human malignancies.

Antimetastatic Effects in Colorectal Cancer

The first indication of the influence of razoxane on the incidence of distant metastases in colorectal cancer came from adjuvant postoperative studies. The reasons to perform controlled adjuvant studies with razoxane in colorectal cancer were the fact that razoxane prevented distant metastases in a variety of animal experiments, and it was one of the few agents showing some activity in advanced colorectal cancer at that time [1].

In a first adjuvant study of Gilbert et al., an interim analysis after 176 patients entered the study (median follow-up 34 months) did not reveal a difference in the incidence of distant metastases between the group treated with razoxane and the group followed-up without further treatment after surgery [1]. However, after a median follow up of 5 years a significant difference became apparent. The authors summarize their results as follows [2]:

> Adjuvant razoxane (125 mg b.d.) given 5 days/week indefinitely following resection of colorectal cancer provided no benefit in terms of survival or recurrence for Dukes' A or B patients when compared to untreated controls. However in Dukes' C patients this treatment reduced the recurrence rate ($p = 0.05$) and possibly increased survival time ($p = 0.08$). Analysis now of the development of metastases in this trial which entered 272 patients over 7 years shows that in the Dukes' C group the incidence of liver metastases in the razoxane-treated patients is only about half that of the untreated patients (18% versus 34%) and that the time to first appearance of the liver metastases is twice as long in the razoxane-treated group as it is in the untreated group (80 weeks versus 40 weeks). It is concluded that the benefit of adjuvant razoxane observed in the Dukes' C patients is due to the antimetastatic activity of the drug in reducing and slowing down the development of hepatic secondaries.

The Austrian Society of Radiooncology (OEGRO) performed a small randomized study on 36 patients with few incompletely resected and mostly inoperable recurrent rectal cancer [3]. The patients were randomly allocated to either radiotherapy alone or to radiotherapy combined with razoxane (125 mg b.d. orally). The combined treatment with razoxane increased the local control rate to radiotherapy alone (39% vs. 8%; $p = 0.05$). The median survival time was not different between the groups (20 months each) for several reasons [3]. No patient with radiation therapy alone but four of 18 patients with the addition of razoxane survived 5 years. A possible explanation for the presumed increase of the long term survival might be the difference seen in the rate of appearance of distant metastases.

The rate of distant metastases during the later course of the disease was 61% in patients with radiotherapy alone and 44% in patients treated with radiotherapy and razoxane despite the fact that no maintenance therapy with razoxane was given, and there was an imbalance in the treatment groups in favour to the radiotherapy-alone group. More patients in the razoxane group were pretreated with chemotherapy for recurrent disease. As a rule, any previous treatment reduces the success rate of a following treatment. In this study, only one patient (7%) in the radiotherapy-alone group had a previous treatment with 5-fluorouracil for 5 days, whereas in the combined treatment group eight of 18 patients (44%) were pretreated by 5-FU-based regimens. In addition, the interval from diagnosis to recurrence was shorter in the razoxane group (12 months vs. 25 months) indicating a more aggressive tumour behavior in the patient group receiving razoxane. A longer latency period between the diagnosis of the primary tumour and the onset of a local recurrence which the authors were not aware of at the time of randomization, seems to be – similar to breast cancer – of definite prognostic significance [4–6].

References

1. Gilbert J, Hellmann K, Evans M, Cassell PG, Stoodley B, Ellis H, Wastell C (1982) Adjuvant oral razoxane (ICRF 159) in resectable colorectal cancer. Cancer Chemother Pharmacol 8: 293–9
2. Hellmann K, Gilbert J, Evans M, Cassell P, Taylor R (1987) Effect of razoxane on metastases from colorectal cancer. Clin Exp Metastasis 5(1):3–8
3. Rhomberg W, Hammer J, Sedlmayer F, Eiter H, Seewald D, Schneider B (2007) Irradiation with and without razoxane in the treatment of incompletely resected or inoperable recurrent rectal cancer. Results of a small randomized multicenter study. Strahlenther Onkol 183:380–4
4. Heide J, Dilcher C, Merten R et al (2001) Lokalrezidiv des Rektumkarzinoms: Tumorresektion, Strahlentherapie oder Radiochemotherapie. Strahlenther Onkol 177(Suppl 1):39, Abstract
5. James RD, Johnson RJ, Eddleston B et al (1983) Prognostic factors in locally recurrent rectal carcinoma treated by radiotherapy. Br J Surg 70:469–72
6. Wong CS, Cummings BJ, Brierley JD et al (1998) Treatment of locally recurrent rectal carcinoma – results and prognostic factors. Int J Radiat Oncol Biol Phys 40:427–35

Antimetastatic Activity in Soft Tissue Sarcomas

In sarcomas of bone and soft tissues, razoxane was given primarily as a radiosensitizer. Used as single agent and in terms of remission induction, no substantial cytotoxic effect was observed in soft tissue sarcomas (STS) (see Section 2.3.1.2). During the last two decades, when razoxane was administered together with radiotherapy, clear progress has been made in controlling inoperable or gross residual STS [1] but distant metastases remained an obstacle to prolonged survival. *A noticeable suppression of distant metastases by razoxane alone was not evident.*

From this background, studies were initiated to extend the radiosensitizing therapy with razoxane by the addition of vindesine, a tubulin affinic drug, to see whether it has any influence on the incidence of metastasis and the application of the drug combination is feasible [2]. One of the reasons for combining razoxane and vindesine came from our experience which showed that vindesine was well tolerated and had no cumulative toxicity even when the drug was given up to 7 years [3], and also because it has already shown antimetastatic activity in cervical cancer. Unpublished pilot series have indicated a dramatic increase of the radioresponsiveness not only in soft tissue sarcomas, but also in some solid tumours, if razoxane is combined with vindesine. A noticable antimetastatic effect of this drug combination emerged slowly and became evident during these studies.

In this context, an extract from an article that appeared in *Clinical & Experimental Metastasis* in 2008 is given below where the strong antimetastatic effect of razoxane combined with vindesine in STS is described in detail.

Combined Vindesine and Razoxane Shows Antimetastatic Activity in Advanced Soft Tissue Sarcomas

Original article: Rhomberg W, Eiter H, Schmid F, Saely Ch (2008) Combined vindesine and razoxane shows antimetastatic activity in advanced soft tissue sarcomas. Clin Exp Metastasis 25(1):75–80

Shortened version

Abstract

Razoxane and vindesine were shown to suppress distant metastasis in animal systems. Both drugs affect the main steps of the metastatic cascade. Therefore, a study was performed to explore the influence of these drugs on the dynamics of metastasis in advanced soft tissue sarcomas (STS).

Twenty-three patients with unresectable ($n = 7$) and oligometastatic STS ($n = 16$) received basic treatment with razoxane and vindesine supported by radiotherapy and occasionally by surgery. Long-term treatment was intended in patients with metastatic disease. The cumulative number of new metastases after 6 and 9 months were determined. Thirty-six patients with comparable stages of STS treated with contemporaneous chemotherapy served as non-randomised, retrospective controls. The prognostic parameters of the groups were comparable.

In patients receiving razoxane and vindesine, the median number of new metastases after 6 months was 0 (range, 0–40) and after 9 months likewise 0 (0–70). The corresponding numbers in the control group were 4.5 (range, 0–40) and 9 (0 to >100) ($p < 0.001$). The progression-free survival at 6 months was 74% in the study group and 23% in the controls. The median survival time from the occurrence of the first metastasis or the time of unresectability was 20+ months (range, 8–120+) vs. 9 months for the controls (range, 2–252) ($p < 0.001$). The combined treatment was associated with a low to moderate toxicity.

Conclusion: Trimodal treatment with razoxane, vindesine and radiotherapy is feasible in patients with unresectable primaries and early metastatic STS. The combination inhibits the development of distant metastases in the majority of patients and prolongs survival.

The reasons for combining razoxane with desacetyl-vinblastine-amide [vindesine (VDS)], a semisynthetic vinca alkaloid, were in part outlined above. A further rationale was the fact that vindesine was shown to be effective in cytotoxic combination therapies in STS [4]. Moreover, VDS has putative radiopotentiating abilities [3, 5] and proven antimetastatic activity in animal systems [6, 7] probably due to its microtubule inhibition which leads to pronounced antiinvasive effects in vitro [8, 9] and inhibits the mitosis. Razoxane by itself is strongly antimetastatic in animal systems [10–12].

Materials and Methods

From 1996 to 2006, 23 patients with advanced adult-type soft tissue sarcomas (STS) received a combined treatment with razoxane and VDS supported by radiotherapy and, in some instances, by surgery. Amongst these patients, 7 had unresectable primary tumours or recurrences without metastases at baseline, and 16 had early metastatic disease, i.e. less than 7 distant metastases. There was no patient selection, all the patients were referred for palliative radiation therapy. Previous treatment with cytotoxic drugs was allowed. Thirty-six patients with comparable age, tumor stages and prognostic features served as non-randomized, retrospective controls.

The Antimetastatic Approach

Conventional cytotoxic chemotherapy is used to induce disease regression or stable disease. In case of progressive disease, the treatment is judged as not effective and will be changed or terminated. In contrast, the antimetastatic approach has the intention to prevent further metastasis – irrespective of the achievement of an objective response of existing lesions. This approach was pursued in our cohort of patients on combined razoxane/VDS treatment: If pre-existing metastases proved resistant to the combination of razoxane/VDS, this therapy was continued ('treatment beyond progression'). In addition, the respective lesions were irradiated and in some cases removed by surgery. In case of only a few new metastases, the razoxane/VDS treatment was likewise continued and local treatment measures were performed again. However, the combination therapy was regarded as ineffective and terminated if more than 5 new metastases appeared within 3 months.

Drug Treatment

The study patients received metronomic chemotherapy with razoxane tablets and small doses of VDS together with concurrent radiotherapy. The treatment was terminated in case of a complete response of unresectable tumours, but continued if metastases were present at the time of patient referral. The protocol required razoxane (Cambridge Laboratories, UK) to be given 5 days before the first irradiation at a dose of 125 mg twice daily by mouth. The drug was continued on radiation days until the end of the radiotherapy. The median overall dose of razoxane was 14 g per patient (range, 7.25–75 g). VDS was given intravenously at weekly doses of 2 mg. The median dose of VDS per patient was 43 mg (16–302).

Three of the 23 patients in the razoxane/VDS group were pre-treated with conventional chemotherapy, i.e. doxorubicin based regimens, and 5 patients received that treatment during the later course of the disease. Further, four patients had received 2–4 doses of mitoxantrone in addition to the razoxane/VDS treatment. This initial treatment variant, however, was discontinued early because of chronic nausea.

Radiation Therapy

External beam radiation therapy (EBRT) was performed with 6 and 25 MeV photons of linear accelerators and conformal planning techniques. Single tumour doses between 170 and 200 cGy were given five times a week at the ICRU (International Commission on Radiation Units) point. The median total dose to unresectable primaries or recurrences was 60 Gy (range, 50–64) and 50 Gy (range, 50–60) to solitary metastases. In case of oligotopic metastases, the average total tumour doses were below 50 Gy. Seven patients received two or more radiation treatments for metastases.

Control Patients

Thirty-six patients with similar age and prognostic features, in particular with similar stages of STS, who received contemporary cytotoxic drugs (doxorubicin based regimens) in addition to radiotherapy served as controls. The control group was selected from 115 patients with adult-type sarcomas who were referred to our department between 1993 and 2004 for adjuvant or palliative radiation therapy.

To be elegible as controls, patients had to have unresectable primaries and/or early metastatic disease with less than 7 distant metastases at the time of referral. This cut-off level was an arbitrary decision. Patients with multiple metastases or patients with complete tumor resections who only received adjuvant radiotherapy, were excluded as controls. For all patients serving as controls, complete clinical follow-up data as well as X-rays, CT and MRT imaging had to be available.

Follow-Up and Evaluation

All patients were followed up until March 2007 or to their death. Abdomino-pelvic and chest CTs were performed every 3 months during the first year. Additional investigations were done dependent on clinical needs. The number of new metastatic foci was counted every 3 months, and the cumulative incidence of new metastases after 6 and 9 months was determined. Further, progression free survival at 6 months and the objective response rate of irradiated tumours was recorded. The survival time was calculated from the occurrence of the first distant metastasis or the time of unresectability of localised sarcomas, and, additionally, from the beginning of the combination therapy (razoxane/ VDS/radiotherapy) in the study cases or from any systemic cytotoxic chemotherapy and/or palliative radiotherapy in the control patients.

Statistical Methods

The Wilcoxon-Gehan statistic was used to compare differences in survival times between the treatment groups. Other between-group differences were tested for statistical significance with the Mann-Whitney U test for continuous variables and with the Chi-squared test for categorical variables, respectively. p values <0.05 were considered significant. All statistical analyses were performed with the software package SPSS 11.0 for Windows.

Results

The main pre-treatment characteristics of our patients including their relevant prognostic parameters are listed in Table 2.10. There were no significant differences between razoxane/VDS treated patients and controls.

Table 2.10 Main clinical characteristics and prognostic factors

	Razoxane + VDS ($n = 23$)	Controls ($n = 36$)
Age in years (range)	64 (31–80)	60 (23–85)
Gender		
Male	12	20
Female	11	16
Time from diagnosis to metastasis in months (range)	11.5 (0–48)	9 (0–240)
Median tumor size at diagnosis in cm (range)	12 (5.5–23)	10 (2–25)
Histological diagnoses		
Liposarcoma	3	7
Malignant fibrous histiocytoma	2	8
Leiomyosarcoma	3	6
Fibrosarcoma	1	2
Angiosarcoma	7	3
Synovial sarcoma	1	3
GIST	1	3
Other rare entities	5	4
Histological grading		
Grade 1	2	1
Grade 2	4	7
Grade 3 + 4	15	24
Unknown	2	4
Conventional chemotherapy, ever given	8/23 (35%)	26/35 (74%)

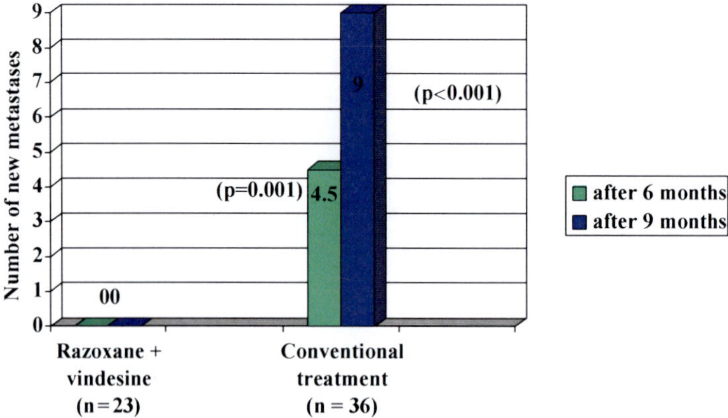

Fig. 2.8 Cumulative incidence of new metastases in soft tissue sarcomas after 6 and 9 months from the start of a palliative treatment (median values)

Development of Metastases

In the razoxane/VDS group the median number of new distant metastases after 6 months was 0 (range, 0–40), and after 9 months likewise 0 (range, 0–70). The corresponding median values for the controls were 4.5 (range, 0–40) and 9 (range, 0 to >100) new metastases after 6 and 9 months, respectively (Fig. 2.8). These differences in the occurrence of new metastases after 6 and 9 months were highly significant ($p = 0.001$ and $p < 0.001$, respectively).

In the subset of patients with unresectable primaries or isolated recurrences, none of the 7 patients treated with razoxane and vindesine, and 9 of 13 control patients developed distant metastases within 9 months ($p = 0.045$). Only after 12 months, 2 of 7 patients of the study group showed a distant spread; both patients had angiosarcomas of the thyroid.

Survival

In 16 oligometastatic patients receiving razoxane and vindesine, the median survival time from the occurrence of the first distant metastasis was 17 months (range 8–120+) vs. 9 months (range 2–252) in the control group ($p = 0.010$). Adding the 7 unresectable localized cases, the median survival was 20+ months (Fig. 2.9). The survival time from the beginning of a systemic drug treatment with or without palliative radiotherapy was 16 months (range 6–120+) in the study patients, and 9 months (range, 2–235) in the controls. The progression-free survival at 6 months was 74% in the study patients and 23% in the controls, respectively ($p < 0.001$).

Among the 7 patients with unresectable primaries or recurrences without metastasis who received the antimetastatic drugs the median survival is as yet 29+ months (range 11.5–46). Six of these 7 patients survived longer than 1 year after the start of the treatment compared to 5 of 13 in the control group ($p = 0.043$).

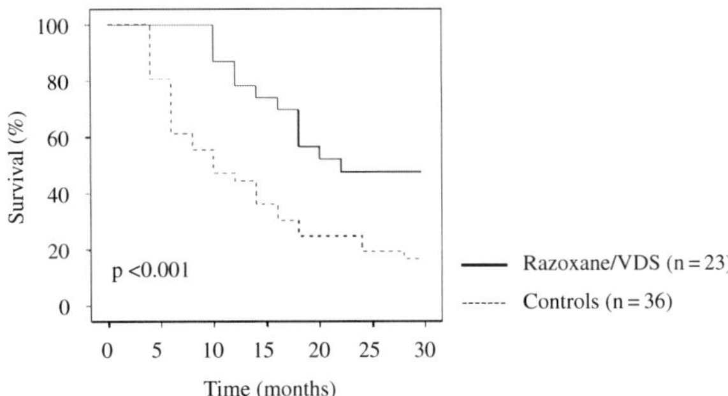

Fig. 2.9 Survival time from the occurrence of first distant metastasis or unresectability of primary/recurrence

Side Effects and Complications (Study Group)

The main side effect of the combined razoxane/VDS/radiotherapy was leukopenia with a nadir at day 16. Leukopenia of grade 3 or 4 was noted in 39% of the patients, no case of neutropenic fever occurred. Pulmonary embolism was seen in 3 patients, one with lethal outcome. Other systemic toxicity included mild to moderate neurotoxicity, diarrhea grade 1–2, and nausea grade 1–2.

Normal tissue reactions to radiotherapy were clearly enhanced by razoxane/VDS. Regional pneumonitis and esophagitis were most frequently observed when parts of the lung were irradiated. Such reactions occurred even after radiation doses of 30 Gy but they were of limited clinical significance.

Discussion

From these data we conclude that the trimodal treatment with razoxane, vindesine and radiotherapy is feasible in patients with unresectable primaries and early metastatic STS. The combination seems to inhibit the development of remote metastases in a majority of patients and prolongs survival.

With the exception of gastrointestinal stromal tumours (GIST) for which survival advantages have been achieved with Imatinib, the median survival in advanced soft tissue sarcomas has not substantially changed for almost three decades irrespective whether doxorubicin alone, CYVADIC or doxorubicin together with ifosfamide has been given [4, 13–15]. The median survival time of advanced STS treated with contemporary chemotherapy ranges between 7 and 12 months [4, 15]; a meta-analysis which included 2,185 patients showed an overall survival time of 51 weeks [14]. In view of the unchanged prognosis of disseminated STS, a comparison of the results of this trial with historical or non-randomized controls seems to be justified, especially if the prognostic features were not different among the compared groups.

Initial tumour size, histological grade, tumour site and lymph node involvement are the major prognostic factors for patients with primary STS [4]. A long disease free interval from diagnosis to first metastasis, low histopathological grade, young age, and absence from liver metastasis are the strongest predictors for a better prognosis in patients with disseminated disease [16, 17], while a doxorubicin based chemotherapy did not affect the survival substantially [18]. All these main prognostic factors did not differ significantly between the razoxane/VDS group and the controls.

The combination of razoxane and vindesine basically represents a combination of an angiogenesis affecting and a tubulin-inhibiting agent although the modes of action of the drugs are overlapping. Both drugs affect the main steps of the metastatic cascade.

The determination of the rate of new metastases for a given period may represent an interesting clinical trial endpoint for the assessment of antiangiogenic substances or, in general, of antimetastatic drug regimens. Presently, no basic data on the incidence and dynamics of metastases in STS are available from the literature. The retrospective evaluation of the metastatic process in the control group proved to be a cumbersome procedure. Numerous inquiries were necessary at different departments. In analyzing CT images and X-rays we had to face some uncertainties, or even the impossibility of counting metastases, especially the lesions of the peritoneum or the pleura. Most precise data on the dynamics of the metastatic process in STS can probably only be obtained by a prospective trial with repeated whole body CT's. CT scans are associated with much higher numbers of visible lesions compared to chest-X-rays. So, there remains some uncertainty, and the figures given on the numbers of metastases in this study must be seen and defined as minimal numbers of detected metastases.

The combination therapy with razoxane, vindesine, and radiotherapy is easy to perform. Patient compliance and tolerance of the drugs was satisfactory. No unsuspected toxicity was observed during long-term treatment, indicating the safety of the treatment. Patients must be warned, however, when giving irradiation to larger lung volumes because of the danger of a pneumonitis.

In summary, the results of this updated pilot study suggest that the treatment has the potential to reduce the propensity of STS for lethal distant metastases. Antimetastatic drug combinations supported by radiotherapy and/or surgery may become a new paradigm for the management of patients with unresectable primaries and oligometastatic STS.

References

1. Rhomberg W, Hassenstein EOM, Gefeller D (1996) Radiotherapy vs. radiotherapy and razoxane in the treatment of soft tissue sarcomas: final results of a randomized study. Int J Radiat Oncol Biol Phys 36:1077–84
2. Rhomberg W, Eiter H, Schmid F, Saely Ch (2007) Razoxane and vindesine in advanced soft tissue sarcomas: impact on metastasis, survival and radiation response. Anticancer Res 27(5B):3609–14
3. Rhomberg W, Eiter H, Soltesz E et al (1990) Long term application of vindesine: toxicity and tolerance. J Cancer Res Clin Oncol 116:651–3
4. De Vita VT Jr, Hellman S, Rosenberg SA (eds) (2001) Cancer principles & practice of oncology. JB Lippincott, Philadelphia, PA, pp 1879–83
5. Storme GA, Schallier DC, De Neve WJ et al (1988) Vinblastine has radiosensitizing activity in limited squamous cell lung cancer. Int J Radiat Oncol Biol Phys 15(Suppl 1):222

 6. Atassi G, Dumont P, Vandendris M (1982) Investigation of the in vivo antiinvasive and antimetastatic effect of desacetyl vinblastine amide sulphate or vindesine. Invasion Metastasis 2:217–31
 7. Mareel MM, Bracke ME, Boghaert ER (1986) Tumor invasion and metastasis: therapeutic implications? Radiother Oncol 6:135–42
 8. Haug IJ, Siebke EM, Grimstad IA et al (1993) Simultaneous assessment of migration and proliferation of murine fibrosarcoma cells, as affected by hydroxyurea, vinblastine, cytochalasin B, razoxane and interferon. Cell Prolif 26:251–61
 9. Mareel MM, Storme GA, De Bruyne GK et al (1982) Vinblastin, vincristine and vindesine: antiinvasive effect on MO4 mouse fibrosarcoma cells in vitro. Eur J Cancer Clin Oncol 18:199–210
10. Hellmann K, Burrage K (1969) Control of malignant metastases by ICRF 159. Nature 224:273–275
11. Baker D, Constable W, Elkon D, Rinehart L (1981) The influence of ICRF 159 and levamisole on the incidence of metastases following local irradiation of a solid tumor. Cancer 48:2179–83
12. Peters LJ (1975) A study of the influence of various diagnostic and therapeutic procedures applied to a murine squamous carcinoma on its metastatic behaviour. Br J Cancer 32(3):355–65
13. Antman K, Crowley J, Balcerzak SP et al (1993) An intergroup phase III randomized study of doxorubicin and dacarbazine with or without ifosfamide and mesna in advanced soft tissue and bone sarcomas. J Clin Oncol 11:1276–85
14. Edmonson JH, Ryan LM, Blum RH et al (1993) Randomized comparison of doxorubicin alone versus ifosfamide plus doxorubicin or mitomycin, doxorubicin, and cisplatin against advanced soft tissue sarcomas. J Clin Oncol 11:1269–75
15. Santoro A, Tursz T, Mouridsen H et al (1995) Doxorubicin versus CYVADIC versus doxorubicin plus ifosfamide in first line treatment of advanced soft tissue sarcomas: a randomized study of the European Organisation for Research and Treatment of Cancer Soft Tissue and Bone Sarcoma Group. J Clin Oncol 13:1537–45
16. Rööser B, Attewell R, Berg NO, Rydholm A (1988) Prognostication in soft tissue sarcoma. A model with four risk factors. Cancer 61:817–23
17. Zagars GK, Ballo MT, Pisters PWT et al (2003) Prognostic factors for disease-specific survival after first relapse of soft-tissue sarcoma: analysis of 402 patients with disease relapse after initial conservative surgery and radiotherapy. Int J Radiat Oncol Biol Phys 57:739–47
18. Komdeur R, Hoekstra HJ, van den Berg E et al (2002) Metastasis in soft tissue sarcomas: prognostic criteria and treatment perspectives. Cancer Metastasis Rev 21:167–83

A Multicenter Phase II Study Confirms the Efficacy and Antimetastatic Activity of the Combined Treatment of Razoxane, Vindesine and Irradiation also in Vascular Soft Tissue Sarcomas

Original article: Rhomberg W, Wink A, Pokrajac B, Eiter H, Hackl A, Pakisch B, Ginestet A, Lukas P, Pötter R (2009) Treatment of vascular soft tissue sarcomas with razoxane, vindesine and radiation. Int J Radiat Oncol Biol Phys 74(1):187–91

A synopsis of this article was already written in Section 2.3.2.1. If there is interest in the background, materials and methods, treatment details, toxicity and some parts of the discussion, the reader is referred to this section. The following data are restricted to the results and some antimetastatic aspects of this work [1].

The ability of razoxane to normalize pathological tumor blood vessels almost demands the drug to be used in the treatment of angiosarcomas. For this reason and in view of the results of the study previously described [2], the Austrian Society of Radiooncology initiated a nonrandomized phase II study to see whether the observations made in soft tissue sarcomas in general can be confirmed especially in angiosarcomas.

Between July 2002 and July 2005 fourteen patients with histologically proven sarcomas of the blood vessels received a combined treatment with concurrent razoxane, vindesine and irradiation. There were 10 male and 4 female patients with a median age of 68 years (range, 47–90 years). Eleven patients presented with angiosarcomas mostly of grade 3 and 4, three patients had malignant hemangiopericytomas. One patient with an angiosarcoma of the scalp was excluded from the treatment analysis because he received erroneously only razoxane and irradiation without vindesine. Thus, the treatment report comprises 13 evaluable patients.

Results

The main pretreatment characteristics, and the tumor response is shown in Table 2.11. All but one patient had no pretreatment with cytotoxic agents; and only 2 patients (# 8 and #9) received any other chemotherapy during the later course of their disease. Measurable disease included 4 unresectable primaries, 4 regrowing gross residual masses after R-2 resection, and one case with distal lung metastases. Three patients presented with microscopic residual disease (R-1 resection, lesions not measurable) and one patient had resection of a scalp angiosarcoma with clear margins.

Response to Radiotherapy

Thirteen patients with sarcomas of the blood vessels (10 angiosarcomas and 3 hemangiopericytomas) were evaluable. The radiation response by histology was as follows: Among 8 patients with measurable angiosarcoma, 6 showed a complete regression of their tumors (75%), one had a partial response and another a minor remission (with tumor measurement by CT after 15 Gy only). One patient with microscopic residuals of an angiosarcoma of the bladder died from repeated hemorrhages 3 months after treatment, and a patient with clear margins after resection of a scalp-tumor remained free of disease for 46+ months.

Three hemangiopericytomas were irradiated: One woman with a large tumor of the pelvis showed a subtotal, partial regression, the other two patients with microscopic residuals were locally controlled for 30 and 63+ months.

Overall, in 8 of 9 patients with gross disease of a vascular soft tissue sarcoma ('89%') major regressions were observed with the trimodal therapy (Table 2.11).

Metastasis

Taking all 13 assessable patients together, only 2 developed new distant metastases within the first 6 months after the start of the combined radiation treatment. Finally, 6 of 13 patients succumbed to distant metastasis, 5 are alive with no evidence of

Table 2.11 Clinical characteristics and treatment outcomes in 13 evaluable patients

Pat. #	Age/sex	Tumor site	Histology	Grade	Irradiated site(s)	Response	Time to DM or regrowth	MTx after RT	Overall survival (months)
1	72 M	Thyroid	AS	G3	Two lung metastases	CR	NED	3 years	120+
2	64 F	Thyroid	AS	G4	Recurr. gross resid.	CR	10 m	6 weeks	11
3	67 M	Thyroid	AS	G4	Recurr. gross resid.	CR	12 m	1 year	16
4	69 M	Thyroid	AS	G4	Recurr. gross resid.	PR	16 m	–[a]	19
5	61 M	Thyroid	AS	G3	Recurr. gross resid.	CR	NED	1 year	27+
6	56 M	Scalp/face	AS	Gx	Adjuvant radiotherapy	LC	NED	–	46+
7	71 F	Scalp/face	AS	G3	Primary tumor, 9 × 14 cm	CR	NED	1 year	41+
8	90 F	Breast	AS	G3	Primary tumor, >10 cm	CR	2 m	–	4
9	57 M	Heart	AS	Gx	Primary tumor, 5 × 3 cm	MR	3 m	–[a]	4.5
10	51 M	Bladder	AS	G3	Microscop. residuals	no LC	–	–	3
11	71 M	Omentium	HPC	G2	Microscop. residuals	LC	30 m	2 months	36
12	47 M	Spine L2	HPC	G2	Microscop. residuals	LC	NED	–	63+
13	77 F	Pelvis	HPC	G3	Primary tumor, 16 × 12 cm	PR	Intercurr. death	–	46

AS, angiosarcoma; CR, complete response; DM, distant metastases; F, female; HPC, hemangiopericytoma; LC, local control; m, months; M, male; MR, minor remission; MTx, maintenance therapy; NED, no evidence of disease; Pat., patient; PR, partial response; RT, radiotherapy; Recurr. gross resid., recurring gross residuals.

[a]Patients received conventional chemotherapy.

disease (NED), one died from hemorrhages without overt disease progression, and one patient died intercurrently.

Among 8 angiosarcoma patients with gross disease, 3 of 4 patients who had maintenance therapy of 1 year or longer are alive with NED whereas 4 of 4 patients without maintenance therapy finally developed distant metastasis and died.

Survival

Taking all patients with gross and microscopic residual vascular sarcomas (angiosarcomas + hemangiopericytomas) together, the median progression-free survival of 12 patients was 21.5+ months (range, 2–120+) and the median overall survival was 23+ months (range, 3–120+).

In the 8 patients with unresectable or metastatic angiosarcomas the median survival time was 17.5 months (range, 4–120+ months), 5 of 8 patients survived 1 year or longer (62.5%). The progression-free survival at 6 months was 75% (6/8 patients).

Comment

In this study a remarkable rate of complete responses was seen in measurable unresectable angiosarcomas, and it seems again that maintenance treatment with razoxane and vindesine could be able to suppress distant metastasis in these tumors.

Data on objective tumor responses to definitive or palliative radiotherapy of angiosarcomas are sparse in the literature. For details, see Section 2.3.2.1. In a study by Mark et al. [3], only 1 of 9 patients (11%) with gross disease treated with radiotherapy with or without chemotherapy was rendered free of disease. Garcia-Schüler et al. observed 6 partial regressions among 13 patients with macroscopic angiosarcomas (46%), and the median progression-free survival was 2.5 months in their series [4]. Thus, a complete response in 6 of 8 patients with macroscopic disease is certainly an outcome that deserves interest.

In patients with unresectable macroscopic angiosarcomas the trimodal treatment led to a median survival time of 17.5 months. If all patients with some form of residual disease are taken together ($n = 12$), the median progression-free survival was 21.5+ months. Abraham et al. described the treatment outcome of 82 patients with angiosarcomas: Of 36 patients with advanced disease, 36% underwent a palliative operation, 78% received radiation, and 58% received chemotherapy. The median survival was 7.3 months [5]. In a larger series of 125 patients treated by Fury et al., the overall 5-year survival was 31%. For unresectable angiosarcoma, no data concerning the effectiveness of radiation was given. Doxorubicin based regimens yielded a progression-free survival (PFS) of 3.7–5.4 months. Paclitaxel led to a PFS of 6.8 months for scalp angiosarcoma and 2.8 months for sites below the clavicle [6].

Unfavourable prognosis of angiosarcomas varies, dependent on the primary site. For instance, angiosarcomas of the thyroid are associated with a dismal prognosis [7, 8], and the median survival was 2.4 months in the largest series reported

[7]. Other unfavourable locations are scalp and face, liver, heart and the skeletal system with 5-year survival rates of 0–15% including all stages. All but one patient of our study had angiosarcomas with locations associated with a bad prognosis.

This drug combination seems to be able to reduce the propensity for distant metastases in angiosarcomas provided that the drugs are continued and given as mainentance treatment. The optimum duration of such a treatment is unknown. In our study, it was observed that in case of a maintenance therapy of 1 year or longer, only 1 of 4 patients with gross disease developed distant metastasis whereas 4 of 4 patients without maintenance therapy developed distant metastasis and died.

The low general toxicity of this multimodal therapy, the outstanding response rate as well as the finding of a reduction of distant metastases deserve further attention for this convenient outpatient-based treatment regimen which might have some curative potential.

References

1. Rhomberg W, Wink A, Pokrajac B, Eiter H, Hackl A, Pakisch B, Ginestet A, Lukas P, Pötter R (2009) Treatment of vascular soft tissue sarcomas with razoxane, vindesine and radiation. Int J Radiat Oncol Biol Phys 74(1):187–91. doi:10.1016/j.ijrobp.2008.06.1492
2. Rhomberg W, Eiter H, Schmid F, Ch Saely (2008) Combined vindesine and razoxane shows antimetastatic activity in advanced soft tissue sarcomas. Clin Exp Metastasis 25(1):75–80
3. Mark RJ, Poen JC, Tran LM et al (1996) Angiosarcoma. A report of 67 patients and a review of the literature. Cancer 77:2400–6
4. Garcia-Schüler H, Jensen A, Röder F et al (2005) Retrospective evaluation of treatment results after radiotherapy of angiosarcomas [Abstract]. Strahlenther Onkol 181(Suppl):65
5. Abraham JA, Hornicek FJ, Kaufmann AM et al (2007) Treatment and outcome of 82 patients with angiosarcoma. Ann Surg Oncol 14:1953–67
6. Fury MG, Antonescu CR, Van Zee KJ et al (2005) A 14-year retrospective review of angiosarcoma: clinical characteristics, prognostic factors, and treatment outcomes with surgery and chemotherapy. Cancer J 11:241–7
7. Ladurner D, Tötsch M, Luze T et al Das maligne Hämangioendotheliom der Schilddrüse. Pathologie, Klinik und Prognose. Wien Klin Wochenschr 102(9):256–9
8. Goh SG, Chuah KL, Goh HK et al (2003) Two cases of epitheloid angiosarcoma involving the thyroid and a brief review of non-Alpine epitheloid angiosarcoma of the thyroid. Arch Pathol Lab Med 127(2):E70–3

2.3.4 Razoxane – A Cytorallentaric Drug

Kurt Hellmann

Slowdown of tumour growth rate – what one might call the 'cytorallentaric' effect (from It. *rallentare*, to slow down) – differs from tumour inhibition by chemotherapy or radiotherapy. Tumour cells are not killed directly, their doubling times are lengthened so that razoxane treated tumours take longer to reach the same weight, volume and dimensions of the control tumours. . . .

Drug development is an unpredictable business; new and unexpected indications as well as new and unexpected adverse reactions may emerge years after a drug was first introduced. No apology is therefore required for considering some results which were obtained two to three decades ago and which show that razoxane is not only experimentally, but also clinically active against angiogenesis and metastasis. Even in what are meant to be exhaustive reviews of tumor angiogenesis (and metastasis) pertinent results with razoxane have not been much in evidence. This might be more readily understood if verification of the received (if simplistic) wisdom that destruction of the tumour neovasculature or prevention of its appearance will destroy cancer, had proven unambiguously successful in the clinic. The reality is, however, that there has yet to be a convincing demonstration from any one of the large number of agents tested for antiangiogenic activity, that clinically effective and worthwhile results can be obtained. A review by Kerbel and Folkman [1] illustrates this very clearly by the paucity of clinical results that they were able to examine now some 31 years after Folkman thought about the therapeutic implications of inhibition of tumour angiogenesis [2]. It also comes some 32 years after we first demonstrated that 'normalization' of the abnormal tumour neovasculature by razoxane prevented metastasis [3] thereby converting a malignant into a quasi-benign tumour – a disease modifying activity – previously unknown.

Apart from pointing out why angiogenesis should work in cancer patients and why it has not – as yet, Kerbel and Folkman also call attention to a study by Rakesh Jain [4]. This 'study' which claims to have discovered a new paradigm for combination therapy –'through normalization of the tumour vasculature with antiangiogenic therapy' is not supported by any experimental or clinical results. The paradigm may not be new [5, 6] but the admission may be a first: that vascular normalization could be significant for cancer therapy. This, and the wide therapeutic potential of normalization of a pathological vasculature is something that – as was stated in a famous Nature paper of 1953 of Crick and Watson – had not escaped our attention, even in 1972, as this and two more comprehensive excellent reviews show [7, 8].

A *Review* by K. Hellmann on *Dynamics of tumor angiogenesis* is of some interest in this connection:

Hellmann K (2003) Dynamics of tumour angiogenesis: effect of razoxane-induced growth rate slowdown. Clin Exp Metastasis 20:95–102

In the last few years there has be an enormous amount of interest in the development, growth and inhibition of blood vessels in both malignant and non-malignant tissues. However, scant attention has been paid to angiodynamics – the rate at which blood vessels proliferate and the consequences of rate variation on their morphological and physiological integrity and character. The character of blood vessels is important in many ways and in none more so than in tumour dissemination and metastasis. We became interested in this aspect of angiogenesis as a result of our investigations into the mechanism of the antimetastatic action of razoxane and its relation to the development of tumour neovasculature.

We discovered the antimetastatic action of razoxane by chance in a random screen [9]. From the clinical perspective, the cancer therapeutic problem is not so much a question of preventing cell division as preventing metastasis. As a result of discussions with Professor K. Karrer (Austrian Cancer Institute, Vienna) we set up in 1966 the first screen for antimetastatic compounds using the Lewis lung carcinoma (3LL) which had exactly the right characteristics for such a screen, but which many believed were unattainable. Nevertheless this screen has subsequently been very widely used to test for antimetastatic drugs.

The first compound to be tested in this screen inhibited all the metastases in all the animals without any significant inhibition of the primary 3LL implant [9]. The compound was razoxane ($\pm$) 1,2-bis-(3,5-dioxo-piperazine-1-yl) propane (ICRF 159; NSC 129,943; ICI 59,118) one of a number of compounds that had shown some activity in our random (US) National Cancer Institute screen to find substances which inhibit standard tumours (S180, Ca755 and L1210) but the decisive nature of the results in the 3LL screen was totally unexpected; results which were later to be repeated by others and in other tumours. At first the findings raised suspicions that they might have been fortuitous and/or that the 3LL metastases were an easy target and that every compound tested would give results similar to razoxane. These fears were unfounded and even today, despite intense efforts in many laboratories, there appears to be no other drug with the same broad-spectrum antimetastatic activity and minimal toxicity of razoxane. That the antimetastatic activity of razoxane might be linked to changes (normalization) of the tumour neovasculature was evident from the initial experiments [3]. Others have pursued their strategy of attempting to neutralize whatever factors may be responsible for developing and maintaining the tumour vasculature. However we have continued to explore the significance of the finding that razoxane, far from destroying the developing tumour neovasculature, normalizes its chaotic structure with important consequences for tumour dissemination and metastasis formation. The continued investigations of the pharmacodynamics of razoxane have clarified to some extent the role of this drug in a number of malignancies and other diseases. It has also become clear that the simplicity of the chemical structure of razoxane (and of its dextro enantiomer, dexrazoxane) stand in sharp contrast to the complexity of their pharmacodynamics. These investigations have taken a considerable time and are far from complete. Consequently, and because publications concerning razoxane are so scattered in time and place, it may be appropriate to assemble some of the results here in one short summary. This will elucidate what can be learned about the influences that can be brought to bear on the development of the tumour neovasculature.

Tumour Angiodynamics

After a meticulously detailed morbid anatomy and histopathology study of some 500 personally autopsied cancer cases, Willis [10] came to the conclusion that 'the structure of the new formed blood vessels in malignant tumours rarely approaches that of normal veins and arteries and THE MORE RAPIDLY GROWING THE TUMOUR THE MORE IMPERFECT IS THE ARCHITECTURE OF THE

VESSELS'. Essentially what Willis said was that there is an inverse dynamic correlation between the rate of tumour growth (which is constantly changing) and the integrity of its neovasculature. Willis went on to say that 'in highly anaplastic growths, carcinomas as well as sarcomas, even the endothelium may be incomplete in places and the vascular channels lined in part by tumour cells'.

Two major consequences flow from these conclusions. First and foremost, since the clinical cancer problem is not the tumour that can be cut out, but the tumour metastases that cannot and since tumour dissemination almost always follows the path of least resistance, ingress and egress of tumour cells through the incomplete vascular channels provides a ready made route for tumour dissemination without the necessity of an initial invasive step, thus greatly enhancing the probability of developing metastases. The incomplete vascular channels may also account for the frequent areas of haemorrhage in and around tumours observed by Willis [10]. Secondly, and perhaps less obviously, if Willis's conclusions are valid and the inverse dynamic correlation between tumour growth and integrity of its neovasculature is true, then the corollary should also hold, namely that slowing down the rate of tumour growth leads to improvement in the developing neovasculature (i.e., the new vessels should appear and behave like normal vessels). This was clearly shown to be the case in the original experiments that investigated the mechanism of the antimetastatic action of razoxane [5, 11, 12]. Doses of this drug which slowed the growth of the Lewis lung carcinoma improved the developing neovasculature to such an extent that it appeared to all intents and purposes to be normal (or normalized). At the same time the metastases from this tumours were dramatically reduced. The antitumour effect on the primary tumour (by weight) was small and should have been seen for what it was: continued growth, but at a slower rate, rather than as a small and in screening terms insignificant tumour 'inhibition'.

Direct and indirect confirmation of the validity of a dynamic correlation between tumour growth rate and the integrity of its neovasculature also comes from other experiments and clinical observations with razoxane. Indirect evidence comes from the effect of razoxane on 15 different tumours in 13 of which growth rates were slowed by the drug, while blood borne pulmonary metastases from 12 of these 13 were greatly reduced (Table 2.12).

While blood borne metastases can readily reach many distant anatomical sites especially lungs, liver, bone and brain, tumour dissemination also frequently occurs via the lymphatics. Heston et al. investigated the effect of razoxane on tumour burden at the inoculation site, at lymphatic drainage sites (primarily axillary, inguinal, and retroperitoneal lymph nodes) and on the number of lung metastases in the hormone refractory rat prostate tumour R3327 MAT-LyLu. The results showed, as the authors comment 'In all categories razoxane exhibited the most dramatic effect' [13]. Although razoxane reduced the tumour volume at the end of the experiment by about 45% and the weight by about 75%, the shape of the tumour growth curves for controls and treated animals were identical: i.e., the razoxane-treated tumours continued to grow but at a reduced rate.

On the question of tumour dissemination, Heston et al. find 'Razoxane suppressed growth at the primary site but was most effective in preventing metastases.

Table 2.12 Razoxane, tumour slowdown, neovasculature normalization and metastasis

Tumour	Slowdown	Vas. Norm.	Metastasis	Lymph	Ref.
3LL	+	+	+		[5,9]
WALKER	+	+			[20]
Hep 3	+	+	+		[21]
ML	±	+	+	+	[39]
Sq, Ca′G′	+	±	+	+	[40]
R3327 MAT-Lu	+		+	+	[13]
PA-III	+		+	+	[14]
B 16	–		–		[41]
DMH	+		+		[42]
S 180	+	–	+		[15]
KHT	+		+	+	[43]
EIPTH Sp1	–	–	–	–	[44]
V2	+	+	+	+	[16]
OSTEO	+		+		[45]
Dukes's C	+		+		[26, 27]
	13/15		12/14		

Blank, not done; +, clear positive evidence; –, clear negative evidence; Vas. norm., normal vasculature.

Of the razoxane treated animals, 68% appeared free of lymphatic metastases and 78% were free of observable lung metastases'. Interesting also are the actual numbers in terms of the wet weight of the lymph nodes and of the number of lung metastatic colonies (Table 2.13).

Razoxane was compared in this series of experiments with diethylstilbestrol (DES), but even the highest doses of this drug, although they reduced the primary tumour sizes by about 40%, did not give a statistically significant reduction of lung metastases. DES was for many years the treatment of choice for treatment of hormone sensitive prostate cancer, but than as now treatment for hormone refractory prostate cancer was unsatisfactory.

There has been no clinical trial of razoxane eiher alone or in combination in this malignancy, even though the results with R3327 MAT-LyLu have been replicated in another prostate adenocarcinoma, PA-III [14]. However, the razoxane effect on the tumour blood vessels has only been examined directly in five tumours and in four evidence of normalization was seen. There have also been studies to determine whether razoxane improves blood flow and/or tumour oxgenation [15–17]. These studies (using the S180, 3LL, V2 rabbit carcinoma or Ehrlich Ca cells) were unable to detect any changes in blood flow, oxygenation or indeed influence on cell

Table 2.13 Effect of razoxane on metastases of the rat R3327 MAT-LyLu prostate cancer

Treatment	Lymphatic mets (total g wet wt.)	Lung mets (number)
Control	7.1 ± 2.1 (11/11)	177 ± 56 (11/11)
Razoxane	0.03 ± 0.02 (3/9)	0.6 ± 0.4 (2/9)

Numbers of animals with metastases are given in parentheses. From Heston [13].

respiration by razoxane. However, it has to be remembered that razoxane would have slowed down the tumour growth rate and consequently tumour demand for oxygen and nutrients. Increased blood flow through a normalized neovasculature would not therefore have been required, particularly if the normalized neovasculature also resulted in a more effective extraction and exchange of substances between the circulating blood and the tumour.

Interestingly and in sharp contrast to the inhibitory action of a variety of anticancer agents, razoxane itself has no inhibitory activity on the respiration of isolated mitochondrial protein and only an inhibition of some 10% or less on the endogenous respiration of Ehrlich ascites cells. Clearly, this is difficult to improve upon [17]. Indirect evidence from clinical observations, in particular from psoriasis, would seem to indicate that razoxane slows down cellular processes that drive proliferation of blood vessels even in non-malignant pathological conditions. Slowdown of tumour growth rate – what one might call the 'cytorallentaric' effect (from It. *rallentare*, to slow down) – differs from tumour inhibition by chemotherapy or radiotherapy. Tumour cells are not killed directly, their doubling times are lengthened so that razoxane treated tumours take longer to reach the same weight, volume and dimensions of the control tumours. The size of razoxane treated tumours can, however, increase slowly even if cells are unable to divide. This is because in cells whose DNA continues to replicate, but which are unable to divide leads to a considerable enlargement in size [18]. Consequently, the tumour may increase in size without any increase in cell numbers. Growth rate slowdown may thus be greater than it might appear when measured by tumour volume, weight or dimension.

Effective treatment with a drug causing tumour growth rate slowdown would therefore be expected to result in a normalized neovasculature and in a reduction or absence of metastasis without necessarily much, if any reduction of the primary tumour size compared with the controls, at least in the short term. All these changes should increase life span, i.e. median survival time of the host, and ultimately it is length of survival (with good quality of life) that is the criterion by which every anticancer drug is (or should be) judged. Effective tumour growth rate slowdown – without adverse side effects – is therefore an avenue worth persuing. At present the avenue is not too crowded.

A largely unrecognised but impressive experimental example which illustrates this antitumour activity was provided by Sandberg and Goldin [19]. They examined the relationship between the antitumour effect (as measured by inhibition of tumour growth) against survival using the 18 most promising anticancer agents of the tens of thousands that had been examined in the screens of the US National Cancer Institute. Amongst the 18 were two drugs that are still the most widely used anticancer agents, Adriamycin (doxorubicin) and cyclophosphamide, but the list also included daunomycin, BCNU, CCNU and razoxane. All 18 compounds were tested against a slow growing breast cancer on which razoxane had no significant activity (T/C 0.69; significant activity requires a T/C of 0.25 or less). For all the other 17 compounds there was an approximately straight line correlation between tumour volume growth inhibition and median survival time: the greater the tumour volume growth inhibition, at optimum doses, the longer was the median survival time. The only exception proved to be razoxane. Despite lack of antitumour activity, measured as volume growth inhibition, razoxane gave the longest median survival time of any of the other putative anticancer agents, thus preventing death where tumour spread is the most likely cause or contributory cause of mortality.

Following treatment with a cytorallentaric drug such as razoxane it is possible to envisage that, if the angiodynamics are at the centre of a chain of events, there would be some six steps with an afferent and an efferent arm.

Step 1. The initiating event in which the drug causing the slowdown interacts with, e.g. ATP to reduce the available energy for cell generation; as a result in Step 2, the metabolic needs of the tumour are reduced, and the size, volume and weight of the tumour remain static. This gives the developing vasculature in Step 3 time to develop normally. Step 4 sees the consequences of Step 1–3. This is the efferent aspect. There may be changes in blood flow and oxygenation, but importantly extraction of nutrients and/or drugs may increase and in Step 5 blood and lymph borne metastases are greatly reduced. As a result in Step 6 median survival time should increase significantly. Thus, the total effect of a cytorallentaric drug is the conversion of a malignant to a quasi-benign tumour.

Normalization of Tumour Neovasculature by Razoxane

The discovery that razoxane could comprehensively reduce 3LL metastases without significant inhibition of the growth of the primary implant [3] (T/C 0.46) (Fig. 2.10) was interesting in a number of ways. Not only did it achieve for the first time the obviously desirable therapeutic goal of preventing tumour spread, but perhaps more interesting was the demonstration that it was possible to delete one malignant characteristic, the most lethal one, without markedly affecting others such as uncontrolled proliferation. Analysis of the mechanism of the antimetastatic action of

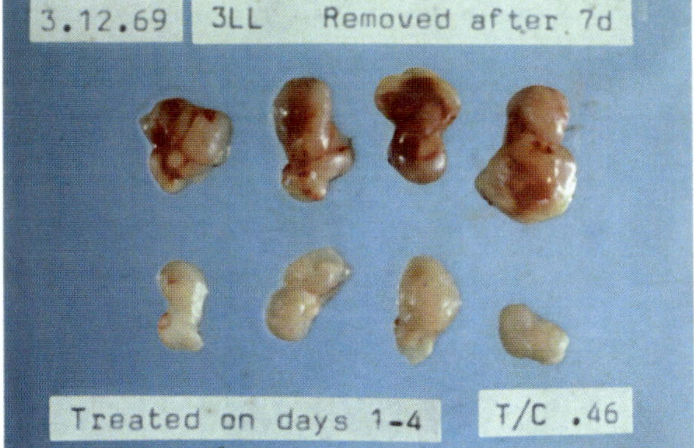

Fig. 2.10 3LL tumour implanted subcutaneously and treated with 30 mg/kg razoxane i.p. daily for 4 days from day of implant. Tumours excised on day 7 and lungs examined 2 weeks later. Normalization of the tumour vasculature reduced haemorrhages and lung metastases (not shown). Precise meaning of the insignificant (T/C 0.46) tumour 'inhibition' is not clear

razoxane revealed that it was due to changes the drug induced in the primary tumour implant. In contrast to the controls, no tumour cells or fragments of them were detectable in the blood of the treated mice [12].

Histological examination of the 3LL tumour implants of animals given razoxane showed that there were marked macroscopic and microscopic differences between the peripheral blood vessels of the control and treated tumours (Fig. 2.11a, b). While those of the controls were haemorrhagic, poorly endothelialized and tortuously sinusoidal, those of the razoxane treated animals showed no evidence of leakiness [5].

These changes in blood vessel morphology were confirmed in another system [20] in which small plastic grids (about 2 cm diameter) were laid on the chorioallantoic membrane (CAM) of 8 day old chick embryos. Suspensions of tumour cells were then carefully dropped into the outer wells formed by intersecting strands of the plastic to form diamond shaped wells. The putative anticancer drug was placed in the middle to allow diffusion to all the outer wells. When cell suspensions of the Walker carcinosarcoma were used, it was found that while razoxane had no apparent influence on the growth of the Walker tumour, there was a dramatic difference between control and treated tumours in the appearance of the blood vessels. There were very few haemorrhages, straight and apparently fully competent blood vessels seen in the treated tumours compared with florid, haemorrhagic tumours and few signs of competent blood vessels visible in the control tumours (Fig. 2.12a, b).

These findings were independently corroborated in other experiments using the CAM assay. Gitterman and Luell [21] implanted the human tumour HEp 3 directly on the CAM of 10-day-old chick embryos and after 10 days examined the lungs for metastases. At all dose levels of razoxane the inhibition of lung metastasis was twofold to threefold greater than the inhibition of the tumour. Moreover, these authors (working in the Research Laboratories at the Merck Institute for Therapeutic Research, Rahway, New Jersey) reported that 'at levels that inhibited metastasis, but not necessarily growth of the primary implant, 5-fluorouracil, razoxane and (a Merck compound) 593A altered the appearance of the primary tumour from

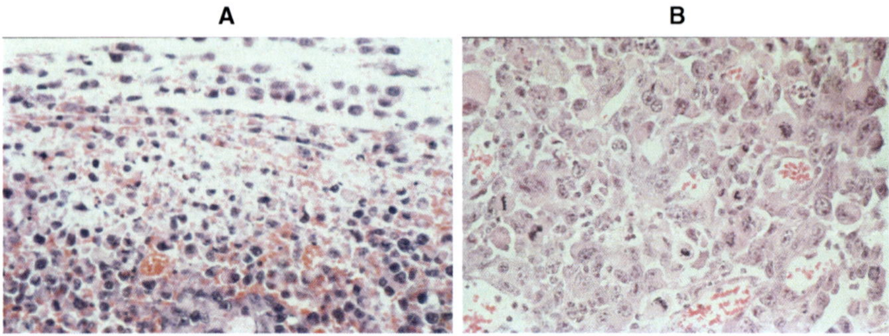

Fig. 2.11 Lewis lung carcinoma without (**a**) and with pretreatment by razoxane (**b**)

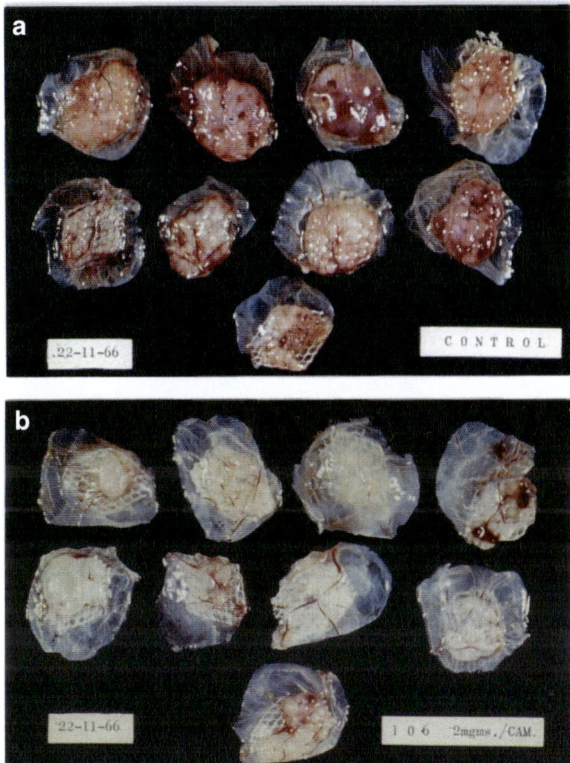

Fig. 2.12 Walker carcinosarcoma grown on the CAM of chick embryos. (**a**) Controls (*above*) showing intense haemorrhagic growth. (**b**) Razoxane 2 mg/CAM resulted in slowing of growth with much reduced areas of haemorrhage

haemorrhagic to non-haemorrhagic. Hadacidin (a glycine derivative) sometimes affected this change, but the nitrosourea BCNU did not'.

Numerous investigations aiming to discover the mechanism of the cytorallentaric effect of razoxane at the cellular level have shown that while the drug was highly effective in blocking cell division in late G2 or early M, it was completely ineffective when given at other points in the cell cycle. Closer analysis found that the block manifested itself by a slowdown at the G2/M border, so that phases of the cycle not normally seen because the cell passes through them very quickly, become visible. Sharpe et al. [22] wrote: 'Mitotic abnormalities were only seen when the drug (i.e. razoxane) was administered during the premitotic and early mitotic (G2/M) period. Mitotic figures were drastically reduced in number and those present showed arrest in early and late prophase or occasionally early metaphase. The latter two stages of mitoses are seldom seen in normal cultures, because cells pass very rapidly through them'. It seems likely that whether razoxane exerts a brake to slowdown the dynamics of the traverse through the cell generation cycle at the border of G2/M or a complete block at this point will depend on the concentration of razoxane and the length of exposure. Recently CDK1 protein expression was measured in K562 cells after incubation in dexrazoxane (the more soluble dextro enantiomer of razoxane) for 24 h and shown to be reduced to less than 50% [23]. Our results suggest that these cells are undergoing two or more S-phases without intervening mitoses, so delaying cellular proliferation.

The influence of dexrazoxane on energy availability was studied by measuring the ADP/ATP ratios in treated cells. This showed a continuous decline of these ratios with increasing concentrations of dexrazoxane. After 24 h incubation there was a 48 ± 12% reduction in the ratio compared to controls, suggesting a block in ATP hydrolysis. The selective cellular slowdown and its consequences for the dynamics of cell division was not as far as could be observed accompanied by any toxic side effects, something that was also noticeable in animal experiments and in clinical trials.

Clinical Impact

There have been a number of small studies, with 20 or fewer patients, to test razoxane as an 'anticancer agent', but only in the leukemias and lymphomas was there a glimmer of activity [24, 25]. In order to see whether the razoxane antimetastatic activity in experimental tumours would translate into clinical therapeutic benefit, a trial was set up in colorectal cancer. Additional trials were also organized to see if the normalization of blood vessels due to razoxane could influence four pathologic conditions, namely psoriasis, Kaposi's sarcoma, Crohn's disease and renal cell cancer in all of which abnormal blood vessels are prominent. Clinical trials aiming to show the effectiveness of antimetastatic drugs have to be very clear about their objectives and the considerable costs involved. It is doubtful if we would have embarked on a trial of razoxane in colorectal cancer if the conditions which prevail today had been in force then. Bureaucratic interference made easy by centralization has reached such proportions that only 'approved' trials can realistically hope to accrue the numbers required to reach statistically valid conclusions.

Razoxane's Antimetastatic Action in Resected Colorectal Cancer

A carefully organized, monitored, randomized, controlled trial recruited 272 patients with resected colorectal cancer [26]. They received razoxane 125 mg bd 5 days/week indefinitely or placebo. Although Dukes' A, B and C patients were entered into the study, only Dukes' C patients are a clearly evaluable group. In Duke's A almost all and in Dukes' B a variable, but unknown number of patients were cured by the surgery. In Dukes' C all had lymph node involvement, but not yet (as far as it was possible to ascertain) liver metastases. The trial followed up patients for 7 years with a median of 5 years. Of the Dukes' C patients who received razoxane, only half as many 7/38 (18%) developed liver secondaries compared with the controls 17/50 (34%). Moreover, the liver secondaries in the razoxane group took twice as long to appear (80 weeks) compared with those in the control group (40 weeks).

A prospective study on the first 126 patients in this trial [27] in which all patients (Dukes' A, B and C) had a liver ultrasound scan every 3 months found that after a median follow-up of 3 years, 28 patients developed liver metastases (12 on razoxane; 16 controls). Time to their recognition was 87 ± 10 weeks for razoxane treated patients and 60 ± 8 weeks for controls ($p < 0.05$). Slowdown of the tumour growth rate must be a possibility to account for this delay [28]. The dynamic aspects: how much slowdown, how much delay, are however crucial. Not only might they affect the angiodynamics and the consequential vascular integrity, but they could also change the response of the tumour to cytotoxics and/or to radiation. Aggressive tumours might then become more manageable and their prognoses no worse than a carcinoma – in situ, if slowdown can be maintained.

Psoriasis

A pilot study in seven patients with severe, intractable psoriasis treated with razoxane immediately showed the high degree of activity the drug has in this disease [29]. Moreover, previous therapy and stage of disease did not influence the activity of razoxane. These findings were borne out and confirmed by other groups [30] who obtained similar results in over 100 patients as well as some 200 unpublished cases. The consensus appears to be that razoxane gives a response rate of between 80 and 90% even in the most refractory cases. An example of response is shown in Fig. 2.13. The more severe the condition, the more effective was the razoxane treatment [31]. Severity of the condition is an indication of the rate of replication of the cell group that provides and provokes the initial and secondary stimulus for psoriasis and any slowdown of their proliferation rate would quickly be seen as clinical improvement. A full discussion of the dynamics of evolution of psoriasis might be very valuable, but this is beyond the scope of this review.

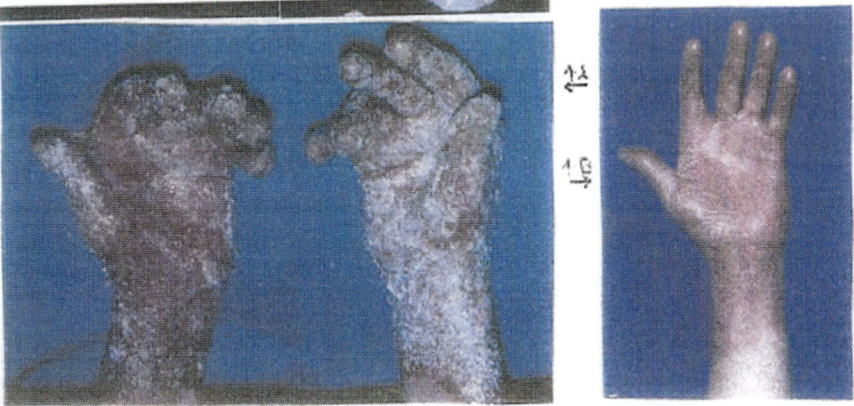

Fig. 2.13 Severe psoriasis. (**a**) Refractory to methotrexate and etretinate. (**b**) Same patient following 3 months of razoxane treatment. Photographs courtesy of Dr. W.A.D. Griffiths

Kaposi's Sarcoma

This malignancy is usually considered to be an haemangiosarcoma. It therefore seemed a possibility that razoxane might have some therapeutic effect. Accordingly a trial took place in Uganda at a time before it was realized that a large proportion of the population was HIV positive. Nevertheless there were some good responses, depending on the histopathology. The overall complete and partial response rate was at least 57%, even if those who received inadequate treatment are included [32]. No information is available on the duration of the responses.

Crohn's Disease

This inflammatory bowel disease was treated with razoxane for three reasons. Firstly because it was believed to be an autoimmune condition like psoriasis; secondly, most of these patients come to surgery at some stage of their disease and thirdly they were all resistant to steroids. This was a small study in nine patients observed for 10 years and treated for the first 6 months with razoxane. None of these patients needed surgery because all reverted to normal bowel habit with normal bowel appearance on sigmoidoscopy [33].

Renal Cell Cancer

The unresectable stage of this angiogenic tumour is one of the most intractable tumours to treat. Until recently, it barely responded to any kind of therapy. Treatment of 40 patients (38 evaluable) with well tolerated doses of razoxane produced an outcome in line with that seen in other tumours where razoxane treated patients with stable disease had significantly improved survival times. In this trial of renal cell cancer, response was classified at 16 weeks. Twenty-seven of thirty-eight patients (71%) progressed before that time. They had a median survival time (MST) of 127 days; while 11/38 (29%) who had stable disease, i.e., they survived longer than 16 weeks, had a MST of 399 days ($p = 0.0026$) without any clinically significant adverse effects [34]. It is not clear, what the molecular target of razoxane was, but there can be little doubt that here too tumour growth rate slowdown was involved possibly together with an antiinflammatory effect.

A clinical trial with a neutralizing antibody (bevacizumab) against the prime suspect in the aetiology of renal cell cancer (VEGF), has recently reported a significant increase in time to progression among 37 patients on this antibody compared with 38 patients on placebo. Toxicity was minimal with hypertension and asymptomatic proteinuria the most prominent problems. There were three partial responses (8%). According to Kerbel and Folkman [1] progressions were rare, but according to Ferrara, disease progression eventually occurred in many patients [35]. We do not have MSTs.

Breast Cancer

The total cumulative dose of the highly active anticancer agent doxorubicin (adriamycin) that can be given to patients is severely restricted by a dose limiting cardiotoxicity. This is of particular importance to patients who are responding to this drug, but have reached a cumulative dose where further doxorubicin treatment will send them into cardiac failure. Since some 50% of breast cancer patients will respond to doxorubicin, measures to reduce its cardio-toxicity are of considerable interest. In 1972, Herman et al. [36] showed that EDTA and subsequently razoxane (which is a cyclized form of EDTA) and its more soluble dextro enantiomer, dexrazoxane, are all highly effective in reducing doxorubicin cardiotoxicity in the isolated dog heart and six mammalian species. Twelve years later several large scale clinical trials were started and also showed dexrazoxane to be highly effective in cancer patients being treated with doxorubicin.

In an attempt to see how long it was possible to delay giving dexrazoxane (DXRz) and still obtain its cardioprotective activity against doxorubicin, Swain et al. [37] gave DXRz only to advanced breast cancer patients who were thought to benefit from more (responders and stable disease), but possibly cardiotoxic doses of doxorubicin. Comparing 102 patients who had received DXRz plus doxorubicin with 99 similar patients who had received placebo plus doxorubicin they found that DXRz had not only been highly effective as a cardioprotectant (3% congestive cardiac failure vs. 22%), but had also nearly doubled median survival time to 882 days vs. 460 days for the placebo group ($p < 0.001$); even after adjustment for disease related prognostic factors. The authors had no explanation for this increase in the median survival time, but it is clear that whatever the molecular target, the result of giving DXRz was an impressive slowdown of tumour growth rate and/or evolution. These data are in line with the results in experimental studies and other clinical conditions, notably colorectal cancer.

Final Comment

The development of razoxane was interrupted in the 1990s by reports that the drug was leukaemogenic. This belief was based largely on findings in patients with refractory psoriasis and bone marrow already compromised by previous treatment with methotrexate, hydroxyurea and PUVA who subsequently also received long term (years) of high dose, chronic neutropenic doses of razoxane. Even in these unfavourable circumstances, none developed acute leukaemia who had received razoxane for less than 1.5 years. Statistical analysis of a group of 301 chemotherapy naïve patients treated with adjuvant razoxane after resection of a colorectal carcinoma and followed for 8 years summated to 1,023 patient years and represents the total patient years potentially at risk of developing acute leukaemia. During this trial only one patient developed acute leukaemia after 2 years on continuous razoxane [38]. The true risk of AML after razoxane therefore is one in 1,023 patient years.

Razoxane has been licensed in the UK since 1977 for the treatment of leukaemia, lymphomas and for use with radiation therapy. It was available on prescription, but was never advertised and thus remains unknown to most oncologists. However, razoxane remains the single most (cost)effective drug against psoriasis in all its manifestations. Dexrazoxane on the other hand is available for the prevention of anthracycline cardiotoxicity in many countries including North America and most of Europe although ironically with gross restrictions in the UK. *No case of leukaemia has been reported following treatment with dexrazoxane and clinical trials for a number of indications are in progress.*

References

1. Kerbel R, Folkman J (2002) Clinical translation of angiogenesis inhibitors. Nature Rev Cancer 2:727–39
2. Folkman J (1971) Tumor angiogenesis: therapeutic implications. N Engl J Med 285:1182–6
3. Salsbury AJ, Burrage K, Hellmann K (1970) Inhibition of metastatic spread by ICRF 159: selective deletion of a malignant characteristic. Br Med J 4:344–6
4. Jain RK (2001) Normalizing tumor vasculature with antiangiogenic therapy: a new paradigm for combination therapy. Nat Med 7:987–9
5. Le Serve AW, Hellmann K (1972) Metastases and the normalization of tumour blood vessels by ICRF 159: a new type of drug action. Br Med J 1:597–601
6. Hellmann K, Rhomberg W (1991) Radiotherapeutic enhancement by razoxane. Cancer Treat Rev 18:225–40
7. Bakowski MT (1976) ICRF 159, (+/–) 1,2 bis (3,5-dioxopiperazin-1-yl) propane, NSC 129943; razoxane. Cancer Treat Rev 3:95–107
8. Hasinoff BB, Hellmann K, Herman EH (1998) Chemical, biological and clinical aspects of dexrazoxane and other bisdioxopiperazines. Curr Med Chem 5:1–28
9. Hellmann K, Burrage K (1969) Control of malignant metastases by ICRF 159. Nature 224:273–5
10. Willis R (1952) The spread of tumours in the human body. Butterworth, London, 115
11. Hellmann K (1971) Serendipity in cancer chemotherapy. New Sci J 21:111–3
12. Salsbury AJ, Burrage K, Hellmann K (1974) Histological analysis of the antimetastatic effect of 1,2-bis (3,5-dioxopiperazin-yl) propane. Cancer Res 34:843–9
13. Heston WDW, Kadmon D, Fair WR (1981) Effect of high dose diethylstilbestrol and ICRF 159 on the growth and metastases of the R3327 MAT-LyLu prostate-derived tumor. Cancer Lett 13:139–45
14. Pollard M, Burleson GR, Luckert PH (1981) Interference with in vivo growth and metastasis of prostate adenocarcinoma (PA-III) by ICRF 159. Prostate 2:1–9
15. Barker Grimshaw M, Davies RW, Hall WS (1981) Combination therapy of a mouse sarcoma using razoxane and electron irradiation. Br J Cancer 44:117–23
16. Young SW, Hollenberg NK, Meinking TL et al The influence of ICRF 159 on lymphatic and hematogenous metastases from rabbit V2 carcinoma. Unpublished report for USPHS Grants GM 18674, GMO1910 and HL20895
17. Gosalvez M, Blanco M, Hunter J et al (1974) Effects of anticancer agents on the respiration of isolated mitochondria and tumor cells. Eur J Cancer 10:567–74
18. Hallowes RC, West DG, Hellmann K (1974) Cumulative cytostatic effect of ICRF 159. Nature 247:487–90
19. Sandberg J, Goldin A (1971) Use of first generation transplants of a slow growing solid tumor for the evaluation of new cancer chemotherapeutic agents. Cancer Chemother Rep 55(3):233–8
20. Hellmann K, Tucker DF (1965) Tumor growth in grids on chorioallantoic membranes of chick embryos. Cancer Chemother Rep 47:73–9

21. Gitterman CO, Luell S (1973) Chemotherapeutic studies on metastasis of human epidermoid carcinoma (HEp 3) in the embyonated chick egg. In: Garattini S, Franchi G (eds) Chemotherapy of cancer dissemination and metastasis. Raven Press, New York, 279–91

22. Sharpe HBA, Field EO, Hellmann K (1970) The mode of action of the cytostatic agent ICRF 159. Nature 226:524–6

23. Hellmann K, Williamson CJ, Sargent JM (2003) Effect of dexrazoxane on the cell cycle and intracellular ATP levels in K562 cells. Br Pharmacol Soc Winter 2003 Meeting; Abstract

24. O'Connell MJ, Begg CB, Silverstein MN et al (1980) Randomized clinical trial comparing two dose regimens of ICRF 159 in refractory malignant lymphoma. Cancer Treat Rep 64:1355–8

25. Flannery EP, Corder MP, Sheehan WW et al (1978) Phase II study of ICRF 159 in non-Hodgkin's lymphomas. Cancer Treat Rep 62:465–7

26. Gilbert JM, Hellmann K, Evans M, Cassell PG, Taylor RH, Stoodley B, Ellis H, Wastell C (1986) Randomised trial of oral adjuvant razoxane (ICRF 159) in resectable colorectal cancer: five year follow-up. Br J Surgery 73:446–50

27. Hellmann K, Gilbert J, Evans M et al (1987) Effect of razoxane on metastases from colorectal cancer. Clin Exp Metastasis 5:3–8

28. Taylor RH, Gilbert JM, Evans M et al (1984) Prospective serial liver ultrasound scanning in resectable colorectal cancer treated with adjuvant razoxane. Clin Exp Metastasis 2:321–31

29. Atherton DJ, Wells RS, Hellmann K (1976) Razoxane (ICRF 159) in psoriasis. Lancet ii:1296

30. Horton JJ, Wells RS (1983) Razoxane: a review of 6 years' therapy in psoriasis. Br J Dermatol 109:669–73

31. Mom A, Aresca S, Fuente G et al (1982) Razoxane in the treatment of psoriatic patients resistant to or intolerant of PUVA, methotrexate and etretinate. Acta Dermatovener 62:357–8

32. Olweny CL, Sikyewunda W, Otim D (1980) Further experience with razoxane (ICRF 159; NSC 129 943) in treating Kaposi's sarcoma. Oncology 37:174–6

33. Kingston RD, Hellmann K (1993) Razoxane for Crohn's colitis and non-specific proctitis. Br J Clin Pract 46:252–5

34. Braybrooke JP, O'Byrne KJ, Propper DJ et al (2000) A phase II study of razoxane, an antiangiogenetic topoisomerase II inhibitor, in renal cancer with assessment of potential surrogate markers of angiogenesis. Clin Cancer Res 6:4697–704

35. Ferrara N (2002) VEGF and the quest for tumour angiogenesis factors. Nat Rev Cancer 2:795–803

36. Herman EH, Mhatre RM, Lee IP et al (1972) Prevention of the cardiotoxic effects of adriamycin and daunomycin in the isolated dog heart. Proc Soc Exp Biol Med 140:234–9

37. Swain SM, Whaley FS, Gerber MC et al (1997) Delayed administration of dexrazoxane provides cardioprotection for patients with advanced breast cancer treated with doxorubicin containing therapy. J Clin Oncol 15:1333–40

38. Kingston RD, Fielding JWL, Palmer MK (1993) An evaluation of the effectiveness and safety of razoxane when used as an adjunct to surgery in colorectal cancer. Int J Colorect Dis 8: 106–10

39. Atherton A (1975) The effect of (+/–) 1,2-bis (3,5-dioxopiperazin-1yl) propane (ICRF 159) on liver metastases from a hamster lymphoma. Eur J Cancer 11:383–8

40. Peters LJ (1975) A study of the influence of various diagnostic and therapeutic procedures applied to a murine squamous carcinoma on its metastatic behaviour. Br J Cancer 32:355–65

41. Hellmann K, Hutchinson GE Unpublished report

42. Gilbert JM, Thompson EM, Slavin et al (1984) Inhibition of experimental colorectal cancer by razoxane (ICRF 159). Br J Surg 71:600–3

43. Baker D, Constable W, Elkon D, Rinehart L (1981) The influence of ICRF 159 and levamisole on the incidence of metastases following local irradiation of a solid tumor. Cancer 48: 2179–83

44. Pimm MV, Baldwin RW (1975) Influence of ICRF 159 and Triton WR 1339 on metastases of a rat epithelioma. Br J Cancer 31:62–7

45. Klenner T, Wingen F, Keppler B et al (1988) Effects on tumour inhibition and survival time of a combination therapy of razoxane (ICRF 159) and two new cytostatica-linked biphosphonates in a transplantable rat osteosarcoma. Clin Exp Metastasis 6(Suppl 1):94

2.3.5 Toxicity of Razoxane

Walter Rhomberg

This part of the monograph deals mainly with the razoxane-induced toxicity seen in the clinic. For preclinical studies of the toxicology of razoxane, and especially dexrazoxane, and further clinical considerations, e.g. for the implications from potential mutagenicity and teratogenicity of the two drugs, the reader is referred to Section 3.3.

Razoxane as Single Agent

If razoxane is given in small daily doses of 125–250 mg as monotherapy, the treatment is very well tolerated. Dose limiting toxicity is bone marrow depression. Leukopenia and anemia of moderate degree are most frequently observed whereas thrombocytopenia is rare. At this dose level the nadir of leukopenia mostly occurrs between day 12 and 16.

Other possible side effects include mild gastrointestinal symptoms such as nausea or diarrhea, weight loss (10% of body weight), skin reactions with maculo-papular rush, hyperpigmentations, and epistaxis. At the nails, bands of hyperpigmentation may occur [1]. Table 2.14 shows the kind and frequency of the most common side effects.

When larger single doses of razoxane are given as in some early clinical studies, e.g. 1,000 mg/m^2 × 3 days as initial dose, nausea and vomiting as well as frequency and degree of leukopenia are more pronounced. Leukopenia with a median white blood cell nadir of 2,000/mm^3 on median day 12 may then be seen in up to 79%, and nausea in 40% of patients [2]. Mild reversible elevations of SGOT and bilirubin are common with larger single doses.

This was also observed when similar doses of *dexrazoxane* (ICRF-187) were administered [3, 4]. Atherton et al. reported lethargy to occur at some time in 21% of their patients with psoriasis [5]. This symptom was usually associated with higher dosage regimens and it tended to occur on the third day of 3-day treatment courses and during the subsequent 1 or 2 days. It was usually accompanied by the more severe degrees of bone marrow depression. Changing to a 2-day treatment course led in every case to its relief [5]. Lethargy was also reported in beagle dogs when *dexrazoxane* was given iv. [3].

Of further interest are vascular side effects as possible correlation to the experimental findings of the strong angiometamorphic effect of razoxane. During the past, we have not observed any serious cardio-vascular side effects, but a few patients, predominantly those with pre-existent atherosclerosis, reported transient

Table 2.14 Main side effects in the monotherapy of razoxane (single doses <300 mg/day)

Side effect	Frequency	Remarks
Bone marrow depression	30–35%	
Leukopenia <3,000/mm^3		Dose limiting toxicity, but rapidly reversible
Anemia (Hb-decrease >2 g%)		
GI-toxicity	15–20%	
Nausea, vomiting (grade 1–2)		
Constipation/diarrhea (grade 1–2)		
Increased salivation or dry mouth		
Various other side effects	5–15%	
Reduction in general condition		
Weight loss (10% of body-weight)		
Exanthema, hyperpigmentation of skin		
Rheumatic syndrome (unknown origin)		Mostly seen in patients with soft tissue sarcoma
Infections		
Rare side effects	<5%	
Watery nose bleeds/discharge		
Angioreactions (e.g., headache)		
Very rare complications	<<1%	
LE-phenomenon, Fournier necrosis		One case, respectively
Leukemia		Problem outlined in Section 2.3.5.3

precordial pain or a deterioration of earlier symptoms related to atherosclerosis. We described two patients who experienced an increase of the frequency of a paroxysmal tachycardia on razoxane [1] but we also observed a favorable angioreaction in a female patient who spontaneously reported a distinct improvement of leg symptoms correlating with regression of her varicose veins on treatment with razoxane.

It is noteworthy that the spectrum of side effects seems to be dependent on the kind of tumor treated. For instance, in soft tissue sarcomas the differential WBC shows frequently a rapidly reversible shift to the left with an increased number of eosinophilic and basophilic leukocytes. In addition, if sarcoma patients were treated with razoxane and radiotherapy, a 'rheumatic syndrome' of unknown origin may occur in 10–15% of the patients probably representing some form of tumor lysis syndrome. Tumour lysis and even rapid liquification of sarcomas and their metastases have been observed even after very low radiation doses on razoxane [1]. In these cases, no hyperuricemia was observed. Such a rheumatic syndrome never occurred in patients with carcinomas. Very rare complications are probably the induction of a LE-phenomenon or a Fournier necrosis by razoxane (unpublished observation of one case).

Beau's lines at the nails, regarded as a result of a sudden interruption of the nail keratin synthesis, were described in a patient with psoriasis treated with razoxane

[6]. However, this phenomenon may also be observed in psoriatic patients not treated with razoxane.

Meticulous observations of haematologic parameters including bone marrow analyses were performed by Atherton et al. during the treatment of patients with psoriasis [5]. Using an intermittent dose regimen of razoxane (for details see Section 2.4.1), the authors noted that – besides the leukopenic reaction – a few giant neutrophils with hypersegmented nuclei were commonly seen in the peripheral blood of patients during treatment. Typically these cells showed an approximately 30% increase in diameter above normal, having between five and eight nuclear lobes, and coarse clumped nuclear chromatin with very fine interlobar connections.

Bone marrow examinations were obtained in four informed volunteers among the psoriatic patients. All marrow examinations were essentially normal at the start of treatment and all showed similar changes after periods of treatment varying between 2 and 40 weeks. The overall cellularity of the bone marrow was markedly reduced; granulopoiesis was depressed; giant metamyelocytes, myelocytes and promyelocytes were present, showing the same coarse clumping of nuclear chromatin observed in the peripheral blood neutrophils. Erythropoiesis was normoblastic; occasional giant late normoblasts were present showing karyorrhexis and prominent basophilic stippling of the cytoplasm. All these patients had normal serum folate and vitamin B-12 levels. In one of these patients there was the opportunity to reexamine the bone marrow 11 months after cessation of a 3-month course of razoxane treatment: The bone marrow had a completely normal appearance, suggesting that the early changes described can be entirely reversible even on prolonged exposure to razoxane [5]. The finding of giant neutrophil leukocyte and erythrocyte precursors in the bone marrow of treated patients and the related finding of giant neutrophils in the peripheral blood are consistent with the specific inhibitory activity of razoxane at the G2/M phase of the cell cycle [5, 7]. DNA synthesis continues normally but nuclear division is inhibited and cell separation [8].

Combined Modality Therapy

Razoxane combined with cytotoxic drugs usually leads to a slightly increased hematotoxicity. For instance, the combination of razoxane and vindesine led to grade 3 and 4 leukopenia in 39% of the patients in a study of soft tissue sarcomas [9]. The nadir of leukopenia was reached on day 16. The leukopenia was rapidly reversible, and there was no case of neutropenic fever. The general tolerance to combined cytotoxic therapies is, as a rule, determined by the specific side effects of the concomitantly given drugs.

Combined with radiotherapy, razoxane leads to a clear increase of normal tissue reactions. The reactions achieved clinical significance in head and neck sarcomas, lung cancer, or if larger fields had to be irradiated. Regional pneumonitis and esophagitis were most frequently observed when parts of the lung were irradiated. Such reactions occurred even after radiation doses of 30 Gy but were of limited clinical significance. However, caution is indicated if larger lung volumes have to be irradiated. The occurrence of pulmonary embolism with razoxane and radiotherapy seems not to be influenced but has yet to be prospectively monitored in specific conditions such as in combined treatments of carcinomas of the bile ducts.

If razoxane is combined with radiation therapy and other cytotoxic drugs, a further enhancement of normal tissue reactions is to be expected. For instance, combined razoxane, vindesine and radiotherapy administered in patients with soft tissue sarcomas led to pronounced local tissue reactions. In the head and neck region and the mediastinum, one has to deal with early mucositis, esophagitis and pneumonitis. In combined modality therapy of angiosarcomas of the thyroid, postoperative impairments added to the problems so that transient nasogastric tube feeding was necessary in some patients. In addition, we observed the rare occurrence of rib necrosis, Fournier necrosis of the gluteal region, severe headache, and alopecia in our series of patients with soft tissue sarcoma (one case for each finding) [9].

The Issue of Leukemia-Induction

In 1981 the first two patients were described who developed secondary acute myelomonocytic leukaemia while taking razoxane for the treatment of malignant disorders [10]. Subsequently, secondary leukemia induction was reported in patients with non-malignant disease during the 1980s when razoxane was investigated as a long term high-dose oral therapy in patients with psoriasis. Until 1987, there were 16 cases of treatment-related leukemias noted in the files of the former ICI (Imperial Chemical Industries, later Zeneca) (personal communication; 26). In these data, two pertinent features have to be taken into consideration:

- First, the median cumulative dose of razoxane in these 16 cases was 160 g (range, 42–450 g), and the mean cumulative dose was 195 g. For comparison, the median dose of razoxane during a radiotherapy course usually varies from 6.5 to 8 g in different studies.
- Six of 16 patients had GI-cancer, 10 suffered from psoriasis. Six out of the ten psoriatic patients had previously received MTX (methotrexate) and HU (hydroxyurea) which both are known to be potentially associated with treatment-related leukemias although to a rather low degree [11].

    In England, hundreds of patients with psoriasis were treated with razoxane from 1975 to about 1984. A 1982 review described razoxane as 'the drug of first choice in the systemic treatment of psoriasis' [12]. In the early 1980s several cases of acute leukaemia were reported after long term administration of high-dose razoxane in patients with psoriasis [10, 13, 14]. The incidence of leukaemias was approximately 1 case per year. The frequency of occurrence of secondary malignancy led to the decision to stop the use of razoxane for this largely benign condition as a primary treatment option. In cases of treatment refractory psoriasis, however, with disabling and potentially life-threatening disease, razoxane and no other treatment may still be a valuable option [8, 15].

What about the adjuvant use of razoxane in malignant tumors? So far, colorectal cancer is the only tumour that has been treated with a prolonged administration of razoxane as a single agent. Todate, 3 razoxane-related acute non-lymphocytic leukemias (ANLL) have indeed been described in colorectal cancer [16, 17] but the occurrence of such an event remained rare and the evidence of an increased incidence of secondary leukemia is not certain in this context. For instance, Gilbert et al. found 3 cases of acute myeloid leukemia among 139 patients with colorectal cancer treated with adjuvant razoxane [16]. This compares with 7 cases of ANLL in 284 patients given adjuvant 5-fluorouracil and semustine [18].

Long term adjuvant treatment with various myelosuppressive drugs does carry a risk of subsequent AML. According to Forbes [19], the leukaemia risk with current adjuvant chemotherapy is likely to be less than a fivefold increase, and the leukemia is predominantly a result of *alkylating agents* that peaks before 10 years. Andersson et al. observed therapy related acute non-lymphocytic leukaemias or preleukaemias in 5 of 71 patients with advanced breast cancer treated with combination chemotherapy consisting of prednimustine, methotrexate, 5-fluoro-uracil, mitoxantrone, and tamoxifen [20]. A case of acute non-lymphocytic leukaemia with inversion of chromosome 16 was described by Baglin et al. in a 48-year-old male who had received oral cyclophosphamide and razoxane in succession. Conventional chemotherapy resulted in complete remission which lasted for 13 months [21].

In the literature, leukaemia induction is also linked to drugs affecting DNA *topoisomerase II* [22] which is – among others – a proven mode of action of razoxane. Most razoxane-related leukaemias belong to type M2 and M3; FAB classification [15]. A review of the published literature on therapy-related acute promyelocytic leukaemia (t-APL) suggests that this is a genuine clinical entity which may be caused by drugs affecting DNA topoisomerase II, and has a prognosis similar to de novo APL [22]. In view of the events of leukemia, Price et al. investigated the frequency of sister chromatid exchanges (SCE) in lymphocytes of 34 patients with colorectal carcinoma treated with razoxane. Their results showed no significant increase in SCE levels in the razoxane group compared with either normal controls or untreated patients [23]. In our own experience which mostly concerns the short term use of low dose razoxane over several weeks or months, we have seen no confirmed case of secondary leukemia in about 1,100 patient years.

Yet, the situation may be different when other derivatives of bis-diketopiperazines are administered, e.g., bimolane in China [24, 25]. Unusual high rates of secondary leukemias were described with this drug [25]. Similar to the few razoxane-induced leukemias, only larger doses of bimolane (mean dose 194 g) were associated with the induction of leukemias [26]. The bimolane-induced leukemias show in part the features of a new subgroup of treatment related leukemias (TRL) characterized by specific rearrangements, whose clinical, hematological and prognostic features may be different from classical TRL characterized by chromosome abnormalities involving absence or deletion of parts of chromosome 5 and/or 7 [26].

If all data were taken together, razoxane doses up to 40 g seem to bear no increased risk for the development of leukemias. The prolonged use of razoxane over many years with cumulative doses between 100 and 450 g may have such a risk. This risk, however, is not excessive, but it makes it necessary to weigh this risk against the threat of the tumour which is treated by long term use of razoxane either in the adjuvant or antimetastatic setting, i.e., the benefit/risk ratio has to be considered under such treatment conditions.

Risk of Secondary Malignant Neoplasms (SMN)

Atherton et al. [5] have discussed the topic in context with the use of razoxane for psoriatic patients. From a theoretical point of view, they considered, the drug's inhibition of the G2/M phase of the cell cycle would not be expected to result in mutagenicity, and this was supported by the absence of any mutagenic effect in the 'Ames test' [27]. It is believed that mutagenicity in this test correlates well with carcinogenicity [28] and razoxane would therefore be predicted to be non-carcinogenic.

Razoxane was tested for its ability to induce primary lung tumours in strain A mice [5, 29], and there was no evidence of carcinogenicity in this test. A small study was performed by the National Cancer Institute in the USA, in which an increased incidence of uterine adenocarcinomas and systemic lymphomata appeared to follow alternate daily intra-peritoneal adminstration of very large doses of razoxane (48 and 96 mg/kg) to rats and mice [30]. Although the route of administration, the administration schedules, and the sex difference in these studies were entirely different from those employed in the patients of Atherton et al., the vague possibility is raised that this drug may have some carcinogenic potential, and this has to be weighed against the likely benefits of therapy in individual patients with benign diseases [5].

In the literature only sporadic descriptions of secondary neoplasms associated with the administration of razoxane are found. In patients receiving razoxane for psoriasis, two cases of epitheliomas were described [31]. Single cases of lymphomas were also noted: A woman with a 20-year history of acral pustular psoriasis developed a cutaneous B-cell lymphoma of the lip following 4.5 years of treatment with razoxane [32], and a 44-year-old man with psoriasis was diagnosed with a mediastinal T-cell lymphoma after having received razoxane for 6-years [33]; and finally few cases of acute leukemias were described which were associated with a long-term administration and high cumulative doses of the drug (see Section 2.3.5.3).

The Toxicity of Dexrazoxane: Are There Differences to Razoxane?

The stereoisomers of a molecule may act in different ways. This is a proven phenomenon, e.g., in the area of antimetastatic activities of razoxane and its related analogues [34]. It remains open whether this applies also to the enantiomers of the molecule, but the question about possible differences in the spectrum of toxicity of razoxane and dexrazoxane is justified. So far no fundamental differences between the two substances have been elucidated. The dose limiting toxicity for both drugs is neutropenia. Some gastrointestinal side effects, skin reactions, or lethargy have likewise been described for razoxane and dexrazoxane.

Differences may chiefly arise from the way of administration of the drugs and their dosages. Dexrazoxane is given iv, and generally at higher single doses compared to razoxane. Side effects associated with the iv-route comprise local venous reactions and anaphylaxis. Pain at the site of injection was frequently seen by Langer (see Section 3.7). Sorensen et al. observed clinical signs of phlebitis in 8 of 64 patients following dexrazoxane therapy, necessitating installation of central venous catheters for therapy continuation [35]. In one patient, the first three courses of the dexrazoxane regimen were associated with pain of injection in direct connection with dexrazoxane. The fourth course was given without a problem via a central venous catheter; however, the fifth course led to a subclavian vein thrombosis [35, 36]. A grade III phlebitis was also reported by Brubaker et al. [37].

So far, one case of anaphylaxis to dexrazoxane was reported [38]. Such a reaction is certainly associated with the iv route of drug administration. Transient elevations of SGOT and/or bilirubin are only seen when higher single doses of dexrazoxane (or razoxane) were given and never at daily metronomic small doses of razoxane. Other uncommon side effects reported on dexrazoxane were the occurrence of a cutaneous and subcutaneous necrosis and panniculitis about 30 cm proximal from the injection site 24 h after the infusion of dexrazoxane into an antecubital vein [36]. We observed one case of Fournier necrosis at the gluteal area in a patient with malignant hemangiopericytoma treated with small doses of razoxane orally and vindesine iv (unpublished). The latency period to this event was several weeks.

Evaluation of the significance of isolated cases of toxicity, however, requires great care with which to judge whether they are just that – isolated cases or something more.

References

1. Rhomberg W (1978) Radiotherapy combined with ICRF 159 (NSC 129943). Int J Radiat Oncol Biol Phys 4:121–6
2. Eagan RT, Carr DT, Coles DT, Rubin J, Frytak S (1976) ICRF 159 versus polychemotherapy in non-small cell lung cancer. Cancer Treat Rep 60:947–8
3. Liesmann J, Belt R, Haas Ch and Hoogstraten B (1981) Phase I evaluation of ICRF-187 (NSC-169780) in patients with advanced malignancy. Cancer 47:1959–62
4. Vats T, Kamen B, Krischer JP (1991) Phase II trial of ICRF-187 in children with solid tumors and acute leukemia. Invest New Drugs 9(4):333–7
5. Atherton DJ, Wells RS, Laurent MR, Williams YF (1980) Razoxane in the treatment of psoriasis. Br J Dermatol 102:307–17
6. Tucker WF, Church RE (1984) Beau's lines after razoxane therapy for psoriasis [letter]. Arch Dermatol 120(9):1140
7. Hallowes RC, West DG, Hellmann K (1974) Cumulative cytostatic effect of ICRF-159. Nature 247:487
8. Griffiths WA (1985) Risk of leukaemia associated with chemotherapy. Br Med J (Clin Res Ed) 290(6467):555
9. Rhomberg W, Eiter H, Schmid F, Saely C (2008) Combined vindesine and razoxane shows antimetastatic activity in advanced soft tissue sarcomas. Clin Exp Metastasis 25:75–80

10. Joshi R, Smith B, Phillips RH, Barrett A (1981) Acute myelomonocytic leukaemia after razoxane therapy. Lancet 2:1343
11. Lim AY, Gaffney K, Scott DG (2005) Methotrexate-induced pancytopenia: serious and under-reported? Our experience of 25 cases in 5 years. Rheumatology 44(8):1051–5
12. Horton JJ, Wells RS (1983) Razoxane: a review of 6 year's therapy in psoriasis. Br J Dermatol 109:669–73
13. Bhavnani M, Wolstenholme RJ (1987) Razoxane and acute promyelocytic leukemia. Lancet 2:1085
14. Caffrey EA, Daker MG, Horton JJ (1985) Acute myeloid leukaemia after treatment with razoxane. Br J Dermatol 113:131–4
15. Lakhani S, Davidson RN, Hiwaizi F, Marsden RA (1984) Razoxane and leukaemia. Lancet 2(8397):288–9
16. Gilbert JM et al (1986) Randomized trial of oral adjuvant razoxane (ICRF 159) in resectable colorectal cancer: fiver year follow-up. Br J Surg 73:446–50
17. Kingston RD, Fielding JW, Palmer MK (1993) An evaluation of the effectiveness and safety of razoxane when used as an adjunct to surgery in colorectal cancer. Report of a controlled randomised study of 603 patients. Int J Colorectal Dis 8(2):106–10
18. Gastrointestinal Tumor Study Group (1984) Adjuvant therapy of colon cancer: results of a prospectively randomized trial. N Engl J Med 310:737–43
19. Forbes JF (1992) Long-term effects of adjuvant chemotherapy in breast cancer. Acta Oncol 31(2):243–50
20. Andersson M, Philip P, Pedersen-Bjergaard J (1990) High risk of therapy-related leukemia and preleukemia after therapy with prednimustine, methotrexate, 5-fluorouracil, mitoxantrone, and tamoxifen for advanced breast cancer. Cancer 65:2460–4
21. Baglin TP, Galvin GP, Pollock A (1987) Therapy-related acute nonlymphocytic leukaemia with inversion of chromosome 16 and a sustained remission. Cancer Genet Cytogenet 27(1):167–9
22. Bhavnani M, Azzawi SA, Yin JA, Lucas GS (1994) Therapy-related acute promyelocytic leukemia. Br J Haematol 86(1):231–2
23. Price CM, Hagger D, Evans M et al (1992) Sister chromatid exchange (SCE) frequency in lymphocytes of patients with colorectal carcinoma treated with razoxane. Cancer Detect Prev 16(4):221–3
24. Ye H, Tu ZH, Li XF (1994) Leukemia associated with bimolane. Zhonghua Nei Ke Za Zhi 33(10):669–71
25. Zhang MH, Wang XY, Gao LS (1993) 140 cases of leukemia caused by bimolane. Chung-Hua-Nei Ko-Tsa-Chih 32(10):668–72
26. Xue Y, Lu D, Guo Y, Lin B (1992) Specific chromosomal translocations and therapy-related leukemia induced by bimolane therapy for psoriasis. Leuk Res 16:1113–23
27. McCann J, Choi E, Yamasaki E, Ames BN (1975) Detection of carcinogens as mutagens in the salmonella/microsome test: assay of 300 chemicals. Proc Natl Acad Sci USA 72:5135
28. Ames BN, Lee FD, Durston WE (1973) An improved bacterial test system for the detection and classification of mutagens and carcinogens. Proc Natl Acad Sci USA 70:782
29. Stoner GD, Shimkin MB, Kniazett AJ, Weisburger JH, Weisburger EK, Gori GB (1973) Test for carcinogenicity of food additive and chemotherapeutic agents by the pulmonary tumour response in strain A mice. Cancer Res 33:3069
30. US Department of Health, Education and Welfare (1978) Bioassay of ICRF-159 for possible carcinogenicity. Department of Health, Education and Welfare, Publication No: NIH 78-1329
31. Horton JJ, MacDonald DM, Wells RS (1983) Epitheliomas in patients receiving razoxane therapy for psoriasis. Br J Dermatol 109:675–8
32. Mallett RB, Langtry JA, Harper JI, Staughton RC (1987) Psoriasis, razoxane and a cutaneous B-cell lymphoma. Br J Dermatol 116(2):243–4
33. Zuiable AG, Aboud H, Coulson I et al (1989) Razoxane and T-cell lymphoma. Br J Dermatol 121(1):149, Letter
34. Zwilling BS, Campolito LB, Reiches NA, George T, Witiak DT (1981) Effect of stereoisomers related to ICRF-159 on metastasis of B16 melanoma. Br J Cancer 44(4):578–83

35. Sorensen B, Bastholt L, Mirza MR et al (1994) The cardioprotector ADR-529 and high-dose epirubicin given in combination with cyclophosphamide, 5-fluorouracil, and tamoxifen: a phase I study in metastatic breast cancer. Cancer Chemother Pharmacol 34:439–43
36. Lossos IS, Ben-Yehuda D (1999) Cutaneous and subcutaneous necrosis following dexrazoxane-CHOP therapy. Ann Pharmacother 33(2):253–4
37. Brubaker LH, Vogel CL, Einhorn LH, Birch R (1986) Treatment of advanced adenocarcinoma of the kidney with ICRF-187: a Southeastern Cancer Study Group trial. Cancer Treat Rep 70:915–6
38. Leoni V, Santini D, Vincenzi B, Grilli C, Onori N, Tonini G (2004) Anaphylaxis to dexrazoxane (ICRF-187) following three previous uncomplicated infusions. Allergy 59:241

2.4 Studies in Non-malignant Diseases

Walter Rhomberg

Razoxane and dexrazoxane have several modes of action which are not restricted to their antineoplastic activity. The most important of them covers the ability to normalize pathologic (tumor) blood vessels, which was first shown in experimental tumour models for razoxane by Le Serve and Hellmann in 1972 [1]. It was reasonable, therefore, to undertake an exploration of razoxane in clinical conditions where an abnormal microvasculature seems to play a role in the pathogenesis of the disease, i.e. in psoriasis and psoriatic arthritis [2–4]. In addition, the reported immunosuppressive activity of razoxane was thought to be of value in diseases with a putative autoimmune pathogenesis such as inflammatory bowel diseases.

Preclinical research has recently indicated that (dex)razoxane augments mitoxantrone-mediated immunosuppressive effects in experimental autoimmune encephalomyelitis. As consequence, the authors suggested clinical trials to be performed in patients with multiple sclerosis to evaluate the unexpected immunosuppressive efficacy of dexrazoxane as add-on treatment [5]. Entirely different modes of action of razoxane seem to be involved in as yet hypothetical treatment options for other neurologic disorders, for instance neurodegenerative diseases like Alzheimer's disease. In Section 3.8 Nigel Greig gives interesting insights into these fascinating new treatment options which have yet to be explored clinically.

2.4.1 Psoriasis and Psoriatic Arthropathy

In 1976, promising initial results with razoxane in the treatment of psoriasis were published by Atherton, Wells and Hellmann [6]. The rationale for this treatment was as follows: In psoriasis, rapid epidermal cell proliferation is associated with a microvasculature resembling that seen in solid malignant tumors [2], and Folkmann has suggested that an analogous reciprocal relationship might exist between the proliferative cells and the microvasculature in such lesions [3]. If this were the case, treatments having an inhibitory effect on the abnormal microvasculature could be of therapeutic benefit.

In 1980, more mature data on the use of razoxane in the treatment of psoriasis was reported by Atherton et al. [7]. A brief synopsis of this work is given below.

Original article: Atherton DJ, Wells RS, Laurent MR, Williams YF (1980) Razoxane in the treatment of psoriasis. Br J Dermatol 102:307–17

(Parts of this material are reproduced with permission of Wiley-Liss, Inc., a subsidiary of John Wiley & Sons, Inc.)

Synopsis

Patients and Methods

A total of thirty-five patients with severe psoriasis were treated with razoxane. Thirty-two were treated principally for incapacitating cutaneous disease which had failed to respond to topical therapy and particularly to dithranol. Minor degrees of arthropathies were common. Further 3 patients were treated primarily for severe arthropathy which has not responded to non-steroidal antiinflammatory drugs.

Sixteen patients had already received other antimitotic drugs, and some had also had photochemotherapy. All of these 16 had received methotrexate for varying periods, but without control of the disease in nine cases, and with the development of adverse effects necessitating cessation of treatment in the other seven.

Razoxane was administered as 125 mg tablets in various intermittent dose schedules. The daily dose was always either 125 or 250 mg three times daily. During maintenance therapy, razoxane was given for either 2 or 3 consecutive days each week, or in a few patients, each fortnight.

Razoxane was not given to patients of either sex who planned to have children within the next 2 years. All patients undertook to use adequate contraception. The patients were all informed of the known adverse effects of the drug and it was explained that its long-term safety remains unconfirmed.

Results

Cutaneous Psoriasis

No patient failed to respond clinically to treatment with razoxane. Regression of psoriasis to mild or minimal residual disease occurred in thirty-one out of the thirty-two (97%) patients receiving treatment primarily for cutaneous disease. Figures 2.2, 2.3, 2.4, and 2.5 in the article (not reprinted in this synopsis) illustrated the amount and kind of responses. Difficulty in obtaining adequate control of cutaneous psoriasis was encountered in only one patient, a 63-year-old woman with unstable erythrodermic psoriasis affecting about 70% of her total skin area. Clinical improvement was accompanied by development of alopecia, nausea, diarrhea and colicky abdominal pain, and subsequently neutropenia, forcing a reduction of dosage. The side effects improved but her psoriasis recurred as the dosage was reduced. Treatment was therefore stopped and has not been reinstituted.

To minimize bone marrow toxicity, the drug was given intermittently rather than continuously. Six slightly different dose schedules were related to different patient groups as to age, sex, performance status and severity of the disease. The aim of treatment was the suppression of disease to a mild level which is acceptable to the patient. The high dosage required to produce complete clearence exposes the patient to an unnecessarily high risk of toxicity.

Psoriatic Arthropathy

Several of the thirty-two patients being treated principally for cutaneous psoriasis had concurrent minor joint symptoms, most commonly in the finger and toe joints. Complete suppression of these symptoms was the rule within the first few weeks after starting treatment. Six patients had more severe degrees of arthritis, and in three of these joint disease constituted the principal indication for treatment with razoxane. Detailed information of these six patients concerning sex, age, duration of psoriasis and arthritis, involved joints, type of psoriasis, previous treatment, duration of treatment and administration schedule of razoxane was documented in a Table of the original article. Early morning stiffness and the Ritchie index improved strongly in 5 of the 6 patients, so did the E.S.R.

Updates of the Initial Results Described in [7]

An update of the above experience was given by Horton and Wells in 1983 [8]. The authors who have been treating patients with psoriasis since 1974 concluded in 1983: Razoxane is an effective drug in the systemic treatment of psoriasis, with an initial response rate of 97%. After continuous therapy for up to 6 years it was found that 72% of patients remained on the drug with a good therapeutic benefit. Razoxane was found to be useful in all forms of cutaneous psoriasis and psoriatic arthropathy. Among the six cases with psoriatic arthropathy all six patients finally experienced a complete response of the disease [8]. The drug did not appear to produce hepatic damage and was therefore regarded as particularly useful in patients intolerant of methotrexate. A late update after 20 years noted that all patients relapsed within 3 months of withdrawal of the drug, and the chronic psoriatic problems remained poorly controlled even on retinoids or topical treatment in a substantial number of patients. Moreover, the authors had to report on four cases of leukemia and the development of multiple cutaneous squamous cell carcinomata in another two patients [9]. Nevertheless, 27% of the former psoriatic patients recorded by Atherton et al. (1980) – after having completed a questionaire many years after treatment – expressed a desire to restart razoxane despite full knowledge of its potential adverse consequences [9].

Additional Treatment Reports and Further Developments

Thirty-six psoriatic patients resistant to PUVA, methotrexate and/or etretinate were treated with razoxane by Mom et al. [10]. Again, the drug was regarded as highly effective in cutaneous and arthropathic psoriasis. Razoxane was well tolerated. Besides some nausea and lethargy, 60% of the patients showed neutropenia, which could be easily controlled [10].

Another treatment report comes from the Westminster Hospital, London, and comprises 15 patients with intractable psoriasis. Eight of these were methotrexate 'failures', two had methotrexate-induced hepatotoxicity and two had not responded to the (at that time new) aromatic retinoid Tigason® (Ro 10–9359). A 2-day regime of razoxane treatment was used. Initially razoxane was given 125 mg 8-hourly, which was increased, in some patients, to 250 mg 8-hourly for 2 days only. The dose was repeated only when the neutrophil count had recovered (>1,500/mm^3). In practice, this was between 7 and 14 days [11]. *Results*: The skin of 14 patients improved significantly; three of them have remained completely clear of the disease and are treated on regular, intermittent, low doses of razoxane. The gross nail deformities of two patients, one with long-standing acropustulosis, improved over 6 months. In four out of five patients with psoriatic arthopathy, the joints improved likewise. The treatment failed in only one patient, a girl aged 22 years, with chronic plaque psoriasis and severe large joint arthopathy. *Comment*: The authors found razoxane to be a very useful drug in the management of severe psoriasis, and made the following points about the regulation of the dose of razoxane: The margin between the therapeutic effect and marrow depression is narrow, and the dose required to clear psoriasis in any individual varies widely. At least 6 weeks treatment may be required before improvement is seen. A regular 2-day regimen of treatment appears to give an optimum therapeutic effect, whereas a 3-day regimen tends to result in a more severe, prolonged neutropenia, with a consequent 'escape' of control of the psoriasis. The neutropenia is dose-dependent, but, in their experience, is always reversible. It can develop within 24 h with a nadir of 3–7 days [11].

Presumably, hundreds of patients with psoriasis were treated with razoxane from 1975 to about 1984 in England. An update in 1983 described razoxane as 'the drug of first choice in the systemic treatment of psoriasis' [8]. However, between 1981 and 1987 some cases of acute leukaemia were reported after long-term administration of razoxane in patients with psoriasis [12–14]. The incidence of leukaemias was approximately 1 case per year.

As a consequence, the drug was stopped for its use in benign clinical conditions but there are still a number of patients with disabling and potentially life-threatening psoriasis who are helped by razoxane and by no other treatment [15, 16]. Illustrative pictures of a disabling form of an acrodermatitis of Hallopeau which is regarded as an unusual variant of localized pustular psoriasis, and its subsequent complete disappearance under razoxane can be seen in a case report of Leigh and Gold [17]. Interestingly to repeat, in an epilogue to this topic Cerio et al. stated that 27% of the former psoriatic patients recorded by Atherton et al. (1980) expressed a desire to restart razoxane despite 'full' knowledge of its potential adverse consequences [9].

Other Skin Disorders

Besides of variants of psoriasis [17], there may be other skin disorders that could respond to razoxane. For instance, pityriasis rubra pilaris is such a disorder of the

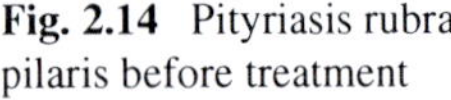

Fig. 2.14 Pityriasis rubra pilaris before treatment

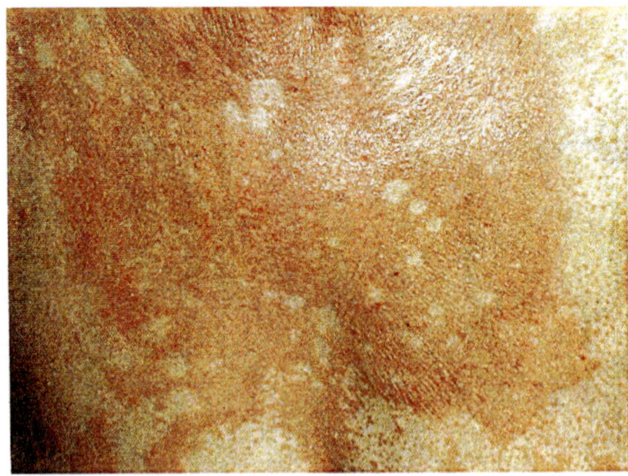

Fig. 2.15 Pityriasis rubra pilaris after treatment with razoxane

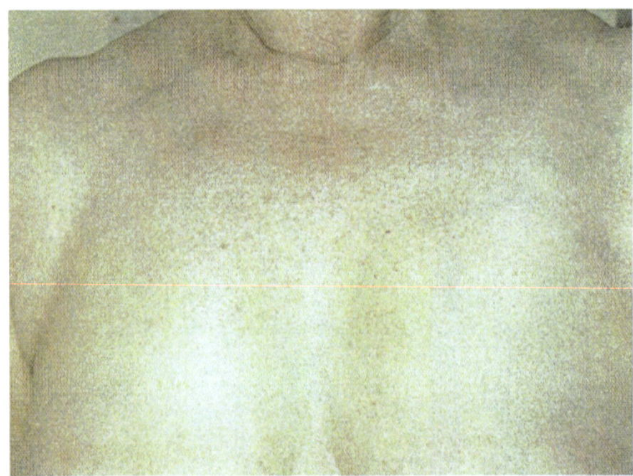

skin that is associated with pathological blood vessel formation. Lun K et al. [18] described a substantial regression of a far advanced case using razoxane. The kind of treatment response is documented in Figs. 2.14 and 2.15.

References

1. Le Serve AW, Hellmann K (1972) Metastases and the normalization of tumour blood vessels by ICRF 159: a new type of drug action. Br Med J 1:597–601
2. Braverman JM, Cohen I, O'Keefe EO (1972) Metabolic and structural studies in a patient with pustular psoriasis (von Zumbusch). Arch Dermatol 105:189
3. Folkman J (1972) Angiogenesis in psoriasis – therapeutic implications. J Investigative Dermatol 59:40

  4. Creamer D, Sullivan D, Bicknell RR, Barker J (2002) Angiogenesis in psoriasis (review). Angiogenesis 5:231–6
  5. Weilbach FX, Chan A, Toyka KV, Gold R (2004) The cardioprotector dexrazoxane augments therapeutic efficacy of mitoxantrone in experimental autoimmune encephalomyelitis. Clin Exp Immunol 135(1):49–55
  6. Atherton DJ, Wells RS, Hellmann K (1979) Razoxane (ICRF-*159*) in psoriasis. Lancet ii:1296
  7. Atherton DJ, Wells RS, Laurent MR, Williams YF (1980) Razoxane in the treatment of psoriasis. Br J Dermatol 102:307–17
  8. Horton JJ, Wells RS (1983) Razoxane: a review of 6 year's therapy in psoriasis. Br J Dermatol 109:669–73
  9. Cerio R, Wells RS, MacDonald DM (2006) The sequelae of razoxane therapy. Br J Dermatol 113(s29):27
 10. Mom A, Aresca S, Fuente G, Pecorini V, Pellerano G, Perez V (1982) Razoxane in the treatment of psoriatic patients resistant to or intolerant of PUVA, methotrexate and etretinate. Acta Derm Venereol 62(4):357–8
 11. Harper JI, Copeman PWM, Straughton RCD (1982) Razoxane regimen for the treatment of psoriasis [Brief communication]. Clin Expl Dermatol 7:581–2
 12. Bhavnani M, Wolstenholme RJ (1987) Razoxane and acute promyelocytic leukemia. Lancet 2:1085
 13. Caffrey EA, Daker MG, Horton JJ (1985) Acute myeloid leukaemia after treatment with razoxane. Br J Dermatol 113:131–4
 14. Joshi R, Smith B, Phillips RH, Barrett AJ (1981) Acute myelomonocytic leukaemia after razoxane therapy. Lancet 2:1343
 15. Griffiths WA (1985) Risk of leukaemia associated with chemotherapy. Br Med J (Clin Res Ed) 290(6467):555
 16. Lakhani S, Davidson RN, Hiwaizi F, Marsden RA (1984) Razoxane and leukaemia. Lancet 2(8397):288–9
 17. Leigh IM, Gold SC (1979) Acrodermatitis of Hallopeau. Br J Dermatol 101(Suppl 17):41–2
 18. Lun K, Walker S, Griffiths WAD, Hellmann K (2001) Razoxane – a novel treatment for pityriasis rubra pilaris. Australasian J Dermatol 42(Suppl 1):A12

2.4.2 Crohn's Disease and Ulcerative Colitis

In 1992, Kingston and Hellmann published a series of 9 cases with Crohn's disease [1]. Oral razoxane 125 mg daily brought active Crohn's disease into remission when used alone and mainly on an out-patient basis in all nine patients studied. Remissions took several months to achieve, but no relapses have occurred on treatment and no surgery has been necessary in any of the patients. The reason to perform the study and some other details are outlined below.

Original article: Kingston RD, Hellmann K (1992) Razoxane for Crohn's colitis and non-specific proctitis. Br J Clin Pract 46:252–5

Synopsis

Background

Some 40% or more of the patients with Crohn's colitis undergo major surgery within 10 years of diagnosis [1]. Many of these patients, often young, with total colonic involvement require panproctocolectomy and a permanent ileostomy. To date (1992), however, there are no drugs of proven effectiveness in the chronic phase of the disease [2–4]. The mainstay of acute therapy is the use of steroids. Other immunosuppressive agents such as azathioprine, 6-mercaptopurine and more recently cyclosporin A have been used either in conjunction with steroids or alone [5–8]. All these agents have a significant toxicity with only occasional beneficial effects [9, 10]. Nevertheless, some studies have suggested that azathioprine might maintain a remission in patients with Crohn's colitis [6] and clinical observations suggest that it has a slow beneficial action, perhaps over months [8].

Because razoxane is an effective experimental immunosuppressive agent and because it has shown considerable activity in 95% of patients with severe psoriasis [11], an apparently auto-immune proliferative disorder, it seemed worthwhile to test the effect of razoxane in Crohn's colitis.

Patients and Methods

During a 5 year period nine patients with Crohn's colitis were treated with razoxane. The criteria for treatment was failure of conventional chemotherapy to control symptoms.

All patients received 125 mg razoxane daily (one tablet) continuously for lengths varying between three and 18 months. Most of the razoxane treatment was given on an out-patient basis. No other treatment was used except oral iron in two patients. Treatment was stopped when patients had been in complete remission for 4–6 months.

The patients with Crohn's disease described in Table 2.15 represent our total experience (no exclusions) during the 5 years (1978–1983) of this study. Two of the patients had been referred for consideration of surgical treatment. Follow-up of all patients was in an inflammatory bowel clinic with serial determination of hematological, chemical and clinical parameters. One patient (P.C.) had a perfectly normal, healthy baby after the treatment with razoxane.

Results

The effects of razoxane on patient's disease parameters is recorded in Table 2.16.

At no time did disease relapse in any patient while on treatment. Three relapses have occurred at 24, 12 and 12 months following cessation of treatment. Patient PC remained well off treatment for 2 years, but then developed episodes of abdominal pain, frequency of bowel action, bleeding, raised ESR, weight loss and further changes of Crohn's colitis on barium enema. At this time she became pregnant and her symptoms improved and remained so. Patient WV, who was the last patient to receive razoxane, responded well to 18 months of treatment but relapsed when treatment was discontinued. Despite counselling with regards to risk of leukemia after prolonged treatment she requested further treatment. This was given again with successful effect on her symptoms and disease parameters. She remained alive and well 2 years later.

Table 2.15   Characteristics of 9 patients with Crohn's disease

Patient	Sex	Age at onset	Extents of disease	Evidence of diagnosis	Previous treatment
PC	F	25	Terminal ileum to descending colon	Histology and radiology	Salazopyrin, local and systemic steroids
EH	M	40	Rectum (previous colectomy)	Histology	Salazopyrin, local steroids
PP	F	59	Descending and sigm. colon	Radiology	Salazopyrin, local and systemic steroids
FA	M	60	Rectum peri-anal	Histology	Salazopyrin, local steroids
GB	M	60	Sigmoid colon rectum	Histology	Salazopyrin, local and systemic steroids
MJ	F	30	Rectum	Histology	Salazopyrin, local steroids
RB	M	74	Rectum	Histology	Salazopyrin, local steroids
DP	F	72	Rectum	Histology	Salazopyrin, local steroids
VW	F	66	Colon and rectum	Histology and radiology	Salazopyrin, local steroids

Table 2.16   Effect of razoxane on disease parameters (shortened version of original table)

Patient	Frequency of bowel action/d	Reduced by months	Blood and excess mucus in stool (months to clear)	Endoscopic appearances (months to return to normal)	ESR (mm/h) Begin	ESR (mm/h) End	Weight (kg) increase during treatment
PC	6 → 1	12	12	6	70	6	12
EH	(ileostomy)	–	3 rectal discharge	9	70	20	12
PP	6 → 1	4	4	4	63	.4	20
FA	8 → 1	4	4	6	64	20	14
GP	5 → 2	9	8	Slightly abnormal	46	10	6
MJ	3 → 1	4	4	Slightly abnormal	51	10	2
RB	4 → 2	3	2	Slightly abnormal	59	52	2
VW	4 → 1	2	1	4	85	36	10

Patient DP, who had rectal and peri-anal Crohn's disease for 15 years, developed a severe anal stenosis on treatment. She also had senile dementia and this led to very difficult management problems due to incontinence. An abdominoperineal resection was performed to aid her management. This proved successful and there were no recurrences of her Crohn's disease until her death 4 years later. Because surgery was performed for reasons other than reactivation of Crohn's colitis DP has been regarded as unassessable.

None of the remaining ten patients has required any surgery, the follow-up times from the commencement of treatment being from 0.9 to 12 years. It is of interest to note the varying times taken to return to normal bowel action, normal sigmoidoscopic appearances, cessation of passage of blood and excess mucus.

All patients gained weight during treatment, the amount varying according to previous weight loss, length of treatment and severity of disease. A considerable fall in ESR was noted in seven patients with Crohn's colitis during treatment (Table 2.16). Three patients with subnormal albumins quickly returned to normal during treatment, six patients never had subnormal albumin levels.

Further evidence of response to treatment has been noted in individual cases. Twelve months after completing treatment patient PC underwent a laparatomy for gynecological reasons and the consultant gynaecological surgeon could not identify any abnormal colon, small bowel or mesentery. The radiological appearances prior to treatment reviewed by several gastroenterologists showed very pronounced Crohn's disease affecting the terminal ileum and colon from caecum to descending colon. Rectal biopsy from the same patient revealed granulomata.

Patient EH, 2 months after completing treatment, underwent an ileorectal reanastomosis successfully and has remained symptom- and apparently disease-free 9 years later.

Patients FA's fistula-in-ano was laid open at the commencement of treatment, healed soundly and has remained healed until the time of his death 4.5 years later.

Patients MJ's multiple perianal and vulval sinuses failed to improve after a resting LIF colostomy but virtually resolved after 12 months treatment with razoxane and have remained healed some 9 years later.

Toxicity and Side Effects

One patient developed leukopenia with WBC below 2,000/mm^3 and platelets below 150,000/mm^3, but both WBC and platelet counts recovered within 2 weeks of cessation of treatment. One other patient only dropped her WBC to 2,900/mm^3 on one occasion. No other side effects were noted. None of the patients have developed any neoplastic disease during the 12 years of the study. The dose of 125 mg daily continuously used in the present series of patients seems to have struck the correct balance between activity and toxicity.

Discussion

It is well recognised that it is difficult to evaluate the effectiveness of drug treatment in Crohn's disease due to its spontaneous remission rate. Nevertheless, the effect on active disease, relapse rate and long term delay or prevention of surgery is a reasonable way of demonstrating some benefit.

Razoxane in this study has been used in patients with active disease uncontrolled by conventional medical treatment. Studies of other cytotoxic agents have often been used in an attempt to prolong remission times or lower the dose of steroids. In the present study razoxane was used alone, principally on an out-patient basis and where possible with the patient at full time work. In this way it has been possible to see clearly the true benefit to the patient.

No patient's disease relapsed on treatment. Furthermore, there is clear evidence symptomatically, sigmoidoscopically, by weight gain and improved haematological and chemical parameters that the active process remitted. Treatment with azathioprine and 6-mercaptopurine have shown that although some patients take several months to go into full remission they may relapse soon after. Four of our patients remain in remission after 10, 9, 7 and 2 years.

Patient compliance was satisfactory, suggesting a lack of significant toxicity and some beneficial action on the disease. None of the patients have developed any neoplastic disease during the 12 years of the study. It should be noted here that there are no reports of any cases of acute leukaemia following continuous low dose razoxane adminstration as used in this study, and that all reported cases of acute leukaemia were associated with being on long-term continuous high dose leukopenic levels of razoxane [12]. The patients in this study took part at a time when the leukaemia risk was not documented and treatment took place on a named patient basis with full discussion with ICI (Imperial Chemical Industries, the pharmaceutical firm that formerly produced and distributed the drug; ed.). In the absence of any serious adverse effect in these patients we believe it is time to reconsider the place of razoxane in the management of Crohn's colitis and the benefit-risk ratio in patients with this often intractable and debilitating disease.

Comment

Seventeen years later, the therapeutic options for Crohn's disease have not been largely extended since the publication of the above article. In recent years, infliximab, a TNF-alpha blocking agent, has shown proven activity in the disease [13]. In a retrospective chart review, 133 patients were identified with records sufficiently detailed to be analyzed. Clinical outcomes were infliximab induction and maintenance responses, defined as the ability to stop and remain off corticosteroids while not requiring additional therapy for active disease.

Hundred-seventeen patients (88%) demonstrated a clinical response to induction; 104 of 117 (89%) were on concomitant immunosuppressive therapy; 80 of 104 on azathioprine/6-mercaptopurine (77%); and 24 of 104 on methotrexate (23%). The mean duration of clinical response was 94 weeks (95% CI 78.8–109.2). The proportion of patients who maintained response at 30 weeks was 83%, at 54 weeks was 64% and at 108 weeks was 45%. Adverse events occurred for 15 of 117 patients (12.8%), consisting of nine infusion reactions, four serum sickness-like reactions, one rash and one infection [13].

Similar data were obtained in a large review from Denmark on 619 patients with Crohn's disease who received a median of 3 infusions of infliximab. A positive clinical response was observed in 82.7% (95% confidence interval, 79.9–85.5) of patients. Infusion reactions were seen in 4.4%. No lymphomas and no increased risk of cancer were observed [11].

In view of these and other recent results, the activity of razoxane in Crohn's colitis still compares favourably with the data reported in the literature, and the statement at the end of the discussion in the work commented on deserves attention again.

References

1. Andrews HA, Lewis P, Allan RN (1989) Prognosis after surgery for colonic Crohn's disease. Br J Surg 76:1184–90
2. The National Co-operative Crohn's Disease Study (1979) Gastroenterology 77:825–944
3. Malchow H, Ewe K, Brandes JW, Goebell H, Ehms H, Sommer H, Jedinsky H (1984) European co-operative Crohn's disease study (ECDS): results of drug treatment. Gastroenterlogy 86:249–66
4. Goldstein F (1980) Reflections on the treatment of Crohn's disease, the NCCRDS Report on randomized clinical trials. J Clin Gastroenterol 2:115–7
5. Parrott NR, Taylor RMR, Venables C, Record CO (1988) Treatment of Crohn's disease in relapse with cyclosporin A. Br J Surg 75:1185–8
6. Willougby JMT, Kumar PJ, Beckett J et al (1971) Controlled trial of azathioprine in Crohn's disease. Lancet ii:944–7
7. Present DH, Korelitz BI, Wesch N, Glass JL, Sachar DB, Pasternack 'BS (1980) Treatment of Crohn's disease with 6-mercaptopurine: a long-term randomised double blind study. New Engl J Med 302:981
8. O'Donoghue DP, Dawson AM, Powell-Tuck J (1978) Double blind trial of azathioprine as maintenance treatment for Crohn's disease. Lancet ii:955–7
9. Lennard-Jones JE, Powell-Tuck J (1979) Drug treatment of inflammatory bowel disease. Clin Gastroenterol 8:187–217
10. Burton I, Korelitz BI (1980) Therapy of inflammatory bowel disease, including use of immunosuppressive agents. Clin Gastroenterol 9:331–50
11. Caspersen S, Elkjaer M, Riis L et al (2008) Infliximab for inflammatory bowel disease in Denmark 1999–2005: clinical outcome and follow up evaluation of malignancy and Mortality. Clin Gastroenterol Hepatol 6(11):1212–7
12. Horton JJ, Caffrey EA, Clark KGA, MacDonald DM, Wells RS, Daker MG (1984) Leukaemia in psoriatic patients treated with razoxane. Br J Dermatol 110:633–4
13. Teshima CW, Thompson A, Dhanoa L, Dieleman LA, Fedorak RN (2009) Long term response rates to infliximab therapy for Crohn's disease in an outpatient cohort. Can J Gastroenterol 23(5):348–52

Chapter 3
Dexrazoxane

Kurt Hellmann and Walter Rhomberg

Abstract   Chapter 3 is concerned essentially with dexrazoxane (ICRF-187) the dextro enantiomer of razoxane (ICRF-159). Although dexrazoxane was initially developed as an anticancer agent (like razoxane) it subsequently emerged as an interesting drug with unexpected properties. Introductory sections first deal with its pharmacology, chemistry and toxicology followed by pharmacokinetics. Dexrazoxane turned out to be a prodrug with iron chelating and topoisomerase II α inhibiting activities. Section 3.4 deals with the identification of dexrazoxane as a protector against the anthracycline-induced cardiotoxicity. Other sections consider the important topics related to the clinical studies of the cardioprotection against the anthracycline toxicity in cancer patients. There is also an original report on the discovery and the outstanding efficacy of dexrazoxane as the only tissue protectant against accidental extravasations from anthracycline chemotherapy. Finally, new data emerging from Greigs Laboratory of Neurosciences (NIH, USA) provide a rationale for the potential clinical use of razoxane/dexrazoxane in specific neurodegenerative conditions such as Alzheimer's or Parkinson's disease.

Keywords Dexrazoxane · Cardioprotective · Cardiotoxicity · Chemoprotective · Topoisomerase II

3.1 Preface

Dexrazoxane (ICRF-187) is the dextro-enantiomer of razoxane (ICRF-159) which first came into focus in the late 1960s as antineoplastic agent. Dexrazoxane is water soluble and can be given iv. Although dexrazoxane was initially developed as

K. Hellmann (✉)
Windleshaw House, Withyham, East Sussex, TN7 4DB, UK

W. Rhomberg (✉)
6700 Bludenz, Unterfeldstrasse 32, Austria
e-mail: walter.rhomberg@gmx.at

K. Hellmann, W. Rhomberg (eds.), *Razoxane and Dexrazoxane – Two Multifunctional Agents*, DOI 10.1007/978-90-481-9168-0_3, © Springer Science+Business Media B.V. 2010

Table 3.1 Developmental timetable of dexrazoxane (ICRF 187)

Razoxane (ICRF 159)	
1964	Chemical synthesis
1969	Discovery of cytostatic and antimetastatic activity. The drug can 'normalize pathological blood vessels' and reduce leakiness and haemorrhages, thereby improving blood vessel function, tumor oxygenation and access of anticancer drugs to the tumor
1970s	Clinical evidence to be a true radiosensitizer
	Discovery of its iron chelating activity, preclinical cardioprotectant
	Effective drug for psoriasis and Crohn's disease
Dexrazoxane (ICRF 187)	
1970s	Emergence as water soluble drug with activities different from razoxane
	Preliminary disprovement of its cytotoxic activity
	Drug confirmed as a cardioprotectant
1980s	First clinical trials demonstrate effectiveness of dexrazoxane as cardioprotectant
	Description of topoisomerase II inhibition
1990s	FDA-accelerated approval of dexrazoxane as a preventive drug against anthracycline-induced cardiotoxicity in May 1995, followed by FDA regular approval in 2002
2000s	Developed as a non-specific tissue protectant to treat accidental extravasations from anthracycline chemotherapy (FDA approval 2007)
	Reappraisal of dexrazoxane's biological activities as a prodrug with iron chelating and topoisomerase II activities for cytoprotection in anthracycline-containing drug therapies and beyond

anticancer agent it emerged as interesting drug with unexpected features. The history and development of the drug is briefly outlined in Table 3.1

The following sections deal with the pharmacology, chemistry and toxicology of the drug as well as with the identification of dexrazoxane as a cardioprotector. In Section 3.6, clinical findings related to the ability of dexrazoxane to protect against the anthracycline-induced cardiotoxicity are discussed in detail. In addition, further potential clinical applications of the drug will be outlined.

3.2 The Pharmacology of Dexrazoxane: Iron Chelating Prodrug and Topoisomerase II Inhibitor

Brian B. Hasinoff

3.2.1 Introduction

Dexrazoxane (Fig. 3.1) is a bisdioxopiperazine and is a ring-closed analog of the iron chelator EDTA. Dexrazoxane has two important biological activities that may

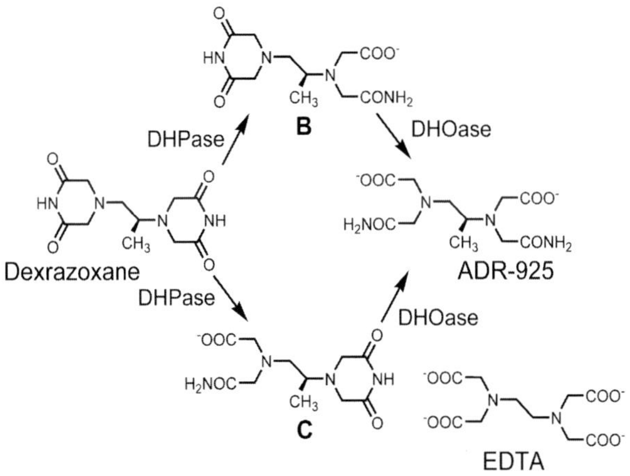

Fig. 3.1 Dexrazoxane, its one ring open metabolites **B** and **C**, and its presumably active iron chelating form ADR-925, and EDTA for comparison. The enzymes dihydropyrimidinase (DHPase) and dihydroorotase (DHOase) metabolize dexrazoxane through the catalysis of the successive ring-opening reactions. DHPase catalyzes only the ring-opening of dexrazoxane and not of **B** and **C**, and DHOase catalyzes only the ring-opening of **B** and **C** and not of dexrazoxane

determine its pharmacology and mode of action. On the one hand dexrazoxane is a prodrug that undergoes hydrolysis to its EDTA-type form (Fig. 3.1) that may chelate iron and prevent doxorubicin-induced oxidative damage to cardiac myocytes. On the other hand the ring-closed form of dexrazoxane (Fig. 3.1) is also a strong catalytic inhibitor of topoisomerase II [1]. The clinical use of the antitumor anthracyclines such as doxorubicin, daunorubicin, epirubicin and idarubicin is limited by a dose-limiting cardiotoxicity [2]. A great deal of evidence has accumulated that this toxicity may be due to iron-dependent oxygen free radical formation on the relatively unprotected cardiac muscle [3]. However, the antitumor activity of doxorubicin is likely a result of its ability to target topoisomerase II [4].

Dexrazoxane is clinically used in the USA, Canada, Europe and in other countries as a cardioprotective agent to reduce doxorubicin induced cardiotoxicity. Additionally, dexrazoxane is also licensed to prevent anthracycline extravasation injury in Europe and the USA [5]. Dexrazoxane, has been shown in both extensive animal studies [6] and in clinical trials [2] to be highly effective in adults and children in reducing doxorubicin-induced cardiotoxicity without affecting the antitumor activity of doxorubicin.

In addition to its ability to reduce doxorubicin-induced cardiotoxicity, dexrazoxane has also been shown in animal studies to reduce or prevent the following: mitoxantrone toxicity, daunorubicin toxicity, bleomycin pulmonary toxicity, isoproterenol-induced cardiotoxicity, acetaminophen-induced hepatotoxicity, alloxan-induced diabetes, and hyperoxia-induced pulmonary damage [7]. It has also recently been shown to prevent cardiac lesions induced by testosterone [8] and ischemia/reperfusion damage in the rat heart [9] and in isolated myocytes [10]. Dexrazoxane has also been shown to suppress mitoxantrone-induced cardiotoxicity in multiple sclerosis patients [11]. Because most of these toxicities likely involve some critical iron-dependent oxygen-radical producing step at some point in the pathogenesis of cellular damage, these results lend support to the hypothesis that dexrazoxane acts by preventing oxygen radical tissue damage.

3.2.2 Chemistry of Dexrazoxane

Dexrazoxane may be considered a neutral pro-drug analog of the tetraacid metal chelator EDTA (Fig. 3.1) which is easily permeable to cells, and which upon hydrolytic metabolism yields its strongly iron chelating diacid diamide form ADR-925 (Fig. 3.1). Dexrazoxane is a bisdioxopiperazine or bis-imide and is the $(+)$-(S)-enantiomer of razoxane. ADR-925 binds Fe^{2+} and Fe^{3+} with formation constants of 10^{10} M^{-1} and $10^{18.2}$ M^{-1}, respectively, [12] which is several orders of magnitude weaker than for binding to EDTA. Nonetheless, it is still a very strong iron chelator. The X-ray crystal structure of the Fe^{3+} complex of ICRF-247, the desmethyl analog of ADR-925, shows that the structure is similar to that of Fe^{3+}-EDTA with 6 coordination positions occupied by the ligand and one by water [12]. We have shown that either ADR-925 or the one ring-open intermediate **B** can either quickly ($t_{1/2}$ ~1 and 6 min, respectively) remove Fe^{3+} from the Fe^{3+}-anthracycline complex, or bind free iron in the cell [10, 13]. The ability to remove iron from its complex with doxorubicin and to chelate free iron in the cell provides a mechanism by which the dexrazoxane hydrolysis product ADR-925 prevents formation of damaging reactive oxygen species.

3.2.2.1 Biochemistry and Pharmacology of Dexrazoxane

Doxorubicin is a quinone that can redox cycle after reductive activation by various reductase enzymes to its free radical semiquinone [7]. The semiquinone radical quickly reacts with oxygen to generate reactive oxygen species such as superoxide anion and hydrogen peroxide [7]. Doxorubicin can also strongly bind iron through its quinone-hydroquinone functional groups and this complex can catalyze formation of the extremely reactive and damaging hydroxyl radical in a redox cycling reaction [7]. Thus, it is this iron-dependent doxorubicin-based oxidative stress that is thought to be responsible for the cardiotoxicity. However, doxorubicin's antitumor activity is likely exclusively due to its ability to bind to DNA and act as a topoisomerase II poison that induces lethal double strand DNA breaks [4, 14], rather than through the formation of reactive oxygen species.

Cardiac mitochondria are a prominent site of injury by doxorubicin. Both doxorubicin and its Fe^{3+} complex have a high affinity for the dianionic phospholipid cardiolipin that is present in high concentration in the inner mitochondrial membrane [15]. Additionally, the cationic doxorubicin, is preferentially taken up by mitochondria because of their negative mitochondrial membrane potential $\Delta\Psi_m$. Doxorubicin has been shown by fluorescence microscopy to localize in the mitochondria of myocytes [16, 17] with doxorubicin-induced dichlorofluorescin oxidation occurring close to the mitochondria [18]. Thus, it would appear that site-specific oxygen radical production by doxorubicin, or its iron complex, may be responsible for doxorubicin-induced mitochondrial damage. Dexrazoxane protects against doxorubicin-induced depolarization of the myocyte mitochondrial membrane as measured with the membrane potential sensing cationic dye JC-1 [17]. It should, however, be noted that the Fe^{3+} complex of ADR-925, like that of EDTA, can be reductively activated and redox cycle to produce hydroxyl radicals [19]. Thus, it is likely that if dexrazoxane protects against doxorubicin-induced oxidative stress in the myocyte, it does so by preventing site-specific iron-based oxidative damage to mitochondria. We also showed that dexrazoxane, **B**, **C** and ADR-925 (Fig. 3.1) entered myocytes and displaced iron from a fluorescence-quenched trapped intracellular iron-calcein complex, a result that provides a mechanistic basis for its cardioprotective effects [10, 13]. We also investigated whether the one-ring open metabolites (Fig. 3.1) of dexrazoxane could protect myocytes from doxorubicin-induced damage [13]. Somewhat surprisingly, metabolite **B** was not able to protect myocytes [13]. Thus, either the anionic metabolites do not have the same access to iron pools in critical cellular compartments, or possibly, and less likely, dexrazoxane protects by some other mechanism such as inhibition of topoisomerase II.

3.2.2.2 Pharmacokinetics and Metabolism of Dexrazoxane

In our studies on the metabolism of dexrazoxane in humans [20] we showed that the one-ring open hydrolysis intermediates **B** and **C** and ADR-925 quickly appeared in the plasma after *i.v.* administration. In patients infused with 1,500 mg/m^2 of dexrazoxane for 15 min and then with etoposide for 90 min, the dexrazoxane metabolite ADR-925 (Fig. 3.1) was detectable in the plasma at the end of the 2 h total infusion period. The ADR-925 level rapidly increased nearly threefold to 29 μM at 15 min post-infusion, and remained nearly constant for 4 h post-infusion [20]. These values compare to a C_{max} for dexrazoxane of 211 μM [20]. In cell and rat models dexrazoxane has been shown to undergo sequential enzymatic hydrolysis (Fig. 3.1) by two tissue zinc hydrolases, dihydropyrimidinase and dihydroorotase, to ADR-925 [21–24]. The fact that dexrazoxane is rapidly metabolized in vivo to its presumably active metal-ion chelating forms is consistent with dexrazoxane acting through the ability of its metal chelating metabolite ADR-925 to prevent iron-based oxidative damage.

3.2.2.3 Tests of Other Iron Chelators as Anthracycline Protective Agents

The ability of other iron chelating drugs, as we and others have shown [25–29], have produced mixed results in their ability to reduce anthracycline-induced cardiotoxicity. Deferiprone is an iron chelator that is used to reduce iron in iron-overloaded patients with β-thalassemia. In a myocyte model it is able to protect against doxorubicin-induced cytotoxicity [30], though in vivo it does not protect against daunorubicin cardiotoxicity [26]. The strong Fe^{2+} chelator Dp44mT, however, does not protect myocytes from doxorubicin-induced damage [31]. The structurally related SIH (salicylaldehyde isonicotinoyl hydrazone) has shown some partial daunorubicin cardioprotective activity in vitro [32], and in vivo at lower doses, though not at higher doses [33]. The clinically used oral iron chelating agent deferasirox (ICL670A) also does not protect myocytes from doxorubicin-induced damage, and in fact increased it [29]. The dexrazoxane analog ICRF-161 with a 3-carbon linker does not inhibit topoisomerase II [25]. While the ICRF-161 rings-opened hydrolysis product displaced Fe^{3+} from its complex with doxorubicin (though not as effectively as ADR-925), and protects against doxorubicin-induced LDH release from cardiomyocytes in vitro, it did not protect against the cardiotoxic effects of doxorubicin in a rat model [25]. Thus, it was concluded that iron chelation alone may not be sufficient for cardioprotection by the bisdioxopiperazines. A series of iron chelating flavonoids also showed good protection in vitro, though they only weakly removed Fe^{3+} from its complex with doxorubicin [28]. In this latter study it was also concluded that iron chelation alone may not be critical for cardioprotective action. However, it does seem clear that some iron chelators, either in vitro or in vivo, have the ability to be cardioprotective. For an iron chelating drug (or prodrug in the case of dexrazoxane) to protect it likely must meet specific requirements. In addition to getting into the cell efficiently and into the right part of the cell, it must be a strong enough iron chelator to remove iron from its complex from doxorubicin, but it must not be so strong that it disrupts the normal iron metabolism of the cell as Dp44mT [29] and deferasirox [29] likely do.

3.2.2.4 Dexrazoxane Inhibition of Topoisomerase II

Dexrazoxane is a strong inhibitor (IC_{50} 10 μM) of the catalytic activity of topoisomerase II [1]. Topoisomerase II alters DNA topology by catalyzing the passing of an intact DNA double helix through a transient reversible double-stranded break made in a second helix and is critical for relieving torsional stress that occurs during replication and transcription and for DNA daughter strand separation during mitosis [14, 34, 35]. Topoisomerase IIα is tightly regulated over the cell cycle and peaks in the G_2/M phase when it is required to separate the DNA daughter strands [35]. Less is known about the highly homologous topoisomerase IIβ isoform which is not tightly regulated during the cell cycle [34]. The X-ray crystal structure of dexrazoxane bound to the N-terminal domain of the highly homologous yeast topoisomerase II has been determined [36, 37] and shows that dexrazoxane binds directly between the two ATP binding sites. Thus, dexrazoxane bridges the two monomers and likely locks the DNA protomers in a closed clamp configuration. As shown in Fig. 3.2 dexrazoxane binds to topoisomerase II in a 'bat wing' type conformation [36]. The dynamic nature of the two topoisomerase II protomers is shown by the fact that in the X-ray structure there is no opening large enough for dexrazoxane to either enter or exit its binding site.

Dexrazoxane does not induce lethal frank DNA strand breaks like the so-called topoisomerase II poisons such as doxorubicin, the other anthracyclines, and other anticancer drugs such as etoposide, amsacrine and mitoxantrone [1, 14]. As such it

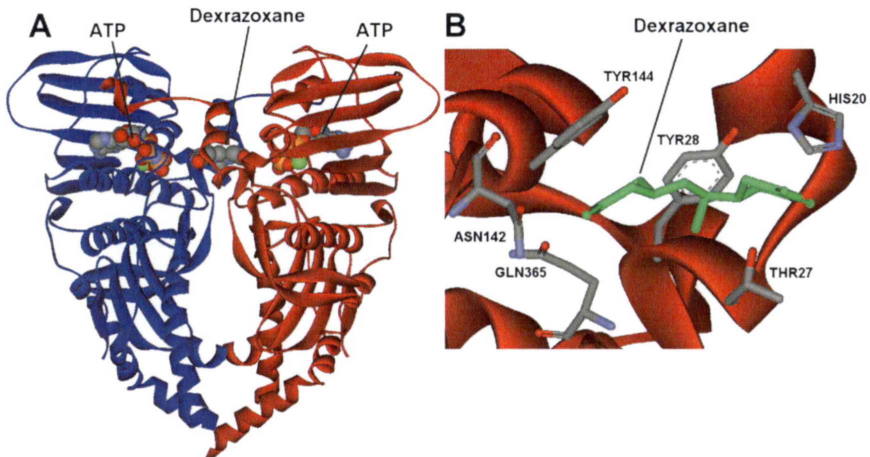

Fig. 3.2 The X-ray structure (Protein Data Bank: 1QZR) of the binding of dexrazoxane to the N-terminal ATP binding region of the yeast topoisomerase IIα dimer (ribbon structures). (**a**) The dexrazoxane, shown in space-filled atoms, binds at the dimer interface close to the two ATP binding sites and locks the molecule into a closed configuration preventing entry of DNA. The 2 ATP molecules are shown on either side of dexrazoxane. (**b**) Close-up of the interaction of dexrazoxane (stick structure) with the amino acid side chains of one of the topoisomerase II chains. The amino acid side chains within 6 Å of dexrazoxane are labeled and shown in stick structure. Hydrogen atoms are omitted for clarity

is termed a catalytic inhibitor to distinguish it from the topoisomerase II poisons. Dexrazoxane blocks the topoisomerase II catalytic cycle after DNA strand passage, but before the hydrolysis of the second ATP [34]. The topoisomerase II poisons act through the increased formation or stabilization of a normally transient topoisomerase II-DNA covalent complex (the 'cleavable complex') [34]. A collision of the DNA replication complex with the cleavable complex is thought to convert the normally transient DNA double strand breaks into permanent double strand breaks which results in inhibition of cell division and induction of apoptosis [14]. An X-ray structure of DNA bound to topoisomerase II displays a large bend in the DNA [38].

Dexrazoxane was initially developed as an antitumor agent and its antitumor activity is likely a result of its ability to inhibit topoisomerase II [1] and prevent DNA daughter strand separation during mitosis and cell division. Dexrazoxane treatment of human leukemia K562 cells induces differentiation and apoptosis [39]. Its mild myelosuppressive effects seen clinically [2] is likely a result of its ability to inhibit topoisomerase II.

Evidence for a role for the topoisomerase II inhibitory effects of dexrazoxane in reducing doxorubicin cardiotoxicity comes from a recent study using H9C2 cardiomyocytes [40]. It was suggested that dexrazoxane antagonizes doxorubicin-induced DNA damage specifically through its interference with topoisomerase IIβ [40]. In addition to antagonizing topoisomerase II cleavage complex formation, dexrazoxane also induces rapid degradation of topoisomerase IIβ, which paralleled the reduction of doxorubicin-induced DNA damage. However, how the prevention of DNA damage in the heart might reduce cardiotoxicity is unknown. The conclusions drawn from this provocative study may have been a consequence of using actively dividing immortal H9C2 cardiomyocytes rather than non-dividing primary myocytes.

It has been observed that dexrazoxane is able to antagonize the growth-inhibitory effects and the DNA-cleaving activity induced by doxorubicin and daunorubicin [41, 42]. This likely occurs because dexrazoxane destabilizes the covalent enzyme intermediate adduct formed between topoisomerase II and DNA (containing a DNA double strand break), and thus reduces lethal frank DNA double strand breaks [43, 44]. While it is possible that the cardioprotective and extravasation protective effects of dexrazoxane may also be due to its ability to inhibit topoisomerase II, it is difficult to envision a mechanism or mechanisms by which this would occur as these tissues are not rapidly proliferating and contain only low levels of topoisomerase $II\alpha$. A possible mechanism involving inhibition of topoisomerase II might involve myocardial remodeling that may be occurring after doxorubicin-induced myocardial tissue damage. Myocardial remodeling is an attempt by the heart to repair damage by initiating hypertrophic processes and new tissue and cell growth [45]. These potentially damaging processes might be inhibited by the topoisomerase II inhibitory effects of dexrazoxane. However, the role, if any, of topoisomerase II in myocardial remodeling is unknown.

3.2.3 Conclusions

The anthracyclines will likely continue to have an important role in the treatment of a variety of neoplastic diseases for some time to come. Anthracycline-induced cardiotoxicity remains a major side effect that limits effective anthracycline dosing in patients. Even though a number of treatment modifications and potential cardioprotective agents have been examined, pretreatment with dexrazoxane has been found to be the most consistently successful [46]. Dexrazoxane cardioprotection has been observed in all reported studies involving anthracycline use in children and adult cancer patients [46]. The concern that dexrazoxane may interfere with the antitumor activity of anthracyclines is based on a single study [47]. An analysis of this study and of numerous other studies indicates that dexrazoxane given at cardioprotective doses does not alter the antitumor efficacy of the anthracyclines [46].

In conclusion, dexrazoxane is highly effective both in reducing anthracycline-induced cardiotoxicity and extravasation injury. It has two important pharmacological activities that could potentially contribute to its efficacy as a cardioprotective drug. It is a prodrug that is hydrolyzed to an iron chelating form, and it is also a strong inhibitor of topoisomerase II. Because of the compelling experimental evidence that doxorubicin-induced cardiotoxicity is due to oxidative stress, and because some other iron chelating drugs have shown some efficacy in reducing doxorubicin-induced damage, at this time the majority of the evidence supports the conclusion that dexrazoxane reduces cardiotoxicity by its rapidly produced metabolite ADR-925 binding free iron and preventing oxidative stress on cardiac tissue. However, it cannot be ruled out that dexrazoxane may also be protective, in part, through inhibition of topoisomerase II.

Acknowledgments This work was supported by the Canadian Institutes of Health Research, the Canada Research Chairs program and a Canada Research Chair in Drug Development for Brian Hasinoff.

References

1. Hasinoff BB, Kuschak TI, Yalowich JC, Creighton AM (1995) A QSAR study comparing the cytotoxicity and DNA topoisomerase II inhibitory effects of bisdioxopiperazine analogs of ICRF-187 (dexrazoxane). Biochem Pharmacol 50:953–8
2. Swain SM, Vici P (2004) The current and future role of dexrazoxane as a cardioprotectant in anthracycline treatment: expert panel review. J Cancer Res Clin Oncol 130:1–7
3. Minotti G, Recalcati S, Menna P, Salvatorelli E, Corna G, Cairo G (2004) Doxorubicin cardiotoxicity and the control of iron metabolism: quinone-dependent and independent mechanisms. Meth Enzymol 378:340–61
4. Gewirtz DA (1999) A critical evaluation of the mechanisms of action proposed for the antitumor effects of the anthracycline antibiotics adriamycin and daunorubicin. Biochem Pharmacol 57:727–41
5. Hasinoff BB (2008) The use of dexrazoxane for the prevention of anthracycline extravasation injury. Expert Opin Investig Drugs 17:217–23
6. Herman EH, Ferrans VJ (1990) Examination of the potential long-lasting protective effect of ICRF-187 against anthracycline-induced chronic cardiomyopathy. Cancer Treat Rev 17:155–60
7. Hasinoff BB, Hellmann K, Herman EH, Ferrans VJ (1998) Chemical, biological and clinical aspects of dexrazoxane and other bisdioxopiperazines. Curr Med Chem 5:1–28
8. Belhani D, Fanton L, Vaillant F, Descotes J, Manati W, Tabib A, Bui-Xuan B, Timour Q (2009) Cardiac lesions induced by testosterone: protective effects of dexrazoxane and trimetazidine. Cardiovasc Toxicol 9:64–9
9. Ramu E, Korach A, Houminer E, Schneider A, Elami A, Schwalb H (2006) Dexrazoxane prevents myocardial ischemia/reperfusion-induced oxidative stress in the rat heart. Cardiovasc Drugs Ther 20:343–8
10. Hasinoff BB (2002) Dexrazoxane (ICRF-187) protects cardiac myocytes against hypoxia-reoxygenation damage. Cardiovasc Toxicol 2:111–8
11. Bernitsas E, Wei W, Mikol DD (2006) Suppression of mitoxantrone cardiotoxicity in multiple sclerosis patients by dexrazoxane. Ann Neurol 59:206–9
12. Diop NK, Vitellaro LK, Arnold P, Shang M, Marusak RA (2000) Iron complexes of the cardioprotective agent dexrazoxane (ICRF-187) and its desmethyl derivative, ICRF-154: solid state structure, solution thermodynamics, and DNA cleavage activity. J Inorg Biochem 78:209–16
13. Hasinoff BB, Schroeder PE, Patel D (2003) The metabolites of the cardioprotective drug dexrazoxane do not protect myocytes from doxorubicin-induced cytotoxicity. Mol Pharmacol 64:670–8
14. Fortune JM, Osheroff N (2000) Topoisomerase II as a target for anticancer drugs: when enzymes stop being nice. Prog Nucleic Acid Res Mol Biol 64:221–53
15. Hasinoff BB, Davey JP (1988) Adriamycin and its iron(III) and copper(II) complexes; glutathione-induced dissociation; cytochrome c oxidase inactivation and protection; binding to cardiolipin. Biochem Pharmacol 37:3663–9
16. Swift LM, Sarvazyan N (2000) Localization of dichlorofluorescin in cardiac myocytes: implications for assessment of oxidative stress. Am J Physiol Heart Circ Physiol 278:H982–90

17. Hasinoff BB, Schnabl KL, Marusak RA, Patel D, Huebner E (2003) Dexrazoxane (ICRF-187) protects cardiac myocytes against doxorubicin by preventing damage to mitochondria. Cardiovasc Toxicol 3:89–99
18. Sarvazyan N (1996) Visualization of doxorubicin-induced oxidative stress in isolated cardiac myocytes. Am J Physiol 271:H2079–85
19. Malisza KL, Hasinoff BB (1996) Hydroxyl radical production by the iron complex of the hydrolysis product of the antioxidant cardioprotective agent ICRF-187 (dexrazoxane). Redox Rep 2:69–73
20. Schroeder PE, Jensen PB, Sehested M, Hofland KF, Langer SW, Hasinoff BB (2003) Metabolism of dexrazoxane (ICRF-187) used as a rescue agent in cancer patients treated with high-dose etoposide. Cancer Chemother Pharmacol 52:167–74
21. Schroeder PE, Davidson JN, Hasinoff BB (2002) Dihydroorotase catalyzes the ring-opening of the hydrolysis intermediates of the cardioprotective drug dexrazoxane (ICRF-187). Drug Metab Dispos 30:1431–5
22. Schroeder PE, Wang GQ, Burczynski FJ, Hasinoff BB (2005) Metabolism of the cardioprotective drug dexrazoxane and one of its metabolites by isolated rat myocytes, hepatocytes and by blood. Drug Metab Dispos 33:719–25
23. Schroeder PE, Hasinoff BB (2005) Metabolism of the one-ring open metabolites of the cardioprotective drug dexrazoxane to its active metal chelating form in the rat. Drug Metab Dispos 33:1367–72
24. Schroeder PE, Patel D, Hasinoff BB (2008) The dihydroorotase inhibitor 5-aminoorotic acid inhibits the metabolism in the rat of the cardioprotective drug dexrazoxane and its one-ring open metabolites. Drug Metab Dispos 36:1780–5
25. Martin E, Thougaard AV, Grauslund M, Jensen PB, Bjorkling F, Hasinoff BB, Sehested M, Jensen LH (2009) Evaluation of the topoisomerase II-inactive bisdioxopiperazine ICRF-161 as a protectant against doxorubicin-induced cardiomyopathy. Toxicology 255:72–9
26. Popelova O, Sterba M, Simunek T, Mazurova Y, Guncova I, Hroch M, Adamcova M, Gersl V (2008) Deferiprone does not protect against chronic anthracycline cardiotoxicity in vivo. J Pharmacol Exp Ther 326: 259–69
27. Simunek T, Sterba M, Popelova O, Kaiserova H, Potacova A, Adamcova M, Mazurova Y, Ponka P, Gersl V (2008) Pyridoxal isonicotinoyl hydrazone (PIH) and its analogs as protectants against anthracycline-induced cardiotoxicity. Hemoglobin 32:207–15
28. Kaiserova H, Simunek T, van der Vijgh WJ, Bast A, Kvasnickova E (2007) Flavonoids as protectors against doxorubicin cardiotoxicity: role of iron chelation, antioxidant activity and inhibition of carbonyl reductase. Biochim Biophys Acta 1772:1065–74
29. Hasinoff BB, Patel D, Wu X (2003) The oral iron chelator ICL670A (deferasirox) does not protect myocytes against doxorubicin. Free Radic Biol Med 35:1469–79
30. Barnabé N, Zastre J, Venkataram S, Hasinoff BB (2002) Deferiprone protects against doxorubicin-induced myocyte cytotoxicity. Free Radic Biol Med 33:266–75
31. Hasinoff BB, Patel D (2009) The iron chelator Dp44mT does not protect myocytes against doxorubicin. J Inorg Biochem 103:1093–101
32. Simunek T, Sterba M, Popelova O, Kaiserova H, Adamcova M, Hroch M, Haskova P, Ponka P, Gersl V (2008) Anthracycline toxicity to cardiomyocytes or cancer cells is differently affected by iron chelation with salicylaldehyde isonicotinoyl hydrazone. Br J Pharmacol 155: 138–48
33. Sterba M, Popelova O, Simunek T, Mazurova Y, Potacova A, Adamcova M, Guncova I, Kaiserova H, Palicka V, Ponka P, Gersl V (2007) Iron chelation-afforded cardioprotection against chronic anthracycline cardiotoxicity: a study of salicylaldehyde isonicotinoyl hydrazone (SIH). Toxicology 235:150–66
34. Nitiss JL (2009) Targeting DNA topoisomerase II in cancer chemotherapy. Nat Rev Cancer 9:338–50
35. Deweese JE, Osheroff N (2009) The DNA cleavage reaction of topoisomerase II: wolf in sheep's clothing. Nucleic Acids Res 37:738–48

36. Classen S, Olland S, Berger JM (2003) Structure of the topoisomerase II ATPase region and its mechanism of inhibition by the chemotherapeutic agent ICRF-187. Proc Natl Acad Sci U S A 100:14510
37. Classen S, Olland S, Berger JM (2003) Structure of the topoisomerase II ATPase region and its mechanism of inhibition by the chemotherapeutic agent ICRF-187. Proc Natl Acad Sci U S A 100:10629–34
38. Dong KC, Berger JM (2007) Structural basis for gate-DNA recognition and bending by type IIA topoisomerases. Nature 450:1201–5
39. Hasinoff BB, Abram ME, Barnabé N, Khelifa T, Allan WP, Yalowich JC (2001) The catalytic DNA topoisomerase II inhibitor dexrazoxane (ICRF-187) induces differentiation and apoptosis in human leukemia K562 cells. Mol Pharmacol 59:453–61
40. Lyu YL, Kerrigan JE, Lin CP, Azarova AM, Tsai YC, Ban Y, Liu LF (2007) Topoisomerase IIβ mediated DNA double-strand breaks: implications in doxorubicin cardiotoxicity and prevention by dexrazoxane. Cancer Res 67:8839–46
41. Hasinoff BB, Yalowich JC, Ling Y, Buss JL (1996) The effect of dexrazoxane (ICRF-187) on doxorubicin- and daunorubicin-mediated growth inhibition of Chinese hamster ovary cells. Anticancer Drugs 7:558–67
42. Jensen PB, Sehested M (1997) DNA topoisomerase II rescue by catalytic inhibitors: a new strategy to improve the antitumor selectivity of etoposide. Biochem Pharmacol 54:755–9
43. Sehested M, Jensen PB, Sorensen BS, Holm B, Friche E, Demant EJF (1993) Antagonistic effect of the cardioprotector (+)-1,2-bis(3,5-dioxopiperazinyl-1-yl)propane (ICRF-187) on DNA breaks and cytotoxicity induced by the topoisomerase II directed drugs daunorubicin and etoposide (VP-16). Biochem Pharmacol 46:389–93
44. Hasinoff BB, Kuschak TI, Creighton AM, Fattman CL, Allan WP, Thampatty P, Yalowich JC (1997) Characterization of a Chinese hamster ovary cell line with acquired resistance to the bisdioxopiperazine dexrazoxane (ICRF-187) catalytic inhibitor of topoisomerase II. Biochem Pharmacol 53:1843–53
45. Swynghedauw B (1999) Molecular mechanisms of myocardial remodeling. Physiol Rev 79:215–62
46. Pouillart P (2004) Evaluating the role of dexrazoxane as a cardioprotectant in cancer patients receiving anthracyclines. Cancer Treat Rev 30:643–50
47. Swain SM, Whaley FS, Gerber MC, Weisberg S, York M, Spicer D, Jones SE, Wadler S, Desai A, Vogel C, Speyer J, Mittelman A, Reddy S, Pendergrass K, Velez-Garcia E, Ewer MS, Bianchine JR, Gams RA (1997) Cardioprotection with dexrazoxane for doxorubicin-containing therapy in advanced breast cancer. J Clin Oncol 15:1318–32

3.3 Toxicology and Pharmacokinetics

Kurt Hellmann

3.3.1 Introduction

Dexrazoxane (ICRF-187; Cardioxane®-Novartis; Zinecard®-Pfizer), the dextro optical isomer of razoxane (ICRF-159; 1,2-di (3,5-dioxo-piperazin-1-yl) propane) protects against anthracycline induced cardiotoxicity in all 6 species in which it has been tested including man (Table 3.2). In early clinical trials dexrazoxane was given at a dose of 1,000 mg/m^2 each time doxorubicin was administered (once every 3 weeks for approximately 6 months) and was well tolerated [1].

Razoxane (trade name Razoxin®) has been on the market in the United Kingdom as an antineoplastic agent since 1977 [2]. One of the disadvantages of razoxane is

Table 3.2 Cardioprotection by ICRF-187 in animal systems

Species	Dosage ICRF-187 mg/kg	Dosage DAUN (A) or DOX (D) mg/kg, or total dose	Result cardiotoxicity	Author	Year
Rabbits	25	3.2 D	+	Herman and Ferrans	1986
Hamster	12.5–200	25 D	+,*	Herman et al.	1985
Mouse	50	4 A	+	Perkins	1982
Guinea-pig	12.5	1.0 A	+	Perkins	1982
Dogs	25	1.75 A	+	Herman et al.	1985
Dogs	12.5	1.0 A	+	Herman and Ferrans	1981
Hamster	100	25 D	+	Herman et al.	1979
Swine	12.5	1.6 A	+	Herman and Ferrans	1983

DAUN, Daunomycin; +, Improvement, reduction; DOX, Doxorubicin; *, death.

its limited solubility in water or saline and it has not therefore been found possible to prepare solutions for parenteral administration. ICRF-187 on the other hand has a solubility in water at least 5 times that of razoxane and this results in the possibility of preparing solutions for parenteral administration [3].

The biological properties of both razoxane (Rz) and dexrazoxane (DXRz) are essentially similar (e.g. antitumour effect, G2/M block of cell division, myelosuppression as dose limiting toxicity) [4–6], and, therefore, a number of studies conducted with razoxane instead of with dexrazoxane are included in this work.

DXRz has been investigated in both men and animals for its toxicological and pharmacological properties. The studies are already published and are summarized here. The report is divided into 8 sections dealing with:

1. Single dose toxicity
2. Repeated dose toxicity
3. Teratogenicity studies
4. Mutagenic potential
5. Oncogenic/carcinogenic potential
6. Pharmacodynamics
7. Pharmacokinetics
8. Local tolerance

3.3.2 Single Dose Toxicity

Studies in Dogs

The toxicity of single doses of ICRF-187 has been studied in Beagle dogs [7, 8]. The drug was given to 4 pairs of dogs (1 female and 1 male per pair) by intravenous infusion. Doses used were 5,000 (250); 10,000 (500); 20,000 (1,000) and 40,000 mg/m^2 (2,000 mg/kg). Because of the large volume of drug solution (up to 2,000 ml/day) required at the maximum tolerated dose of ICRF-187 which could be used (10 mg/ml), infusion rates were between 3.8 and 4.9 ml/min. Constant infusion rates were maintained with a continuous infusion pump which was connected by a cannula to the saphenous vein. The dogs were restrained in slings during infusion. According to the schedule one dog was necropsied at day 8 and one at day 45.

At the highest dose used, the male treated with 40,000 mg/m^2 (2,000 mg/kg) was found to be moribund 3 days after treatment but the female survived to the end of the observation period. Clinical signs of toxicity during and after treatment were vomiting, diarrhea, anorexia, lethargy, weight loss and hypothermia. Even at this dose the female showed signs of recovery from the toxic effects on day 5. All the animals, however, at all concentrations reduced their food consumption, and consequently weight loss was seen. No deaths occurred at the other concentrations used and 20,000 mg/m^2 (1,000 mg/kg) is therefore the highest tolerated toxic dose.

Changes were seen in the hematology and in serum chemistry values on day 2 in both dogs treated with 20,000 mg/m^2 (1,000 mg/kg). The haematological changes seen were an increase in neutrophils, and lymphopenia. The blood chemistry showed an increased value for BSP, SGOT, SGPT, and prothrombin time. At day 4 were also seen markedly elevated BUN values, creatinine, SGOT, SGPT, total and direct bilirubin and prothrombin time. Also on day 4, the female dog had anemia, leukopenia, hypoglycaemia, proteinuria and increased values for liver function tests. Furthermore serum indices of hepatic dysfunction in this dog remained elevated throughout most of the 45 day observation time. At the lowest dose administered, i.e. 5,000 mg/m^2 (250 mg/kg) only moderate neutrocytosis and lymphopenia occurred with a delayed increase in urinary protein in the dog held for observation. At this dose only very slight increases in SGOT values were observed. The haematology was essentially the same as control values for both male and female dogs with possibly slight reductions, though of doubtful significance in red blood cell count, haematocrit and haemoglobin values.

Histopathology showed that microscopic lesions were less severe in the dogs held for 45 days at all doses. At the lethal dose, however, the dog which had died had generalised haemorrhage of tissues, necrosis of lymphoid tissues, small intestinal mucosa and renal proximal convoluted tubules; cytoplasmic vacuolation of hepatocytes, pancreatic acinar cells and lining cells of proximal convoluted tubules. There was bone marrow hypoplasia with an increased myeloid erythroid ratio. Histopathological changes 45 days after treatment with the lethal dose were hepatocellular cytoplasmic vacuolation and focal necrosis, hyperplasia, haemorrhage, histiocytosis, erythrophagocytosis and haemosiderosis in lymphoid tissue. Only the dogs necropsied on day 8 had mild focal necrosis of lymphoid tissue and the lining cells of the proximal convoluted tubules.

A clear dose response effect was seen with doses between 5,000 mg/m^2 (250 mg/kg) and 40,000 mg/m^2 (2,000 mg/kg). Toxic effects were reversible at all doses studied below 40,000 mg/m^2 (2,000 mg/kg).

Studies in Other Species

Given to *mice*, intravenous doses of 3,000 mg/m^2 (150 mg/kg) – the only dose examined – showed no toxicity and no lethality.

In *humans*, doses of 1,000 mg/m^2 as a single dose are well tolerated with no evidence of any toxicity (see also Section 2.3.5).

Summary

Available evidence from the experiments and the results quoted above indicates that large single doses of ICRF-187 (5,000 mg/m^2, i.e. 250 mg/kg) can be given without encountering much toxicity. It is also clear from the results, particularly at the lower doses, that any microscopic evidence of tissue damage is reversible. There are two kinds of toxicity, one related to the inhibition of cell division by ICRF-187, and the renal and hepatic toxicity which could be due to the very large volumes administered, and the fact that the dogs had to remain in slings for hours. All of which may have contributed to the nonspecific toxicity seen.

However, all the work on single doses of ICRF-187 toxicity has been primarily done in one mammalian species. It is worth noting, therefore, that the racemic compound ICRF-159 was evaluated not only in Beagle dogs, but in Rhesus monkeys as well [9]. Both ICRF-159 and ICRF-187 have already been investigated in human beings for some years, and although ICRF-159 is given by mouth and ICRF-187 by intravenous infusion, the toxicological profile (with the exception of signs of transient hepatic toxicity with ICRF-187) in man has been broadly similar. The results obtained in the studies with ICRF-159 in dogs were similar to those now seen with ICRF-187. While one cannot be absolutely certain, the probability is, that the results with ICRF-187 in another species such as Rhesus monkeys would also have been similar to that seen with ICRF-159.

Clinical implications. It is clear that single large doses of ICRF-187 are well tolerated. Maximum tolerated (MTD) doses have been worked out in a phase I escalation study [10]. At the MTD there was little hepatic or renal toxicity and what there was, was reversible. Dose limiting toxicity was nearly always due to bone marrow depression.

3.3.3 Repeated Dose Toxicity

Repeated dose toxicity has been studied in 7 pairs of Beagle dogs (1 male and 1 female for each pair) [7, 8]. One pair received the control solution (nonpyrogenic sterile 0.9% sodium chloride), i.e. the vehicle in which the ICRF-187 was made up. These control animals received 25 ml/kg. Six pairs of dogs were treated once daily for 5 consecutive days at a level of 500, 250, 125, 62.5, 31.3 or 15.6 mg/kg/day. This corresponds to 10,000, 5,000, 2,500, 1,250, 625 or 312.5 mg/m^2/day. Dogs treated with 500 mg/kg died after the 4th or 5th dose. At lower doses of 250 mg/kg the males were autopsied on day 6 as per protocol while the females died on day 8. At other doses the protocol called for one animal of the pair to be autopsied at day 6 and the other at day 52.

The clinical signs of toxicity seen in both these groups were anorexia, vomiting, haematemesis, weight loss and loss of blood per rectum. The male treated with the highest dose level was hypothermic 3 h after the end of each day's treatment. The temperatures being at least 2°F below the normal temperature for Beagle dogs (100–102°F).

Anorexia and weight loss occurred at all doses down to 31.3 mg/kg. Reduced food intake during the 5 days treatment was observed even after infusions of the vehicle alone, and this would indicate that the administration of large doses of fluid to these animals on a repeated daily basis is not well tolerated.

Haematological changes were noted at all doses even down to 15.6 mg/kg though changes at the lower doses were only slight to moderate changes in circulating blood cells. There were also slight changes in the levels of alkaline phophatase, SGOT and SGPT. At values above 31.3 mg/kg more severe changes, usually leukopenia, thrombocytopenia, increased BSP retention and increased SGOT and SGPT levels were noted. Potassium and calcium values were only slightly changed.

At the higher levels of 500 and 250 mg/kg there were haemorrhages in the gastrointestinal tract, in the adrenal cortex and urinary bladder as well as in lymphoid tissues, in some cases going on to necrosis. The mucosa of the intestinal tract also showed signs of necrosis with bone marrow demonstrating hypoplasia. Most of these effects can be seen at an early stage by day 6 and appear to be irreversible at the higher doses. At the lower doses the effects are only observable in the early stages but not later on and therefore appear to be reversible.

Multiple Dose Study with Rest Periods

Three pairs of dogs were used in this study [7, 8] and the treatment schedule was 5 consecutive daily infusions for 3 treatment periods with intervening 9 day rest periods. The dose levels were 125, 31.3 or 15.6 mg/kg. The dogs were autopsied at day 34 and 45. Clinical signs of toxicity at the highest level was anorexia and weight loss with the female showing greater reduction in food intake than the male. At the highest dose level there were also changes in serum chemistry observable during the treatment periods whereas the nadir in circulating blood cells occurred after the end of dosing. SGOT and SGPT levels showed the greatest increases during the first series of treatments. Depressions of white blood cells and platelet counts were most prominent after the first treatment period. The maximum decrease in the red cell count occurred at the end of the third series of infusions.

Microscopic lesions were not present in dogs autopsied 45 days after treatment at any of the doses. This would indicate that with this schedule of treatment and these doses histopathological toxicity is entirely reversible, because some histopathological changes were seen in dogs autopsied 24 h after the last dose.

Summary and Conclusions

In summary, the toxic effects of ICRF-187 in dogs were most prominent on mitotically active tissues, i.e. bone marrow, lymphoid tissue and gastrointestinal mucosa.

Similar results have been previously described after ICRF-159 treatment in both dogs and monkeys [9]. Additionally however, there appear to be hepatic and renal toxicites after i.v. infusions of ICRF-187. Although these organ systems were not affected by ICRF-159, it has to be remembered that the ICRF-159 was given by mouth and ICRF-187 by means of large intravenous infusion volumes.

The importance of the rest period between a series of 5 daily doses allowed a total dose of 1,875 mg/kg to be given without lethality. Phase I studies have shown that leukopenia is the dose limiting toxicity when patients were given ICRF-187 in doses ranging from 200 to 1,500 mg/m^2 i.v. daily for 5 days. Leukopenia began at 800 mg/m^2. Non-myelosuppressive toxicity included reversible increases of liver enzymes, mild nausea and vomiting, low grade fever, alopecia and pain at injection site. It is also worth noting that although ICRF-187 is a chelating agent, serum calcium, potassium and sodium concentrations were not affected by treatment.

Clinical implications. The clinical implications of the preclinical toxicity studies are that the doses of ICRF-187 which it was thought to give (1,000 mg/m^2 once every 3 weeks) were entirely safe.

3.3.4 Teratogenicity Studies

There do not appear to have been any preclinical studies on the influence of ICRF-187 on reproduction. However, the effects on reproduction of the racemic compound ICRF-159 have been closely studied in mice, rats and rabbits. While it is always possible that even where two optical isomers which display for the most part similar biological activities may still have some activity where they behave differently, it is unlikely to be in a situation (reproduction) which depends fundamentally on a biological function (normal cell reproduction) which both isomers are known to inhibit in an identical manner.

Studies on reproduction with ICRF-159 [11] showed that when this drug was given to pregnant Balb/c mice, Sprague Dawley rats, Wistar rats or New Zealand white rabbits, the effect was either embryolethal or a malformation rate of approximately 7.5%. Some foetuses exhibited retarded development. In all 3 species, the time at which they were maximally sensitive to ICRF-159 was early in gestation between days 6–10 in mice, 6–8 in rats, and 7–10 in rabbits.

In those rats where malformations were found after ICRF-159 given on days 6–8 of the pregnancy, malformations which formed 7.3% of the offspring in the case of Sprague Dawley rats and 7.2% in the case of the Wistar rats, showed a variety of different effects ranging from omphalocele to hydronephrosis or, and these were the commonest malformations, rib or sternum abnormalities. In some cases more than one abnormality was seen in the same foetus and this was probably due to administering the drug on 3 consecutive days during organogenesis.

From a detailed consideration of these results and the cytotoxic effect of ICRF-159 it seems highly improbable that the effects of ICRF-187, dose for dose, would be different from those obtained with ICRF-159. ICRF-159 does not appear to be able to cross the blood-brain barrier and it would be surprising if it were able to cross the placental barrier.

A detailed analysis of the effects of varying dosages of ICRF-159 given orally on specific days of pregnancy [11] has shown that the number of resorptions as well as litter size and foetal weights, are directly proportional to the dosage given. Furthermore, the number of implantations falls proportionally to the drug administered when more than 7.5 mg/kg is given.

Comment and Clinical Implications

Clearly, direct evidence of the effect of ICRF-187 on reproduction would be more informative, but having regard to the identical biological behaviour of ICRF-187 and its racemate ICRF-159, it is difficult to believe that the outcome of a study on the teratogenicity of ICRF-187 similar to that carried out by Duke [11] on ICRF-159 would have given different results.

It would be advisable to exclude women in the first trimester of pregnancy from receiving ICRF-187. Women of child bearing age should be instructed not to become pregnant whilst receiving ICRF-187. These warnings and restrictions are included on the package inserts. Moreover, it must be remembered that dexrazoxane is part of a treatment with antineoplastic drugs for advanced cancer.

3.3.5 Mutagenicity

No specific studies of mutagenicity by ICRF-187 have yet been undertaken. However, studies on the racemate ICRF-159 have shown that in salmonella/ microsome assay doses up to 5,000 μg/plate were negative [12]. On the other hand, in the mouse micronucleus test, ICRF-159 in the very large doses of 200 and 400 mg/kg given i.p. was not only cytotoxic to the bone marrow (which limited the analysis) but showed a fivefold increase in micronucleated polychromated erythrocytes compared to control values. In the Chinese hamster metaphase assay, ICRF-159 (up to 500 mg/kg orally) induced abnormal chromosome condensation and an increase in structural chromosome aberrations (sevenfold compared to control value) as well as an increase in the number of polyploid cells (eightfold compared to control value).

Thus, the mutagenic effect of ICRF-159 was restricted to eukaryotic organisms and was associated with specific chromosomal changes. Deficiency of evidence is noted in so far that it would be preferable to have some information on the mutagenicity of ICRF-187 at therapeutic doses.

Clinical implications. It would be difficult to judge, however, what additional hazard, if any, ICRF-187 would have when given together with another drug such as doxorubicin which itself is strongly mutagenic. The likelihood that ICRF-187 would change the situation or present a hazard to the patient, particularly when used in the relatively low therapeutic doses, is small. It would seem much more likely, and indeed there is evidence for this, that ICRF-159 may not only prevent the doxorubicin cardiotoxicity, but also the doxorubicin carcinogenicity (Hellmann K, Finch M, unpublished).

3.3.6 Carcinogenicity

No long or short term animal studies to test whether ICRF-187 is carcinogenic have been done. However, studies done on ICRF-159 at the United States National Cancer Institute to test whether ICRF-159 is carcinogenic in animals were published many years ago in the Federal Register [13].

These studies were done over a period of 1.5 years with 1 year of drug administration and an additional 6 months of observation. The investigators used continuous intraperitoneal administration of relativly high doses of ICRF-159 in a dose of either 48 or 96 mg/kg for the rats, and 40 or 80 mg/kg for the mice. The drug was given on 3 days/week continuously for 52 weeks. The study employed Sprague-Dawley rats and B6C3F mice of both sexes with 35 animals per group for the drug injection groups, but untreated controls and vehicle controls consisted of 10 rats and 15 mice of each sex. All surviving rats were killed between 81 and 86 weeks and all surviving mice at 86 weeks.

Mean body weights were depressed in rats and mice administered ICRF-159, and mortality was dose related among male and female rats and male mice. The high mortality among the male rats may have been associated with inflammatory lesions observed in the lungs, the liver, and the pleural and peritoneal cavities. Sufficient number of female rats and of both male and female mice were at risk for development of late-appearing tumors. In the male rats, time-adjusted analysis of the incidence of tumors was used for determining statistical significance.

At autopsy, careful examination of all the organs showed that in the mice lymphomas were produced, though when compared with the incidence of spontaneous lymphomas in the controls, the difference was marginally greater in the ICRF-159 treated animals.

In the rats however, while the males had no detectable tumours, 8 of 32 of the females developed uterine adenocarcinomas. The sex difference in susceptibility is not readily explicable. However, there are reasons which could possibly account for this development of tumours which have to be taken into consideration before the drug can be labelled carcinogenic. These reasons were not taken into account by the NCI who felt that the drug was 'marginally' carcinogenic (Weissburger, personal communication).

The reasons which may account for the apparent carcinogenicity are firstly that at the dose of ICRF-159 employed throughout these experiments, i.e. 40–96 mg/kg, the solubility is low particularly in saline and there may have been some ICRF-159 precipitate deposited in the peritoneal cavity. The minute particles could quite easily enter the fallopian tubes from the peritoneal cavity and by this means reach the uterus. The possible irritant effect of small insoluble deposits has therefore to be taken into account. Another possibility is that the continuous administration of a drug which can prevent cell division non-selectively may be responsible for interfering with the repair of uterine blood vessels and the mucosa which could result eventually in aberrant repair and therefore transformation to a malignant state. A third possibility would be that the carcinogenic effect seen in the females is a combination of both the above reasons. A further objection to the claim that ICRF-159 is carcinogenic is that the animals had to be treated virtually throughout their life span with large doses of ICRF-159 in order to obtain the small number of tumors which have been seen. Since it is not anticipated to give ICRF-187 at a very high dose or for very long, even these eratic results cannot be taken as evidence that there may be some carcinogenic problems with ICRF-187.

Prolonged continued use of high doses of almost any cytotoxic drug may produce tumours in animals and problems in patients, and several cases of acute leukaemia have been observed with ICRF-159 when taken continuously for more than 18 months. In all these cases leukopenia was a constant feature, but with doses that do not have this effect, the chances of developing any form of blood dyscrasia would be expected to disappear. Moreover, the acute leukaemias which have been seen following prolonged use of ICRF-159 have been of the myeloid type and this series of cells are the most sensitive to any drug which inhibits cell division.

Clinical implications. Since ICRF-187 (dexrazoxane) is given together with doxorubicin that is itself a powerful mutagen and carcinogen it does not seem relevant, at present, to investigate the carcinogenic potential of ICRF-187 further.

3.3.7 Pharmacodynamics

ICRF-187 has no known pharmacologic activity on the cardiovascular, neurological, respiratory, urinary or gastrointestinal systems at therapeutic doses. On the other hand, ICRF-187 has a number of well documented pharmacodynamic effects which are potentially of considerable therapeutic interest. Thus, it has:

1. A broad spectrum inhibitory effect on cell division. This cytostatic effect is most marked on lymphoid, haematopoietic, gonadal and mucosal tissues, but the degree of activity depends on dosage and length of exposure to the drug.
2. It blocks cell division at the end of G2/M – a point at which most tumour cells become more radiosensitive.
3. ICRF-187 (like ICRF-159) potentiates the activity of some other cytotoxics. This is because most of those agents block cell division at other parts of the cell cycle. The result is a better therapeutic index, i.e. the activity is potentiated to a greater degree than the toxicity [4, 14].
4. ICRF-187 (as well as ICRF-159) are potent suppressors of tissue damage caused by several anticancer drugs [4, 14, 15] especially the anthracyclines doxorubicin and daunorubicin. These clinically useful drugs have as their dose limiting toxicity an insidious chronic cardiac toxicity which ultimately causes congestive cardiac failure with high mortality.

ICRF-187 has been studied in a number of laboratories as a cardioprotector and has been found to give effective protection against the cardiotoxic effects produced by doxorubicin and daunorubicin. This protection is obtained without interfering with the anticancer activity of these drugs. The cardioprotective effect of ICRF-187 has been clearly demonstrated in a number of mammalian species [4], an overview of which is given in Table 3.2.

There are two distinct phases of tissue damage caused by the anthracyclines. The acute phase causes an indiscriminate destructive effect especially in the subcutaneous tissues when doxorubicin accidentally extravasates. Experimentally, more precise damage can be seen in the small gut which can lead to complete sloughing off of the mucosa to leave a denuded base. This understandably leads to rapid death owing to the inability of the animal to feed [15]. The acute damage to the gut produced by daunorubicin can be prevented by ICRF-159 and the mortality drastically reduced [15].

Doxorubicin-induced cardiotoxicity may also have an acute component which can be picked up by changes in the EKG's and which can be prevented by ICRF-159 [16]. The chronic phase is seen in the slow development of anthracycline cardiotoxicity which is the result of a dose related cumulative exposure of the myocardium to doxorubicin resulting inexorably in cardiac failure. This can be largely prevented by ICRF-187 (or ICRF-159).

It must be stressed, however, that an optimal dosage scheme has yet to be discovered. It is also worth noting that most of the studies have employed single doses of ICRF-187 at some time close to the injection of doxorubicin, and there may be scope for further optimization by a more comprehensive experimental design. There is also the possibility of further bio-chemical elucidation of the mechanisms involved in the cardioprotection by ICRF-187 which may in turn lead to a better utilization of the drug.

In this connection it is known that doxorubicin can produce considerable nephrotoxicity in rats resulting in severe proteinuria. This toxic effect can also be prevented to a large extent by the prior administration of ICRF-159 (Hellmann, Finch, unpublished). It may be that the kidney and the cardioprotective effect have similar basic biochemical mechanisms and that this tissue protection extends to other anticancer drugs.

Summary

ICRF-187 protects against the dose limiting cardiotoxicity of doxorubicin in a variety of species including humans. It does so without adding any toxicity of its own or interfering with the anticancer activity of doxorubicin.

3.3.8 Pharmacokinetics

This area is dealt with in depth in the Section 3.2.

In supplementation to this section, it may be added that children seem to have a higher tolerance to ICRF-187 than adults. In a phase I study, ICRF-187 was given as a 2-h iv infusion daily for 3 days in 46 evaluable pediatric patients. The maximum tolerated dose was 3,500 mg/m^2/day $\times$ 3 based on changes in hepatic function and coagulation abnormalities encountered when larger dosages were administered. Pharmacokinetic analysis showed that the children have a larger volume of distribution per kilogram of body weight in the central compartment and total body and a more rapid totalbody clearence than adults. These parameters can explain only part of the increased tolerance of children to ICRF-187 [17].

3.3.9 Local Tolerance

No definitive preclinical studies have been undertaken to examine the local effects of ICRF-187. Formal phase I studies in man showed first no untoward local effects of the injection of ICRF-187 [10]. Recently however, reports on pain at the site of injection and venous thrombosis have appeared in the literature (see Section 2.3.5). Experimentally, intraperitoneal injections were without evidence of local intolerance.

References

1. Speyer J, Green MD, Kramer E et al (1988) Protective effect of the bispiperazinedione ICRF-187 against doxorubicin-induced cardiac toxicity in women with advanced breast cancer. N Engl J Med 319(12):745–52
2. Bakowski MT (1976) ICRF 159, (+/–) 1,2 bis (3,5-dioxopiperazin-1-yl) propane, NSC 129943; razoxane. Cancer Treat Rev 3:95–107 (Review)
3. Repta AJ, Baltezor MJ, Hansal PC (1976) Utilization of an enantiomer as a solution of a pharmaceutical problem: application to solubilization of 1,2-di (4-piperazine-2,6-dione)propane. J Pharma Sci 65(2):238–42
4. Herman EH, Witiak DT, Hellmann K, Waravdeker VS (1982) Biological properties of ICRF-159 and related bis(dioxopiperazine) compounds. Adv Pharmacol Chemother 19:249–90
5. Creighton AM, Hellmann K, Whitecross S (1969) Antitumor activity in a series of bis-diketopiperazines. Nature 222:384–5
6. Hellmann K (1975) Chemother Congress Proceedings, London, Plenum Press
7. Levine BS, Henry MC, Port CD, Rosen E (1980) Preclinical toxicologic evaluation of ICRF-187 in dogs. Cancer Treat Rep 64(12):1211–5
8. Final Report (7th Oct 1977), IIT Research Institute (unpublished)
9. Gralla EJ, Coleman GL, Jonas AM (1974) Preclinical toxicology studies with ICRF-159 (NSC-129943) – a new antineoplastic drug. Cancer Chemother Rep 35(1):1–7
10. Vogel CL, Gorowski E, Davila E et al (1987) Phase I clinical trial and pharmacokinetics of weekly ICRF-187 (NSC 169780) infusion in patients with solid tumors. Invest New Drugs 5(2):187–98
11. Duke DI (1975) Prenatal effects of the cancer chemotherapeutic drug ICRF-159 in mice, rats, and rabbits. Teratology 11(1):119–26
12. Albanese R, Watkins PA (1985) The mutagenic activity of razoxane (ICRF-159): an anticancer agent. Br J Cancer 52(5):725–31
13. U.S. Department of Health, Education and Welfare (1978) Bioassay of ICRF-159 for possible carcinogenicity. Department of Health, Education and Welfare, Washington, Publication Number: NIH 78-1329
14. Poster DS, Penta JS, Bruno S, MacDonald JS (1981) ICRF-187 in clinical oncology. Cancer Clin Trials 4(2):143–6
15. Wang G, Finch MD, Trevan D, Hellmann K (1981) Reduction of daunomycin toxicity by razoxane. Br J Cancer 43(6):871–77
16. Tian Hu S, Brändle E, Zbinden G (1983) Inhibition of cardiotoxic, nephrotoxic and neurotoxic effects of doxorubicin by ICRF-159. Pharmacology 26(4):210–20

17. Holcenberg JS, Tutsch KD, Earhart RH et al (1986) Phase I study of ICRF-187 in pediatric cancer patients and comparison of its pharmacokinetics in children and adults. Cancer Treat Rep 70(6):703–9

3.4 Identification of Dexrazoxane as a Cardioprotector

Eugene H. Herman

This review is dedicated to the memory of Dr. Victor Ferrans who was a superb cardiac pathologist and served as an inspirational mentor for numerous cardiotoxicity studies.

Anthracyclines such as doxorubicin and daunorubicin were introduced into clinical use over 40 years ago. These agents were found to possess a broad spectrum of antitumor activity and as a result have been widely used against both hematologic and solid neoplasms. Cardiotoxicity is one of the most important dose-limiting toxicities that is associated with clinical use of anthracycline-type antitumor agents. The cardiotoxic effects of anthracyclines were not detected in preclinical studies but was first reported during early clinical trials. Attempts to ameliorate anthracycline cardiotoxicity were directed toward: (1) decreasing myocardial concentration of the agents, (2) developing less cardiotoxic analogues and (3) identifying other compounds which will block or overcome the adverse myocardial effects of anthracyclines. Studies exploring the possibility that certain agents might exert cardioprotective activity could only be initiated once animal models of anthracycline cardiotoxicity were developed.

The characteristics of anthracycline cardiotoxicity described in early clinical studies included hypotension, tachycardia with or without arrythmia and congestive heart failure. Initial experimental studies sought to determine whether these types of cardiac alterations could be reproduced in animals. Early experimental studies showed that injections of high doses of daunorubicin could induce acute cardiac arrhythmias in Syrian Golden hamsters and Rhesus monkeys. Subsequent investigations determined that pretreatment with certain types of adrenergic blocking agents either prevented or increased the arrhythmic dose of daunorubicin thus indicating that this acute toxicity might in part be mediated by some component of the sympathetic nervous system.

Isolated perfused hearts were used as a model to identify potential direct cardiotoxic effects of anthracyclines. In this model both daunorubicin and doxorubicin were found to cause a significant increase in coronary perfusion pressure (increased vascular resistance). Pretreatment with agents previously been found to prevent or suppress daunorubicin-induced arrhythmias in hamsters and monkeys had no effect on the changes in perfusion pressure. However, increases in coronary perfusion pressure were found to be negligible or absent when the hearts were exposed to ethylendiaminetetraacetic acid (EDTA). At that time two EDTA-related Imperial Cancer Research Fund compounds (ICRF 159 'razoxane' and 187 'dexrazoxane') were being evaluated in the National Cancer Center program for antitumor activity. Surprisingly, pretreatment with either razoxane or dexrazoxane attenuated anthracycline-induced coronary pressure increases in the isolated heart preparation.

The information gained from the isolated heart experiments began to be explored in the hamster model given high intravenous (iv) doses of daunorubicin (the same high iv doses of doxorubicin caused lethality during injection). In these studies animals were pretreated with a single dose of razoxane or dexrazoxane (intraperitoneal (ip)) 30 min prior to daunorubicin. Pretreatment with razoxane or dexrazoxane resulted in an increase in the numbers of hamsters surviving as well as a decrease in the toxicity observed in animals given daunorubicin alone. Pretreated hamsters had less severe cardiac alterations than did animals given daunorubicin alone. Alterations in the heart and non cardiac tissues such as the kidney and liver did not appear to be sufficiently severe to account for the high mortality. However, severe gastrointestinal damage was noted in hamsters given daunorubicin alone. These lesions were much less severe in animals pretreated with razoxane or dexrazoxane.

The reduction in acute high dose anthracycline toxicity provided the first evidence that razoxane and dexrazoxane can exert in vivo protective activity. However, the cardiac alterations that limit clinical use were induced by treatment regimens which involved chronic administration of low anthracycline doses. The first attempt to reduce chronic anthracycline cardiac toxicity was reported in rabbits given daunorubicin at 3 week intervals (rabbits were unable to tolerate comparable doses of doxorubicin) with or without dexrazoxane (because of increased water solubility, dexrazoxane was used exclusively in chronic anthracycline studies). After 5 treatment periods, myocardial lesions similar to those found in the hearts of patients with anthracycline cardiotoxicity were observed. These lesions were present in the hearts of all rabbits given daunorubicin alone. When rabbits were pretreated with dexrazoxane myocardial alterations were either absent or significantly less severe than those in the animals receiving daunorubicin alone. These studies showed that cardioprotective activity could be demonstrated up to a period shortly after the final dose of daunorubicin.

The persistence of dexrazoxane cardioprotective activity was addressed in another rabbit study. In this study, rabbits were treated with 3.2 mg/kg daunorubicin, with or without 25 mg/kg dexrazoxane once every 3 weeks over a period of 18 weeks (6 doses). Three months after the last dose, the hearts of all 7 rabbits receiving daunorubicin alone had myocardial alterations of varying severity. In contrast, the hearts from five of seven rabbits pretreated with dexrazoxane were normal and the hearts from the other two rabbits in the group showed only minimal changes. The results of these studies indicate that dexrazoxane provides prolonged protection against anthracycline-induced cardiomyopathy, rather than causing only a delay in the appearance of cardiac alterations.

The observations in rabbits led to efforts to determine whether dexrazoxane would also protect against doxorubicin cardiotoxicity. This question was addressed in beagle dogs that were treated weekly with doxorubicin (1 mg/kg; cumulative dose

15 mg/kg) or 1.75 mg/kg doxorubicin every 3 weeks (cumulative dose 14.4 mg/kg) either alone or 30 min after dexrazoxane (12.5 mg/kg). In each of these three models, both the incidence and severity of the myocardial lesions were significantly attenuated in animals given doxorubicin in combination with dexrazoxane. In a fifth animal model, chronic weekly treatment with doxorubicin (1 mg/kg $\times$ 12 weeks) to male spontaneously hypertensive rats (SHR) and genetically related Wistar-Kyoto rats (WKY) caused cardiomyopathy and nephropathy. The cardiotoxic effects of doxorubicin were more severe in SHR (confirming observations of increased doxorubicin sensitivity in hypertensive patients). Pretreatment with dexrazoxane significantly decreased the severity of the myocardial lesions in both SHR and WKY rats.

Studies were undertaken to determine whether the schedule of dexrazoxane treatment is an important factor in exerting cardioprotective activity. Beagle dogs were given either doxorubicin (1.75 mg/kg) alone, simultaneously with dexrazoxane (35 mg/kg) or followed 2 h later by dexrazoxane at 3 week intervals. Examination of the myocardium after 7 treatments revealed that in comparison with animals given doxorubicin alone, the incidence and severity of doxorubicin-induced myocardial alterations were significantly reduced in the two groups treated with dexrazoxane and doxorubicin. However, the cardioprotection was significantly better in dogs given the dexrazoxane simultaneously with doxorubicin (in this instance the degree of protection was similar to that observed when dexrazoxane is given 30 min prior to doxorubicin) than in those where dexrazoxane was given 2 h after doxorubicin. Thus, the timing of the administration of the two agents is an important factor in maximizing the degree of cardioprotection.

The dog model was used to determine whether the use of dexrazoxane influences the tolerated cumulative dose of doxorubicin. In this study Beagle dogs were treated, at 3–4 week intervals, with doxorubicin (1.75 mg/kg) either alone or 15 min after dexrazoxane (25 mg/kg). Animals ($N = 8$) administered doxorubicin alone tolerated only seven to eight doses (cumulative doses of 12.25–14 mg/kg) and had severe myocardial alterations. Both the incidence and the severity of cardiac lesions were significantly attenuated in the animals given the combination of dexrazoxane and doxorubicin. Of the five animals that received 20–25 doses of doxorubicin (cumulative dose of 35 and 43.75 mg/kg), four had no myocardial lesions and one had minimal alterations. Of the three dogs that received 52.5 mg/kg doxorubicin (30 doses), one had minimal lesions and two had lesions of moderate severity. This study showed that dexrazoxane provided significant protection when given with doxorubicin and made it possible to give cumulative doses of the latter drug, which otherwise would have been lethal.

These and other experimental studies served as an impetus for the eventual evaluation of dexrazoxane as a protector against doxorubicin cardiotoxicity in human patients. In 1984 the first clinical trial (women with breast cancer) of dexrazoxane was initiated at New York University. This trial confirmed the cardioprotectant activity of dexrazoxane and showed that its use neither increased noncardiac toxicities nor altered the antitumor activity of the chemotherapy medication. Further trials which began in 1988, also in women with breast cancer, confirmed that dexrazoxane exerted significant cardioprotective activity as measured by non-invasive testing and congestive heart failure. At present dexrazoxane is the sole clinically approved agent used to attenuate doxorubicin cardiotoxicity in human patients.

More detailed information regarding the investigations cited in this review and other studies related to dexrazoxane cardioprotection can be found in [1–4].

References

1. Herman EH, Ferrans VJ (1998) Preclinical animal models of cardiac protection from anthracycline-induced cardiotoxicity. Semin Oncol 25(Suppl 10):15–21
2. Hasinoff BB, Hellmann K, Herman EH, Ferrans VJ (1998) Chemical, biological and clinical aspects of dexrazoxane and other bisdiketopiperazines. Curr Med Chem 5(1):1–28
3. Herman EH, Ferrans VJ (1998) Animal models of anthracycline cardiotoxicity: Basic mechanisms and cardioprotective activity. Prog Pediatr Cardiol 8:49–58
4. Herman EH, Ferrans VJ and Sanchez J (1992) Methods of reducing anthracycline cardiotoxicity. In: Muggia F, Green MD, Speyer J (eds) Cancer treatment and the heart. Johns Hopkins Press, Baltimore, pp 114–69

3.5 Dexrazoxane as Antitumour Agent

W. Rhomberg

The following contributions deal mainly with the protection from anthracycline-induced cardiotoxicity and the various aspects of tissue protection by DXRz. For the purpose of completeness, however, a short note on what has been found on DXRz as antitumor agent is preceding these articles.

The clinical development of dexrazoxane (DXRz; ICRF-187) was started off by using it as an antineoplastic agent [1]. Phase I trials with the agent began already in 1979 when first hints of antitumor activity were noted.

Dexrazoxane as Cytotoxic Agent

In a phase I pharmacokinetic trial, dexrazoxane was given as 96-h continuous infusion to 37 patients with advanced malignancies [2]. DXRz plasma clearence (ss) and elimination $t(1/2)$ were 7.2 ± 1.6 l/h/m(2) and 2.0 ± 0.8 h, respectively. In this study, no data on tumour responses were given, but it was stated that the infusion can be safely administered.

In 1981, a phase I evaluation of the drug was performed by Liesmann et al. in patients with advanced malignancy using a dose schedule of 200 mg/m^2 daily for 5 days every 3 weeks. The tumours included 3 soft tissue sarcomas, 6 carcinomas of the lung, 3 colorectal carcinomas, and 6 others. Among these 18 patients, no objective remissions were achieved, and even no stabilization of disease was observed [3].

Some years later, a phase II trial of DXRz was done in patients with AIDS related *Kaposi's sarcoma*. Thirteen patients entered on this study, eight patients had received prior chemotherapy for AIDS-Kaposi-Sarcoma. The dose of ICRF-187 was

1,000 mg/m^2 IV daily for 3 days every 3 weeks. There were no complete responses, and one partial response lasting 6 months was seen [4]. Toxicity was significant, and of the first 5 patients treated, 3 out of 5 had grade III or IV neutropenia. Because of this, subsequent patients received 800 mg/m^2 IV days 1–3 if previously untreated. Activity of DXRz in this disease seems to be much lower compared with the results achieved by Olweny et al. with oral razoxane in endemic Kaposi's sarcoma (see Section 2.3.1.2)

A larger phase II trial of DXRz was initiated by Vats et al. in 31 children with *solid tumours* and 35 with acute *leukaemia* [5]. Among the solid tumours there were 12 Ewing's sarcomas, 2 neuroblastomas, 3 lymphomas and 14 other solid tumors. DXRz was administered at a dose of 3 g/m^2/day for 3 days as a 4 h infusion each day. In patients with leukaemia (22 ALL, 13 AML), no objective response was seen in the bone marrow although 7 patients had a decrease in peripheral blast count. There were no objective responses seen in patients with a solid tumour. Due to the rarity of the pediatric solid tumours statistical significance can not be attributed to response in all the solid tumours except *Ewings's sarcoma* where no response was seen in 12 patients.

DXRz was well tolerated at this dose level, the major toxicity was hematopoietic depression. Significant but rare toxicites included moderate to severe nausea and vomiting, and elevation of bilirubin and transaminases. The toxicity was graded according to Pediatric Oncology Group criteria as grade I (mild), grade II (moderate), grade III (severe), and grade IV (life-threatening). The authors concluded: Although inactive in the current study, DXRz might be more active in another schedule [5].

In a phase II trial of the Southeastern Cancer Study Group, no responses were seen among 40 patients with advanced *renal cell carcinomas* when DXRz was given at a dose of 3.8 g/m^2/week iv for four doses, followed by a 2-week rest, followed by a second course with a 25% increase in dose if there had been no grade 2 neutrophil toxicity [6].

Dexrazoxane was also administered in *non-small cell lung cancer*. In a study of Natale et al. [7], ICRF-187 at a dose of 1,250 mg/m^2 daily over 3 days, repeated every 3 weeks, was given to 29 patients with measurable non-small-cell lung cancer (NSCLC). There were 23 males and 6 females. Among 25 assessable patients including 15 who had not received prior chemotherapy, no objective responses were seen. ICRF-187 seems not to be active in NSCLC.

Indirect drug testing occurred in a study of 10 patients with malignant *mesotheliomas* when DXRz was used as cardioprotector together with high-dose doxorubicin and GM-CSF [8]. Doxorubicin is the most widely studied agent for the treatment of malignant mesothelioma, and it is associated with a moderate response rate of approximately 17% when given in conventional doses. There were no objective responses observed in this study, and moreover, it was stated that high-dose doxorubicin administered with DXRz is unacceptably toxic in this patient population. The median survival of the ten patients was 4.8 months from study entry, with 40% of patients surviving 6 months and 20% surviving 1 year [8].

Although dexrazoxane was not systemalically tested in other major tumor entities as yet, the drug seems to be even less active than razoxane in terms of remission induction in human malignant tumours.

Dexrazoxane as Radiosensitizing Agent

Dexrazoxane has been successfully used as cardioprotective drug in the treatment with anthracyclines for many years but it has not at all been tested as radiosensitizing agent in human malignancies. However, in view of the presently limited availability of razoxane-tablets which are unreplacable as a radiosensitizing agent, it is strongly recommended to test the (+) enantiomer dexrazoxane (Zinecard®, Cardioxane®) as radiosensitizer and/or antimetastatic drug. In practice, this clearly requires the availability of an oral form of dexrazoxane in order to give daily metronomic doses of the drug.

In preclinical testing, dexrazoxane was shown to exhibit radiopotentiating activity when given together with etoposide [9]. Radiotherapy and anthracyclines with DXRz (here given as an anthracycline cardioprotector) seem to be well tolerated, and hitherto no unusual interference has been reported [10, 11].

Antimetastatic Activity

Dexrazoxane has not been tested clinically for its antimetastatic activity. In view of the antimetastatic activity of razoxane, when given together with vinca-alkaloids, and the limited availability of razoxane tablets (at least at present) it would be of some clinical value to be able to substitute oral DXRz for razoxane, and to test this drug for antimetastatic activity.

This view is strengthened since there is a known relation between the antimetastatic activity of razoxane and its antiangiogenic activity which is directed against the development of pathological tumour blood vessels, and furthermore, since it was recently shown that dexrazoxane was also strongly antiangiogenic if administered in small repeated doses [12].

Summary and Comment

Dexrazoxane given as single agent shows essentially no cytotoxic activity in the screening trials so far performed, but in contrast to razoxane (the racemic form of the molecule), it has as yet not been tested for potential radiosensitizing or antimetastatic activities.

Little is known about its interaction with various chemotherapeutic agents but dexrazoxane demonstrated profound synergistic cytotoxic effects in leukemic cell lines in vitro together with daunorubicin and cytosine arabinoside [13] suggesting that DXRz is not antagonistic to drugs used in the treatment of leukaemia. The present clinical experience confirms that DXRz given as cardioprotectant does not interfere with the antileukemic effect of doxorubicin [14]. Treatment data on breast cancer and soft tissue sarcomas support this view of a non-interference with contemporary drug regimens. There is only one unconfirmed report on a reduced response rate in breast cancer if DXRz was added to anthracycline based therapy, and this report is also not supported by 2 independent meta-analyses.

References

1. Von Hoff DD (1998) Phase I trials of dexrazoxane and other potential applications for the agent. Semin Oncol 25(4 Suppl 10):31–6
2. Tetef ML, Synold TW, Chow W et al (2001) Phase I trial of 96-hour-infusion of dexrazoxane in patients with advanced malignancies. Clin Cancer Res 7(6):1569–76
3. Liesmann J, Belt R, Haas Ch, Hoogstraten B (1981) Phase I evaluation of ICRF-187 (NSC-169780) in patients with advanced malignancy. Cancer 47:1959–62
4. Chachoua A, Green M, Wernz J, Muggia F (1989) Phase II trial of ICRF-187 in patients with acquired immune deficiency related Kaposi's sarcoma. Invest New Drugs 7:327–31
5. Vats T, Kamen B, Krischer JP (1991) Phase II trial of ICRF-187 in children with solid tumours and acute leukaemia. Invest New Drugs 9:333–7
6. Brubaker LH, Vogel CL, Einhorn LH, Birch R (1986) Treatment of advanced adenocarcinoma of the kidney with ICRF-187: a Southeastern Cancer Study Group trial. Cancer Treat Rep 70(7):915–6
7. Natale RB, Wheeler RH, Liepman MK, Sauder A, Bricker L (1983) Phase II trial of ICRF-187 in non-small cell lung cancer. Cancer Treat Rep 67(3):311–3
8. Kosty MP, Herndon JE, Vogelzang NJ, Kindler HL, Green MR (2001) High-dose doxorubicin, dexrazoxane, and GM-CSF in malignant mesothelioma: a phase II study-Cancer and Leukemia Group B 9631. Lung Cancer 34(2):289–95
9. Hofland KF, Thougaard AV, Dejligbjerg M et al (2005) Combining etoposide and dexrazoxane synergizes with radiotherapy and improves survival in mice with central nervous system tumors. Clin Cancer Res 11:6722–9
10. D'Adamo DR, Anderson SE, Albritton K et al (2005) Phase II study of doxorubicin and bevacizumab for patients with metastatic soft-tissue sarcomas. J Clin Oncol 23:7135–42
11. De Giorgi U, Giannini M, Frassinetti L et al (2006) Feasibility of radiotherapy after high-dose dense chemotherapy with epirubicin, preceded by dexrazoxane, and paclitaxel for patients with high-risk stage II-III breast cancer. Int J Radiat Oncol Biol Phys 65(4):1165–9
12. Maloney SL, Sullivan DC, Suchting S, Herbert JMJ, Rabai EM, Nagy Z, Barker J, Sundar S, Bicknell R (2009) Induction of thrombospondin-1 partially mediates the antiangiogenic activity of dexrazoxane. Br J Cancer 101:957–66
13. Budman DR, Calabro A, Kreis W (2001) In vitro effects of dexrazoxane (Zinecard) and classical acute leukemia therapy: time to consider expanded clinical trials? Leukemia 15(10):1517–20
14. Moghrabi A, Levy DE, Asselin B et al (2007) Results of the Dana-Farber Cancer Institute ALL Consortium Protocol 95-01 for children with acute lymphoblastic leukemia. Blood 109(3):896–904

3.6 Protection Against Anthracycline-Induced Cardiotoxicity. Clinical Aspects

Clinical trials with toxicity protectors are complex. They must include sufficient dose-limiting events for study, and have to assess both the toxicity of the basic treatment and the protector, and antitumor effects. The first paper in this section is a report and critical analysis by R. Rubens on two pivotal studies showing the efficacy of dexrazoxane as cardioprotector. This article is part of records that were used to get dexrazoxane licenced. It is followed by a comment of K. Hellmann on the important 'Swain-trials' in patients with breast cancer, and finally by an update of the clinical trials of dexrazoxane for prevention of anthracycline-induced cardiotoxicity until 2010, written by R. Jones.

3.6.1 Two Pivotal Studies of Dexrazoxane as Cardioprotector: A Report Including Pharmacology and Safety Issues

R. Rubens

Problem Statement

Anthracyclines are among the most utilized antitumour drugs worldwide. The anthracycline doxorubicin has a wide spectrum of activity and is used with a high degree of efficacy in many human cancers [1, 2].

Many patients needing systemic chemotherapy for cancer receive doxorubicin at some time during their clinical course. An obstacle to the use of anthracyclines, however, is a cumulative dose-limiting cardiotoxicity [3–7]. This toxicity is characterized by diffuse myocardial injury which can result in a dose related irreversible cardiomyopathy. The probability that clinical evidence of congestive heart failure will develop when cumulative doses of doxorubicin rise above 450 mg/m^2 is 10% [4, 8]. A progressive increase in the frequency of this complication is noted with higher cumulative doses of doxorubicin: a 20% probability at $550–600 \text{ mg/m}^2$ and a 41% probability at cumulative doses $>600 \text{ mg/m}^2$ [4, 8].

Other factors that increase the risk of cardiomyopathy are an age of more than 65 years, prior mediastinal irradiation, preexisting cardiac disease, hypertension and previous chemotherapy with alkylating agents [3].

Because of this dose related cardiotoxicity the usual clinical practice is to terminate doxorubicin treatment at an empirically derived cumulative dose of $500–550 \text{ mg/m}^2$. This limit should be 400 mg/m^2 in patients with known risk factors. Since patients with preexisting cardiac disease are at high risk of developing cardiotoxicity even at lower doses, it is recommended not to start doxorubicin in these patients.

Discontinuing or withholding an (effective) drug therapy at an arbitrary level of 550 mg/m^2 has therapeutic implications. For example, von Hoff et al. reported in a retrospective analysis of 3,941 patients with different types of cancer, that doxorubicin chemotherapy was withdrawn in 185 patients, because the total cumulative dose of 550 mg/m^2 was reached [3]. Of these 185 patients 19% were in complete remission from their disease, 28% were in partial remission and 39% were classified as having stable disease. It is possible that these remissions might be prolonged with additional doxorubicin.

Therefore, several attempts have been made to ameliorate, prevent or at least delay the onset of anthracycline related cardiotoxicity. One approach is to alter the dosing schedule of doxorubicin from once every 3 weeks to weekly regimens or slow infusions. Lowering the peak blood levels of the drug clearly appears to offer a means of reducing cardiac toxicity. In a number of studies less clinical cardiac toxicity was claimed with no loss of antitumour effect, however, conflicting data were reported by others [8–11]. A disadvantage of the described regimens is that these may be less amenable to the out-patient setting. Continuous infusion necessitates indwelling catheters and pumps which can be uncomfortable and expensive and increase the risk of bacterial sepsis.

Another approach to the problem has been attempts to develop anthracycline analogues that retain the antitumour effect but do not cause cardiac toxicity. A number of analogues have shown promise but none have clearly demonstrated equivalent antitumour activity without cardiotoxicity [7]. To date more than 16 such analogues have been described. With 4'-epi-adriamycin (epirubicin), which is one of the most promising analogues, cardiotoxicity appears at a higher dose, but is still observed [12]. The higher dose is required to obtain the same degree of antitumour activity as doxorubicin.

A third approach is the development of cardioprotective agents. This became of interest when it was demonstrated that cardiotoxicity and antitumour activity of anthracyclines were mediated through separate mechanisms [13, 14]. Of the different compounds only the chelating agent ICRF-187, which is the subject of this review, has been shown to provide clinically significant cardioprotection against anthracycline induced cardiotoxicity.

The Clinical Development Program for ICRF-187

ICRF-187 was originally synthesized as a possible antitumour agent. It is the more soluble (+) enantiomer of the racemic mixture known as ICRF-159 or razoxane [15]. Razoxane was marketed by ICI Pharmaceuticals (in the 1970s–1990s – later by Zeneca) for the treatment of lymphomas, including mycosis fungoides, acute leukemias and Kaposi's sarcoma and in combination with radiotherapy for the treatment of all forms of soft tissue, chondro- and osteosarcomas [16]. Razoxane is administered orally. ICRF-187 demonstrated substantial antitiumour activity in different preclinical models, and a number of phase I cancer clinical trials were performed with the drug. An evaluation of the toxicity profile of the drug with different doses and treatment schedules is given in Report EC-900702 [17], and pharmacologic aspects of these initial phase I cancer clinical studies are described in Report EC-900701 [18].

Phase II cancer clinical trials with ICRF-187 were performed in non-small cell lung cancer, head and neck cancer, renal cancer, and AIDS related Kaposi's sarcoma. The drug, however, demonstrated only marginal antitumour activity. An evaluation of patients, activity and toxicity is provided in Report EC-900702 [17].

The clinical development program for dexrazoxane (Cardioxane®, ICRF-187) as a cardio-protector against anthracycline induced cardiotoxicity comprised two randomized phase III studies which are further outlined below. The trials involved women with advanced breast cancer, requiring doxorubicin chemotherapy. ICRF-187 for both trials was supplied by the National Cancer Institute, Bethesda, Md, USA.

The formulation of ICRF-187 that was marketed by EuroCetus B.V., Amsterdam, The Netherlands, for the prevention of doxorubicin induced cardiotoxicity (Cardioxane®), consists of the hydrochloride form of the drug. After reconstitution, and dilution in infusion fluids, the two formulations contain the same amount of unprotonated ICRF-187. Neither of the two formulations contains excipients or preservatives.

A single intravenous dose bioequivalence study in patients with advanced breast cancer comparing Cardioxane® with NCI-ICRF-187 (National Cancer Institute, USA) demonstrated that the two formulations were bioequivalent with respect to the investigated pharmacokinetic parameters elimination rate constant, Kel, half life, $t1/2$, mean residence time, MRT, area under the curve, AUC (0–48) and area under the curve extrapolated to infinity, AUCinf [19]. The design of the study was such that the study had sufficient sensitivity to detect potentially clinically significant differences in pharmacokinetic parameters. Based on the results, the two formulations Cardioxane® and NCI-ICRF-187 (Zinecard®, Pfizer), may be considered as therapeutically equivalent.

Clinical Pharmacology

The early clinical investigations provide pharmacodynamic and pharmacokinetic data on the use of ICRF-187 in patients with various types of cancer [17, 18]. All the studies were previously published in the literature. As the administration of an investigational cytotoxic drug to normal volunteers was considered to constitute a risk without conceivable benefit, no studies were performed in healthy subjects.

Pharmacodynamics

ICRF-187 is an analogue of ethylene diamine-tetraacetic acid (EDTA). Its mechanism of action as a cardioprotector is not fully elucidated, but it is postulated that ICRF-187 exerts its cardioprotective effect by chelating iron [20]. ICRF-187 is membrane permeable, and may act through its intracellularly formed ring opened hydrolysis product ICRF-198, which has an EDTA-type structure and is a strong chelator of iron and copper [21]. ICRF-198 itself is too polar to cross cell membranes. The uptake and subsequent hydrolysis of ICRF-187 to ICRF-198 in the myocardium protects the cellular components by reducing the amount of available iron and preventing the formation of free oxygen radicals forming metal ion-doxorubicin complexes [21]. In man, ICRF-187 causes an increase in urinary clearance of iron and zinc by an average factor of ten [22, 23].

Differences in intracellular metabolism and/or uptake of ICRF-187 by tumor cells and normal myocardial cells may explain the selective protection against doxorubicin-induced cardiotoxicity, without affecting the doxorubicin antitumor activity. In patients receiving ICRF-187, bone marrow studies show that the drug produces G2/M accumulation in the marrow [24]. The S-phase cells are depleted. The role of chelation in the antitumour effect of ICRF-187 remains unclear.

The dose and schedule of administration are important for the therapeutic effect of ICRF-187 as cardioprotector. Dosages of ICRF-187 as cardioprotector are considerably lower when compared to dosages of ICRF-187 used in the anticancer setting. In animal studies, cardiac protection was achieved with an ICRF-187 to doxorubicin ratio of 10–15: 1, and when the drug was administered 30 min before the anthracyclines [25–28]. Based on these studies a dose of 1,000 mg/m^2 of ICRF-187 was initially selected for clinical cardioprotective use, combined with the commonly used doxorubicin dose of 50 mg/m^2, though now 500 mg/m^2, i.e. 10 times the doxorubicin dose, is considered more appropriate. This dose is 4–5 times lower than the maximum tolerated dose in humans, as demonstrated in phase I clinical studies.

Pharmacokinetics

After a single intravenous injection of 14-C labelled ICRF-187 to mice and dogs (12.5 and 100 mg/kg body weight respectively), tissue distribution and elimination of ICRF-187 was rapid [29, 30]. A plasma $t1/2$ alpha of 3 min is observed in both dogs and mice. In dogs maximum concentrations of 100 and 31 μg/g tissue were seen in liver and kidney respectively within 2 h. The cardiac concentration did not reach these high levels, but gradually increased to 6 μg/g tissue during the same period. Thin layer chromatography and autoradiography of liver, kidney, plasma and urine from treated dogs revealed the presence of large quantities of unchanged parent compound and hydrolysis product.

Based on a pharmocokinetic study in 3 adult rhesus monkeys it can be considered unlikely that ICRF-187 will reach levels of clinical significance in the cerebrospinal fluid (CSF) of man. After 2 h iv infusions of 300 mg/m^2 14-C labelled ICRF-187, levels of CSF radioactivity in rhesus monkeys reached approximately 10% of plasma levels near the end of the 2 h infusion period [31].

A number of pharmacokinetic studies are evaluated that have been performed in adult and pediatric cancer patients. Different treatment schedules were used and dosages ranged from 60 to 7,400 mg/m^2.

Pharmacokinetics of ICRF-187 are independent of dose and schedule of administration, but do show some inter-individual variation. The reason for this is most likely the heterogenicity in the patient population treated with ICRF-187. There is no evidence for any difference in pharmacokinetics beteen women and men.

Serum pharmacokinetics follow a two compartment model [32, 33]. Mcan $t1/2$ alpha values are approximately 0.25 h (range 0.1–1.5 h) and $t1/2$ ß values 2.3 h (range 1.5–5.1 h). One group of investigators suggest the existence of a third 'deep' tissue compartment in which the drug is localized [33]. Although ICRF-187 is excreted in the bile, there is no evidence for an active excretion mechanism [32]. Apparently the drug passively equilibrates between serum and bile. ICRF-187 does not concentrate in pleural fluid [32].

A comparative study in children and adults demonstrates that children (age < 21 years) have a larger central volume of distribution per kilogram of bodyweight and a more rapid total body clearance than adults [34]. These results may in part explain the difference in toxicity patterns and maximum tolerated dose between adults and children. Distribution and elimination half-lives in children do not differ from values in adult patients.

There is no significant serum protein binding of ICRF-187; less than 2% of the drug is protein bound [32]. Total urinary drug recovery of unchanged drug is averaging 48% (range, 25–62%). In patients with low creatinine clearence drug bioavailability may be increased [32, 33].

In a study by Hochster et al. the pharmacokinetics of ICRF-187 were studied when the drug was given 30 min before the administration of doxorubicin [35]. ICRF-187 was given at four different dose levels (60, 300, 600 and 900 mg/m^2), to 4 male and 9 female cancer patients (3 patients per dose level), concomitant with a doxorubicin dose of 60 mg/m^2. The distribution half-life, $t1/2$ alpha, of ICRF-187 was 0.4 h and $t1/2$ ß 2.8 h. Total body clearence of the drug was 4.8 ml/min/kg. The data are consistent with those from studies in which ICRF-187 was used as a single agent, and demonstrate that doxorubicin does not influence the pharmacokinetics of ICRF-187.

Comment

The studies reported from clinical and preclinical investigations indicate similar qualitative and quantitative pharmacokinetics in man and the animal species studied.

Pharmacokinetics of ICRF-187 follow a two compartment model and are independent of dose and schedule, but do show some inter-individual variation. This variation may in part be explained by the heterogeneous characteristics of the patient population studied. At present no data are available on the metabolism of ICRF-187. The dosage schedule of ICRF-187 as a cardioprotector, 30 min before doxorubicin is based on the pharmacokinetic data. In view of the half-life of ICRF-187, a significantly longer waiting period before administering doxorubicin, could potentially lead to a decreased efficacy of ICRF-187 as a cardioprotector. The proposed mechanism of action requires the presence of ICRF-187 in the intracellular compartments of the myocardium, before doxorubicin can cause damage.

Since ICRF-187 is primarily excreted through the kidneys, renal impairment may increase the drug availability. At the dose of ICRF-187 for the cardioprotective indication (500 mg/m^2), an increase in plasma levels could enhance the myelosuppressive effect of the chemotherapy regimen. Special attention, therefore, should be given to patients with compromised renal function.

Clinical Trials (Efficacy of ICRF-187 as a Cardioprotector)

This section will primarily review the results of the controlled, randomized phase III studies demonstrating the efficacy and safety of ICRF-187 as a cardioprotector against doxorubicin induced cardiotoxicity [36, 37]. The database includes the 212 patients enrolled in two randomized trials. The protocols that were used by both study centers were virtually identical.

The first study was initiated in New York. The principal data were described in a report from the NY-University and is related to trial NYU 83–05 [36]. The other study (AVL 86-01) was initiated in Amsterdam as a European confirmatory study [37].

Trial NYU 83-05

Materials and Methods

Evaluation of Study Design and Assessment Procedures

Female patients with advanced breast cancer were randomized to receive a standard regimen of chemotherapy containing 50 mg/m^2 iv doxorubicin, 500 mg/m^2 iv cyclophosphamide and 500 mg/m^2 iv 5-FU, or the same chemotherapy regimen with concomitant administration of 1,000 mg/m^2 iv ICRF-187 given 30 min before chemotherapy. Patients were stratified according to cardiac risk factors and prior adjuvant chemotherapy.

Dose modifications for haematological toxicity, hepatic dysfunction and modifications for other toxicities as described in the protocol were in line with good clinical practice.

The primary endpoint of the study was the development of cardiac toxicity, which was either the development of clinical cardiac failure, a left ventricular ejection fraction (LVEF) on rest gated pool scan of less than 45%, a fall in resting LVEF of 20% over 2 successive courses of therapy, or a Billingham biopsy grade of 2 or above. Cardiac function under stress was not evaluated.

Each end point of cardiotoxicity was examined separately. The guidelines for cardiac monitoring of the patients and the cardiologic criteria for stopping treatment were in agreement with common clinical practice and published in the medical literature [6, 8]. One could argue whether a New York Heart Association (NYH Ass) grade I congestive heart failure is sufficient reason for removal from study, the more so as the criteria provided by the New York Heart Association may be subject to varying interpretation, limiting the accuracy and reproducibility. However, only two patients on the control arm were removed solely due to NYH Ass grade I cardiotoxicity. The five other patients on the control arm with NYH Ass grade I, all also had an abnormal LVEF, justifying treatment discontinuation.

Each patient underwent a detailed cardiac history and physical examination upon enrollment into the trial and at the intervals specified in the protocol, including 12-lead ECG's and chest radiographs obtained by standard methods. The analysis of the ECG's was detailed, including rate, rhythm, ectopics, conduction disturbance, chamber enlargement, evidence of myocardial infarction and repolarization abnormalities.

Emphasis was placed on the elimination of investigator bias. The chest radiographs and ambulatory ECG's were read in a blinded fashion. The histories and physical examinations were obtained by cardiac co-investigators who were blinded as to the treatment group.

The nuclear cardiac gated pool scans were used to assess cardiac function at baseline and when cumulative doxorubicin doses reached 300 and 450 mg/m^2 and each 100 mg/m^2 increment thereafter. All scans were performed in the same laboratory by the same group of cardiologists and nuclear medicine physicians. The patients were scanned at rest. The scanning procedure involved the injection of 99-Tc-labelled serum albumin into the antecubital vein. The patient was connected to an ECG and after a time for equilibration of the isotope, the patient was imaged in different views with a gamma scintillation counter. The gamma scintillation data and simultaneous ECG were fed into a computer where the entire cardiac cycle was analysed. Images of the central circulation were acquired during several hundred cardiac cycles at end diastole and end systole as determined from the ECG. By measuring changes in chamber volume the ejection fraction and ejection rate could be ascertained. Films of the images abtained were reviewed by the cardiologists, who were blinded to the patient's treatment group, to determine changes in ventricular contraction.

Endomyocardial specimens were obtained in the cardiac catheterization laboratory whenever the cumulative dose of doxorubicin reached 450 mg/m^2. In view of the invasive nature of the procedure biopsy was performed with separate informed consent as approved by the institutional review board. A number of patients did not consent to this procedure.

Right heart catheterization was performed with a triple linear thermodilution catheter, inserted through the internal jugular or femoral vein. With the position of the catheter confirmed fluoroscopically, five biopsy specimens of the right ventricular septum were obtained with a no. 9 French Standford endomyocardial bioptome and placed immediately in glutaraldehyde. An adequate biopsy sample was required to at least 2 mm in its greatest dimension.

Histologic sections were prepared and stained with hematoxylin and eosin and also prepared for electron microscopy. Doxorubicin-induced changes were assessed by the pathologist according to the Billingham scale. Interpretation of a minimum of 4 samples was done blinded as to the treatment group.

As defined by Billingham, a score of less than 2.5 is associated with a low risk of cardiac failure (<10%), while a score of 2.5 is associated with a 10–25% risk of developing cardiac failure [38, 39]. For this study a score of 2 was the endpoint in terms of cardiac pathology.

Comment: In general, the study design and cardiac assessment procedures as described in the protocol, assured a standardized, reproducible acquisition and analysis of the data. By assessing cardiotoxicity by separate teams, blinded to treatment assignment, investigator bias was eliminated. The study was well organized and reported in sufficient detail to allow objective evaluation.

Demographics

A total of 150 patients entered the study, 74 patients were randomized to receive chemotherapy only (control arm) and 76 patients to receive chemotherapy and concomitant ICRF-187 (ICRF-187 arm). For the two treatment arms, there were no significant imbalances for strata according to cardiac risk factors, age, weight, menstrual status, performance status and prior hormonal, cytotoxic, radiotherapeutic or surgical treatments. No imbalances existed between patients with locally advanced disease or those with metastatic disease. Sites of metastatic lesions were equally distributed. It shold be noted that 1 patient on the control arm and 3 patients on the ICRF-187 arm had a prior infarction while 5 patients on the control arm and 8 on the ICRF-187 arm had abnormalities in baseline ECG. However, these imbalances would weigh against ICRF-187, and are acceptable in this study.

Treatment

At cut-off date September 15, 1989 15 patients were still on study: 4 on the control arm and 11 on the ICRF-187 arm.

Considering all patients, the patients on the ICRF-187 arm received significantly more cycles of treatment (mean 11.0 vs. 8.6, $p = 0.009$) and this is also reflected in the cumulative dose of doxorubicin received (mean 502 vs. 397 mg/m^2, $p = 0.015$). The percentage of the total projected cumulative dose of doxorubicin is the same on the control and ICRF-187 arm.

However, the percentage of the projected cumulative dose of fluorouracil and cyclophosphamide was somewhat lower on the ICRF-187 arm (mean 80.8% vs. 87.6%, $p = 0.012$). This probably relates to ICRF-187 adding to the myelosuppressive effect of the chemotherapy regimen, necessitating dose reductions. Dose reductions were, according to the protocol, preferentially made for fluorouracil and cyclophosphamide only. However, the 7% difference in projected cumulative dose for fluorouracil and cyclophosphamide is of doubtful significance. Moreover, upon completion of treatment the total cumulative dose of these drugs is higher for the patients receiving ICRF-187, since they could receive more cycles of therapy. It seems unlikely that the slightly lower percentage of projected dose of fluorouracil and cyclophosphamide is clinically relevant. The equivalent antitumour efficacy in both treatment arms supports this view.

Results

Cardioprotection

A total of 35 patients developed cardiac toxicity while on therapy: 30 patients on the control arm and 5 patients on the ICRF-187 arm ($p < 0.0001$).

Of the 30 patients developing cardiotoxicity on the control arm, 25 patients had a LVEF <45% and/or a fall from baseline LVEF >20%. Fourteen of them also developed clinical signs of cardiotoxicity (NYH Ass >1). Of the 5 patients developing cardiotoxicity on the ICRF-187 arm, 4 patients had LVEF abnormalities of whom 1 patient had NYH Ass grade 2 clinical cardiotoxicity.

A total of 18 patients (24%) on the control arm developed clinical signs of cardiotoxicity during treatment. Of these patients, 9 patients (12%) had NYH Ass grade 3–4 congestive heart failure (CHF). Only 2 patients on ICRF-187 had clinical signs of cardiotoxicity (NYH Ass grade 2).

Long-term follow up of the patients who had come off study was very limited. For a variety of reasons such as death form progressive disease, intercurrent clinical progression or other medical problems, patients were lost for cardiac follow up. The data that were reported, included two patients without signs of clinical cardiac toxicity on the control arm at the time of their removal from study, who developed cardiotoxicity 1 and 3 months afterwards. Furthermore, the toxicity scores of 4 patients on the control arm who had developed mild NYH Ass grade 1 clinical cardiotoxicity subsequently worsened at follow up. No such delayed toxicity was observed in patients on the ICRF-187 arm. As a result, a total of 14 patients (19%) on the control arm developed NYH Ass grade 3–4 CHF and 3 patients (4%) had grade 2 CHF.

The two patients developing CHF on ICRF-187 had grade 2 toxicity. Since in particular NYH Ass grade 3–4 clinical cardiotoxicity markedly limits the physical performance status of patients, both the on study and follow up data indicate an improved quality of life in terms of physical activity for patients receiving concomitant ICRF-187.

A comparison of the change in LVEF per dose level, calculated for each patient as the difference from her own baseline value, is given in Table 3.3. The comparison shows a significant difference in fall of LVEF from baseline up to cumulative doses of 699 mg/m^2 of doxorubicin. Comparison at higher doses are not relevant as only 1 patient on the control arm received >699 mg/m^2 doxorubicin. The results are graphically presented in Fig. 3.3.

Endomyocardial biopsies were performed in 30 patients (16 on the control and 14 on the ICRF-187 arm), at a median cumulative doxoubicin dose of 452 mg/m^2. Two of the biopsies both in patients on the control arm, yielded tissue inadequate for evaluation.

Table 3.3 LVEF fall from baseline (comparison per dose level). Mean fall from baseline in EF units (number in brackets: LVEF fall as % of baseline value)

Dose range	mg/m^2 doxorubicin	N	Control	N	ICRF-187	N	All	p
Level 1	<400	57	5.5 (8.0%)	54	2.0 (2.4%)	111	3.4 (5.3%)	0.023
2	400–499	43	14.7 (20.7%)	38	2.7 (3.2%)	81	9.1 (12.5%)	0.0001
3	500–599	16	15.4 (22.8%)	33	1.9 (2.0%)	49	6.3 (8.8%)	0.0001
4	600–699	6	16.3 (24.2%)	23	0.6 (0.1%)	29	3.8 (5.1%)	0.0002
5	700–799	1	3.0 (4.3%)	17	4.7 (5.9%)	18	4.6 (5.8%)	
6	800–899	–	–	10	2.2 (2.2%)	10	2.2 (2.2%)	
7	900–999	–	–	8	2.1 (2.3%)	8	2.1 (2.3%)	
8	1,000–1,099	–	–	7	+0.4 (+1.8%)	7	+0.4 (+1.8%)	
9	1,100–1,199	–	–	3	3.0 (4.2%)	3	3.0 (4.2%)	

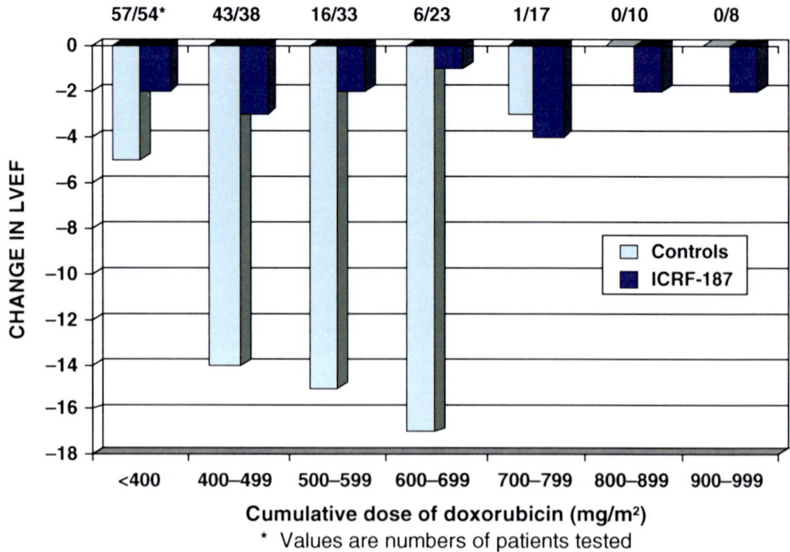

Fig. 3.3 LVEF fall from baseline per dose level doxorubicin control vs. ICRF-187

Billingham biopsy scores are given in Table 3.4. A chi squared test used to determine the significance of the differences in the severity of the cardiomyopathy scores between the two treatment groups, showed that these were significantly higher on the control arm ($p = 0.002$).

However, interpretation of the biopsy results is limited by the fact that only 22% of the patients on the control arm and 18% of patients on the ICRF-187 arm consented to this invasive procedure. In addition, patients with major signs of heart failure (NYH Ass grade 3 and 4, all on the control arm) were often unable to undergo biopsy.

The characteristics of the patients who underwent cardiac biopsy do not show evidence of a selection bias.

Table 3.4 Cardiac biopsy results (Billingham score)

	Control	ICRF-187	All	p
# Patients evaluated	16	14	30	
Median dose at Biopsy	452 (300–550)	491 (425–1,000)	471 (300–1,000)	N.S.
Billingham score: 0	4 (25%)	8 (54%)	12 (40%)	$p = 0.002$[a]
0.5	1 (6%)	0	1 (3%)	
1.0	3 (19%)	6 (46%)	9 (30%)	
1.5	0	0	0	
2.0	6 (37%)	0	6 (20%)	
N.E.	2 (13%)	0	2 (7%)	

[a]Note: X^2 for linear trend: (Mantel-Haenszel X^2).

Role of Cardiac Factors on the Development of Cardiac Toxicity

A subanalysis of the data to investigate the impact of preexisting cardiac risk factors (as defined in the study protocol), demonstrated that ICRF-187 cardioprotection was not limited to either patients with or without cardiac risk factors. Patients with cardiac risk factors are considered to be at high risk for developing doxorubicin induced cardiotoxicity, and it is recommended not to start treatment with doxorubicin without frequent, careful monitoring of cardiac function.

Based on the assumption in the medical literature that prior infarction, angina, rheumatic heart disease or abnormal baseline ECG are strong predictors of cardiotoxicity, when compared with age > 65 years and diabetes mellitus, a subanalysis of the patients with these preexisting cardiac risk factors was performed. All 5 patients on the control arm with these risk factors developed cardiotoxicity, whereas none of the 7 patients on the ICRF-187 arm with these risk factors developed cardiotoxicity. These findings, although limited by small patients numbers, suggest that ICRF-187 may reduce the impact of preexisting cardiac disease as a cofactor of increased risk of doxorubicin induced cardiotoxicity.

Antitumour Effect

ICRF-187 did not reduce the antitumour activity of the chemotherapy regimen. The observed rates of objective response and the time to disease progression were similar to those observed by other investigators in patients with metastatic breast cancer who were treated with the same regimen, including patients in whom adjuvant therapy failed [40].

Although progression free survival and actual survival are longer on the ICRF-187 arm this was not statistically significant. A therapeutic advantage would be expected only in the patients who received a dose of doxorubicin that was greater than the usual dose for stopping treatment, 450–550 mg/m^2. As the patients had metastatic disease, this dose was reached just after the median time to disease progression. Larger trials would be required to test for improved survival. It should be noted, however, that the present study was not designed to demonstrate such an effect of ICRF-187.

Reasons for Patient Removal from Study

Of the patients not receiving ICRF-187, 30 patients (43%) were removed due to the development of cardiotoxicity. Twenty-three patients (33%) were removed due to disease progression. The majority of patients on the ICRF-187 arm (42 patients; 65%) was removed due to disease progression, while only 5 patients (8%) had to come off study because they developed cardiotoxicity. Presumably, the reason for the differences in proportions with 'disease progression' is that the ICRF-187 patients could stay longer on treatment, and have more chance of reaching the end-point of progression. Treatment of 14% of the patients on the ICRF-187 arm was terminated at their own request, whereas 3% of patients on the control arm refused further treatment. No statement is made in the study reports on this difference between the treatment arms.

Comment: In general the data presented in the study report provide significant evidence for the clinical efficacy of ICRF-187 as a cardioprotector against doxorubicin induced cardiotoxicity in women with advanced breast cancer.

Report AVL 86-01

Materials and Methods

Evaluation of Study Design and Assessment Procedures

The second controlled randomized phase III trial was designed as a European confirmatory trial of the above study. The same protocol was used in both study centers. In view of the invasive nature of the cardiac biopsy procedure, it was left out in the assessment of doxorubicin induced cardiotoxicity.

Demographics

A total of 62 woman with advanced breast cancer were entered the study, 27 on the control arm and 35 on the ICRF-187 arm. For the two treatment arms there were no significant imbalances for age, weight, menopausal status, oestrogen receptor status, performance status and prior hormonal, cytotoxic, radiotherapeutic or surgical treatments.

The majority of patients had metastatic disease: 27 patients on the control and 33 patients on the ICRF-187 arm. Most patients had 1 preexisting cardiac risk factor (prior radiation to the mediastinum).

Treatment

At December 2, 1989, 12 patients were still on study: 5 on the control arm and 7 on the ICRF-187 arm.

Considering all patients, both patient groups received an equivalent number of treatment cycles (mean 6.6 vs. 6.0; N.S.) and equivalent cumulative doses of doxorubicin (mean 321 mg/m^2 vs. 295 mg/m^2; N.S.). The percentage of total projected cumulative dose of doxorubicin is the same on the control and ICRF-187 arms and is approximately 100% (95.6% vs. 97.7%; N.S.).

An imperfection of the data presented is that the percentages of projected cumulative dose of fluorouracil and cyclophosphamide are not included.

Results

Cardioprotection

The extent of cardioprotection by ICRF-187 was documented by the assessment of LVEF. A comparison of the mean fall in LVEF from baseline per dose level, showed that patients on the control arm had a seven to eightfold higher fall in LVEF at a cumulative doxorubicin dose of 400–499 mg/m^2 and a threefold higher fall at 500–599 mg/m^2.

However, the interpretation of the results is limited by the fact that cardiac function of a considerable number of patients on the control arm was not evaluated by assessment of LVEF, whereas all patients on the ICRF-187 arm were evaluated by LVEF. Nevertheless, the results provide supportive evidence for the cardioprotective properties of ICRF-187.

Only one patient on the control arm and none of the patients of the ICRF-187 arm had a fall from baseline LVEF to below 45%. None of the patients was removed from study due to cardiotoxicity. The majority of patients had disease progression causing removal from study, before any signs of cardiac failure were evident.

Antitumour Effect

That ICRF-187 did not reduce the antitumour activity of the fluorouracil, doxorubicin, cyclophosphamide chemotherapy is in agreement with the data of the New York study. Data on the progression free survival are not included in the report. Considering the fact that after a median of 5 cycles of chemotherapy the majority of patients on both treatment arms was removed from study due to disease progression, and considering the poor survival of 458+ days, the relatively young patient group (median age 45 years) obviously had a poor prognosis when compared to patients in the NYU study.

Comment: In general, the data presented in this study report are less powerful and less detailed than the data provided in report NYU 83 – 05, and because of its lower number of patients the study is statistically less robust than the first study. None of the patients on both treatment arms was removed from study due to cardiotoxicity. Compared to the patients in the New York study (NYU 83-05), the patients in this study had a poor prognosis and a majority of patients was removed after a median of 5 cycles of chemotherapy due to disease progression, before any signs of clinical cardiotoxicity were apparent. This obviously limits the cardiac evaluation of the study. Nevertheless the data on LVEF indicate that patients on the ICRF-187 arm had a better cardiac performance compared to patients on the control arm. The report, therefore, provides supportive evidence for the clinical efficacy of ICRF-187 as a cardioprotector against doxorubicin induced cardiotoxicity.

Safety of ICRF-187

Patient Population

Clinical experience obtained so far with ICRF-187 is described below and includes besides the cardioprotective clinical studies, a number of different phase I/II cancer clinical trials [17]. Including these patients, safety data are available for 486 patients. Of these, 212 patients received the proposed clinical cardioprotective dose of ICRF-187 (1,000 and 500 mg/m^2) in combination with doxorubicin (50 mg/m^2). Because the two randomized cardio-protective studies had a similar design, and monitored the same safety parameters, the safety data from these studies have been pooled for the following discussion. Similarly, the safety data of the phase I and II cancer clinical trials were also pooled.

Adverse Events in Cancer Clinical Trials

All reported phase I and II cancer clinical trials were published in the literature. The principal side effects of ICRF-187 in phase I studies, in a total of 167 patients with different types of cancer was transient, mild to moderate leukopenia, with nadir counts on day 12 and recovery within day 21. Mild and transient thrombocytopenia, nausea, vomiting, alopecia and transient elevations in SGOT, SGPT and/or bilirubin are noted at the MTD for different treatment schedules.

Other reported toxicities of ICRF-187 include malaise, low grade fever, increased urinary clearance of iron and zinc, anaemia, abnormal clotting, transient elevation of serum triglycerides and amylase and transient decrease in serum calcium.

It appears that a higher total dose per treatment cycle can be administered on a weekly $\times$ 4 schedule (1 h infusion; recommended doses 3,800–7,420 mg/m^2 weekly for 4 weeks), when compared to other schedules (daily for 3 days, daily for 5 days or 48 h continuous infusion every 21–28 days).

Unlike the dose-limiting myelosuppression observed in adult patients treated with high-dose ICRF-187, there was no consistent pattern of dose-dependent myelo-suppression in pediatric patients. Instead, the dose limiting toxicity was hepatic dysfunction, characterized by transient elevations in bilirubin and serum transaminase levels. These elevations were noted after the 2nd or 3rd dose and usually returned to within normal limits within 2 weeks. At a dose level of 3,000 mg/m^2 daily $\times$ 3, life-treatening hepatic dysfunction was observed accompanied by abnormal blood clotting. This was observed in 3 of 46 patients.

Toxicity patterns of ICRF-187 in phase II cancer clinical studies were similar as reported for the phase I studies. Because treatment intensity was less, however, toxicity patterns were less pronounced. In none of the phase II studies ICRF-187 demonstrated substantial antitumour efficacy, and the drug was not further developed as an anticancer agent.

Adverse Events in Cardioprotection Trials

The proposed clinical cardioprotective dose of ICRF-187 is 500 mg/m^2 in combination with 50 mg/m^2 of doxorubicin. This dose is 6 times lower than the MTD in humans.

The data presented in Report NYU 83-05 and AVL 86-01 demonstrate that concomitant administration of ICRF-187 and fluorouracil, doxorubicin and cyclophosphamide chemotherapy, did not increase the incidence of any clinical noncardiac toxic manifestations caused by the chemotherapy regimen [36, 37]. The incidence of mucositis, infection, fever, alopecia, nausea and vomiting was similar on the two arms of the study. Toxicities were scored by ECOG toxicity grading and reported giving the worst scores observed during any cycle of treatment. ECOG toxicity and response criteria are included in the report.

A total of 8 deaths related to chemotherapy treatment were observed in the two randomized cardioprotection trials: 5 on the control arm (3 due to sepsis, 1 due to disease progression and 1 due infection without positive blood cultures) and 3 on the ICRF-187 arm (1 due to intracerebral hemorrhage, 1 due to sepsis and 1 due to a Lysteria infection). ICRF-187 did not add to the incidence of chemotherapy related deaths. No deaths related to ICRF-187 were reported.

Laboratory Evaluations of Safety

Haematology. A marginal but statistically significant lower, median nadir white blood cell count (WBC) was noted on cycle 2 of treatment in patients receiving ICRF-187. Similarly, in cycle 1, the platelets showed significantly lower mean and median nadir counts at cycle 1. No differences appeared for either hematocrit or haemoglobin.

These findings suggest a small, but definite, suppressing effect of ICRF-187 on the bone marrow for the white cell and magakaryocytic series, but not on erythropoiesis. However, this interpretation is not confirmed by the analysis of haematological toxicity according to ECOG grading: no significant difference is observed in the WBC and Hb toxicity grading during the first two cycles. This implies that any bone marrow suppressive effects are truly minimal.

Biochemistry. Biochemistry results including SGOT, LDH, bilirubin, alk. phosphatase, BUN and s-creatinine, classified according to ECOG toxicity grading, were not significantly different on the two treatment arms. At a dose of 500 mg/m^2, ICRF-187 does not add to the hepatic or renal toxicity of the given chemotherapeutic regimen. Most patients had grade 0 toxicity. ECOG grade 3 toxicities in liver function tests, seen on both arms of the study, were related to liver metastases (SGOT, LDH, bilirubin) and/or bone metastases (alk. phosphatase).

Other Safety Aspects of ICRF-187

In a mouse micronucleus assay of ICRF-187, a 1,000 mg/kg oral dose caused an increase in the incidence of micronuclei in polychromatic erythrocytes in males and females (see Section III-D of the dossier). To date no secondary malignancies have been reported following therapy with ICRF-187. However, the racemic mixture ICRF-159 was reported to be associated with the development of AML and skin epitheliomas in patients receiving the drug for a prolonged period of time.

The use of ICRF-187 as a cardioprotector is, however, restricted to cancer patients that are being treated with doxorubicin (containing) chemotherapy. It should be noted that the majority of cytostatic drugs, including the anthracyclines used in the cancer setting have mutagenic, teratogenic and carcinogenic potential.

Like other cytotoxic drugs, ICRF-187 should be administered under the supervision of a qualified physician experienced in the use of cancer chemotherapeutic agents, who is aware of the associated risks.

Extrapolation of Preclinical Data on Toxicology

Repeated dose toxicity studies performed in rats, rabbits and dogs with doses of Cardioxane (ICRF-187) administered more frequently and at higher doses than the anticipated human dosing regimen, have revealed reversible toxic effects.

The toxicity of ICRF-187 has been shown to be manifest in animals in mitotically active tissues, e.g. bone marrow, lymphoid tissue, testes and gastrointestinal mucosa, with the drug administration schedule the primary factor in the degree of tissue alteration.

In phase I cancer studies the degree of toxicity was also schedule dependent. The dose limiting toxicity of ICRF-187 in adults was myelosuppression. There was no evidence for toxic effects on lymphoid tissue, testes and gastrointestinal mucosa.

No clinical signs of toxicity due to ICRF-187 were observed in clinical cardio-protection trials, in which ICRF-187 was given at a dose fo 500 mg/m^2 once every 3 weeks.

Prescribing Information

The prescribing information includes a complete listing of the adverse events considered to be related to the administration of ICRF-187.

Based on case reports of liver dysfunction after high dose ICRF-187 in phase I cancer clinical trials, it is recommended to perform routine liver function tests in patients with initial liver function abnormalities. Since renal dysfunction may increase bioavailability of ICRF-187, patients with initial renal dysfunction should be monitored carefully for haematological toxicity.

ICRF-187 should not be administered to fertile persons of either sex not practicing effective contraception. Animal reproduction studies have not been conducted with ICRF-187. It is not known whether ICRF-187 can cause foetal damage when administered to pregnant women or whether it can affect reproduction capacity. There is no information available on the excretion of ICRF-187 in human milk, or its effect on lactation.

Risk Benefit Conclusions/Quality of Life

One obstacle to the use of anthracyclines in cancer chemotherapy is its irreversible cumulative dose-limiting cardiotoxicity. Data from 4 controlled randomized clinical trials in patients with advanced breast cancer demonstrate that ICRF-187 offers significant protection against cardiac toxicity caused by doxorubicin, without adversely affecting the antitumour effect of a chemotherapy regimen consisting of fluorouracil, doxorubicin and cyclophosphamide.

Nineteen percent of the patients studied on the control arm, not receiving ICRF-187, experienced major treatment related cardiac dysfunction, resulting in a limitation of physical performance status (NYH Ass grade 3 and 4). In patients with grade 3 toxicity ordinary physical activity causes fatigue, palpitation, dyspnoea or anginal pain. In patients with grade 4 cardiotoxicity, symptoms of cardiac insufficiency or anginal syndrome may be present even at rest. If any physical activity is undertaken, discomfort is increased. An additional 4% of patients on the control arm had cardiac failure slightly limiting their physical activity (NYH Ass grade 2). These patients are comfortable at rest, but ordinary physical activity results in fatigue, palpitation, dyspnoea, or anginal pain. Thus, 23% of patients on the control arm had cardiac dysfunction, adversely affecting quality of life in terms of limitation of physical activity. Only two patients on the ICRF-187 arm developed clinical cardiac toxicity (NYH Ass grade 2). The data demonstrate that patients on the ICRF-187 arm had a significantly better cardiac performance and resultant better quality of life in terms of physical abilities, compared to patients on the control arm.

ICRF-187 cardioprotection occurs irrespective of cardiac risk factors. This opens additional chemotherapeutic options to patients with preexisting cardiac disease who are at high risk for developing cardiotoxicity for whom doxorubicin is not normally recommended.

By reducing the risk of developing cardiotoxicity, patients receiving ICRF-187, who are responding to doxorubicin or who might benefit from it, may safely receive higher cumulative doses of the anticancer drug, potentially increasing the antitumour efficacy of the chemotherapy regimen.

Dose dependent chronic cardiotoxicity is a toxicity of all anthracyclines. The anthracycline induced cardiotoxicity is thought to result form the generation of oxygen free radicals, while intra-DNA intercalation and topoisomerase II – induced DNA cleavage are considered to constitute the mechanism of tumour cell killing.

In the clinical studies reviewed here, doxorubicin was used as a 'model-drug' to study the cardioprotection of ICRF-187. Considering the mechanism of cardioprotection of ICRF-187, reducing the availability of intracellular iron and thereby preventing the formation of oxygen free radicals, the drug is of potential therapeutic value for use with other anthracyclines, when it would be expected to reduce the risk of patients developing cardiotoxicity.

The clinical cardioprotective dose of ICRF-187 is 500 mg/m^2 in combination with 50 mg/m^2 of doxorubicin. ICRF-187 should be given as a short continuous infusion over 15 min, 30 min before doxorubicin administration. At this dose of ICRF-187, the drug offers significant protection without adding significantly to the non-cardiac toxicity of the regimen. The only added toxicity occurring after concomitant use of ICRF-187 is a small, but definite accentuation of leukopenia and thrombocytopenia. These toxicities are manageable and do not form a limitation to the use of ICRF-187.

References

1. Tan C, Etcubanas E, Wollner N et al (1973) Adriamycin – an antitumour antibiotic in treatment of neoplastic disease. Cancer 32:9–17
2. Benjamin RS, Wierik PH, Bachur NR (1974) Adriamycin chemotherapy – efficacy, safety and pharmacological basis of an intermittent single high-dose schedule. Cancer 33:19–27
3. Von Hoff DD, Layard MW, Basa P et al (1979) Risk factors for doxorubicin-induced congestive heart failure. Ann Int Med 91:710–7

4. Unverferth DV, Magorien RD, Leier CV, Balcerzak SP (1983) Doxorubicin cardiotoxicity. Cancer Treat Rev 9:149–64
5. Praga C, Beretta G, Vigo PL et al (1979) Adriamycin cardiotoxicity: survey of 1273 patients. Cancer Treat Rep 63:827–34
6. Schwartz RG, Mc Kenzie WD, Alexander J et al (1987) Congestive heart failure and left ventricular dysfunction complicating doxorubicin therapy: seven-year experience using serial radionuclide angiocardiography. Am J Med 82:1109–18
7. Weiss RB, Sarosy G, Clagett-Carr K, Russo M, Leyland-Jones B (1986) Anthracycline analogs: the past, present, and future. Cancer Chemother Pharmacol 18:185–97
8. Braunwald E (1988) Heart disese – a textbook of cardiovascular medicine, 3rd edn. WB Saunders Company, Philadelphia, pp 1748–51
9. Lokich J, Bothe A, Zipoli T et al (1983) Constant infusion schedule for adriamycin: a phase I – II clincial trial of a 30 day schedule by ambulatory pump delivery system. J Clin Oncol 1:24–8
10. Speyer JL, Green MD, Dublin N et al (1985) Prospective evaluation of cardiotoxicity during a six-hour doxorubicin infusion regimen in women with adenocarcinoma of the breast. Am J Med 78:555–63
11. Green MD, Speyer JL, Wernz J et al (1983) Prolonged infusion chemotherapy with adriamycin in combination with 5-FU and cytoxan in an attempt to reduce cardiotoxicity. Proc Am Soc Clin Oncol 2:26
12. Green MD, Speyer JL, Muggia FM (1982) A phase I/II study of 4′epidoxorubicin administered as a 6 hour infusion. Proc Am Soc Clin Oncol 1:20
13. Myers CE, McGuire WP, Liss RH, Ifrim I, Grotzinger K, Young RC (1977) Adriamycin: The role of lipid peroxydation in cardiac toxicity and tumour response. Science 197:165–7
14. Doroshow JH, Reeves J (1980) Anthracycline enchanced oxygen radical formation in the heart. Proc Am Assoc Cancer Res 21:266
15. Bakowski MT (1976) ICRF 159, (+/–) 1,2 bis (3,5-dioxopiperazin-1-yl) propane, NSC 129943; razoxane. Cancer Treat Rev 3:95–107
16. Package Insert Razoxin® – Imperial Chemical Industries PLC (Pharmaceutical Division), Macclesfield
17. Clinical Experience with ICRF-187. EuroCetus Research Report 900702. EuroCetus B.V., 1990 (Section IV B2)
18. Clinical Pharmacology of ICRF-187. EuroCetus Research Report 900701. EuroCetus B.V., 1990 (Section IV A)
19. A single intravenous dose bioequivalence study comparing Cardioxane (EuroCetus B.V., The Netherlands) with NCI-ICRF-187 (National Cancer Institute, USA). EuroCetus Research Report 900903, EuroCetus B.V., 1990
20. Herman EH, Ferrans VJ (1987) Amelioration of anthracycline cardiotoxicity by ICRF-187 and other compounds. Cancer Treat Rev 14:225–9
21. Hasinoff BB (1989) The interaction of the cardioprotective agent ICRF-187; its hydrolysis product (ICRF-198) and other chelating agents with the Fe (III) and Cu (II) complexes of adriamycin. Agents Actions 26:378–85
22. Von Hoff DD, Howser D, Lewis BG et al (1981) Phase I study of ICRF-187 using a daily for 3 days schedule. Cancer Treat Rep 65:249–52
23. Eastland G (1987) ICRF-187. Drugs Future 12:1017–20
24. Poster DS, Penta JS, Bruno S, MacDonald JS (1981) ICRF-187 in clinical oncology. Cancer Clin Trials 4:143–6
25. Herman EH, Ferrans VJ (1981) Reduction of chronic doxorubicin cardiotoxicity in dogs by pretreatment with ICRF-187. Cancer Res 41:3436–40
26. Herman EH, Ferrans VJ, Jordan W, Ardalan B (1981) Reduction of chronic daunorubicin cardiotoxicity by ICRF-187 in rabbits. Res Comm Chem Pathol Pharmacol 31:85–97
27. Herman EH, El-Hage A, Witiak DT (1982) Protection against acute daunorubicin toxicity by ICRF-187 and related compounds. Fed Proc 41:1477

28. Wang G, Finch MD, Trevan D, Hellmann K (1981) Reduction of daunomycin toxicity by razoxane. Br J Cancer 43:871

29. McPherson E, Archilla R, Schein PS, Tew KD, Mhatre RM (1984) Preclinical pharmacokinetics, disposition and metabolism of ICRF-187. Proc Am Assoc Cancer Res 25:1436

30. Mhatre RM, Tew KD, van Hennik MB, Waravdekar VS, Schein PS (1983) Absorption, distribution and pharmacokinetics of ICRF-187 in dogs. Proc Am Assoc Cancer Res 24:1147

31. Von Hoff DD, Soares N, Gormley P, Poplack DG (1980) Pharmacokinetics of ICRF-187 in the cerebrospinal fluid of subhuman primates. Cancer Treat Rep 64:734–6

32. Earhart RH, Tutsch KD, Koeller JM et al (1982) Pharmacokinetics of (+)-1,2 di(3,5-dioxopiperazin-1-yl) propane intravenous infusion in adult cancer patients. Cancer Res 42:5255–61

33. Vogel CL, Gorowsk E, Davila E et al (1987) Phase I clinical trial and pharmacokinetics of weekly ICRF-187 (NSC-169780) infusion in patients with solid tumours. Invest New Drugs 5:187–98

34. Holcenberg JS, Tutsch KD, Earhart RH, Ungerleider RS, Kamen BA, Pratt CB, Gribble TJ, Glaubiger DL (1986) Phase I study of ICRF-187 in pediatric cancer patients and comparison of its pharmacokinetics in children and adults. Cancer Treat Rep 70:703–9

35. Hochster H, Speyer J, Liebes L et al (1990) Phase I and pharmacokinetic study of escalating ADR-529 (ICRF-187) with doxorubicin. Proc Am Soc Clin Oncol 9:82

36. A randomized trial of cardioprotection with ICRF-187 in patients with advanced breast cancer receiving 5-fluorouracil, doxorubicin and cyclophosphamide. EuroCetus Research Report NYU 83-05. EuroCetus B.V., 1990 (Section IV B1)

37. A randomized trial of cardioprotection with ICRF-187 in patients with advanced breast cancer receiving 5-fluorouracil, doxorubicin and cyclophosphamide. EuroCetus Research Report AVL 86-01. EuroCetus B.V., 1990 (Section IV B1)

38. Legha SS, Benjamin RS, Mackay B et al (1982) Reduction of doxorubicin cardiotoxicity by prolonged continuous intravenous infusions. Ann Int Med 96:133–9

39. Bristow MR, Mason JW, Billingham ME, Daniels JR (1981) Dose-effect and structure-function relationships in doxorubicin cardiomyopathy. Am Heart J 102:709–18

40. Calabresi P, Schein PS, Rosenberg SA (eds) (1985) Medical oncology – basic principles and clinical management of cancer. MacMillan Publishing Company, New York

3.6.2 *Comments on the Definitive Trials of Dexrazoxane Protection against Anthracycline Cardiotoxicity: The Swain Trails*

Kurt Hellmann

Speyer et al. [1] undertook the first randomized, controlled clinical trial to determine whether DXRz would permit doses of doxorubicin greater than that limited by the conventional maximum tolerated dose to be given to women with advanced breast cancer. These investigators found that doses of doxorubicin in excess of the maximum tolerated dose of 450 mg/m^2 could indeed be given without any evidence of cardiotoxicity if DXRz (ratio 20:1) had been given 30 min beforehand. These results formed the basis of six further randomized clinical trials, three of which were monitored by the Food and Drug Administration.

Analysis of the largest of the Swain trials (88001) [2] appeared to show that the addition of DXRz to the 5-Fluorouracil/DOX/cyclophosphamide (FAC) regimen had reduced the response rate in women with advanced breast cancer from 96 of 152 patients (63%) to 67 of 141 patients (48%). This highly significant result ($p = 0.007$) was however not in line with time to progression in the DXRz group nor with trial

88006. Nevertheless, the finding led to a crucial decision: The protocol of 88001 and a smaller study (88006) would be amended so that patients would only receive DXRz in addition to the FAC regimen after they had already received 300 mg/m^2 of DOX and were thought to benefit from further doses of DOX. This decision and amendment selected doxorubicin responders and excluded the non-responders largely from the trials.

The trial (88001 and 88006 combined) involved 1,008 patients and was carefully monitored. It deserves the closest scrutiny.

Comments on the Swain Trials (88001 and 88006) – The Definitive Trials [2, 3]

1. Trial 88001 involved dexrazoxane primarily in relation to doxorubicin (DOX), only secondarily in relation to breast cancer, and it is therefore of interest for all patients where DOX is used.
2. Dexrazoxane was highly effective in preventing DOX-induced congestive heart failure (CHF), but CHF is only the tip of the cardiotoxic iceberg. It is highly likely, therefore, that DXRz prevented myocardial damage by DOX, which might not have been evident, because of cardiac compensation. This subclinical damage may only become clinically obvious when stress is laid on the cardiovascular system through, for example, pregnancy, weight lifting, serious infection, diabetes, or the development of high blood pressure [4].
3. Delaying the administration of DXRz until after 300 mg/m^2 DOX had been given to those patients who might respond to DOX was clearly wise, if DXRz interfered with responses as judged by the size of the tumors and as apparently indicated in study 88001. However, study 88001 response results were skewed by an unusually high response rate in the placebo group, which were even out of line with the placebo response rate in 88006 and with those in other trials using identical doses of identical drugs in identical trial situations [1]. In these trials, placebo responses (i.e., FAC and placebo) were practically identical with those in 88006. Moreover, time to progression, which in the interim analysis had seemed also to be significantly shorter in the DXRz-treated patient group, had on analysis of the mature data become essentially identical for the two groups ($p = 0.23$). Survival also was identical for placebo- and DXRz-treated groups ($p = 0.88$) [2, 3].

   There is then no evidence from this very large trial that there is any 'interference' from the addition of DXRz in the most important responses to the patient: time to progression of disease and survival. Quite the contrary, as was shown by analysis of the survival data of the breast cancer patients receiving DXRz only after 300 mg/m^2 of DOX had been given [3, 5] where the median survival time had increased.
4. If it is true that all (or most) of the patients who received DOX in excess of 300 mg/m^2 were responders, then this is a particularly valuable group in which to compare the effect of DXRz in addition to FAC treatment to determine whether DXRz had any additional effect in this patient group apart from its cardioprotection. Analysis of such a group who received only placebo plus FAC before and after 300 mg/m^2 DOX compared with a similar group who received placebo with FAC before and placebo/DEX after 300 mg/m^2 of DOX revealed

that survival was doubled at 3 years [3] and was still statistically significantly improved at 5 years with a median survival time of 948 days [5], which is one of the best results achieved in advanced breast cancer.

The improved survival of responders might not have been seen in any analysis that included all the patients (responders and nonresponders) because nonresponders account for some 50% of patients and calculation of the median survival time might then remain unchanged, reflecting essentially the survival time of the nonresponders. If this analysis is correct, it would also imply that DXRz is not able to convert intrinsic refractory tumors into responsive tumors even though in preclinical tests Rz potentiated each of the three drugs of the FAC combination when individually tested against a responsive malignancy (Tables 3.5 and 3.6) [6].

Perhaps the most important supplementary conclusion that can be drawn from Swain et al.'s results [3], in addition to their main conclusions, is that DXRz seems to be unable to convert nonresponding tumors to responsive ones, but keeps responders responding.

5. It was an important observation that neutropenia followed by the addition of DXRz to FAC had no clinical sequelae, there were no infections, no fevers, no death, and no necessity to change or delay administration of FAC. The only side effect of DXRz was temporary pain at the site of injection.

6. This trial also showed that the addition of DXRz to the FAC combination had a marked influence on the severity of the gastrointestinal toxicity of the FAC regimen. Remarkably, there was a reduction of toxicity at all intestinal levels. Dysphagia was reduced from 5 to 0%, nausea and vomiting were reduced ($p = 0.02$ and $p = 0.003$, respectively), and stomatitis and esophagitis also showed signs of reduced toxicity ($p = 0.07$ and $p = 0.08$, respectively). These results are

Table 3.5 L1210 Leukemia: summary of experience with ICRF-159 (Rz) combinations

Drug 1	Observed ILS	Drug 2	Observed ILS	Combination	Calculated ILS	Observed ILS
CTX (2)	162.4[a]	Rz (2)	19.0	CTX + Rz (7)	181.4	211.7[b]
5-FU (2)	87.9	Rz (2)	19.0	5-FU + Rz (7)	106.9	148.1[c]
DOX (2)	36.6	Rz (2)	19.0	DOX + Rz (7)	57.6	161.5[d]
MTX (2)	69.9	Rz (2)	46.4	MTX + Rz (7)	116.3	105.4[e]
VCR (2)	22.1	Rz (2)	46.4	VCR + Rz (7)	68.5	85.7
DTIC (2)	25.0	Rz (2)	46.4	DTIC + Rz (7)	71.4	122.9
HXM (1)	1.6	Rz (2)	42.2	HMX + Rz (7)	43.8	84.4

Note: numbers in parentheses are in the numbers of groups involved in observed ILS. Except for the MTX/Rz combination, all observed results were more than additive. This degree of overall enhanced activity has not been seen with any other drugs in combination with a selection of standard chemotherapeutic agents (Wampler [6]).

ILS, increase in life span; CTX, cyclophosphamide; 5-FU, 5-fluorouracil; DOX, doxorubicin; MTX, mitoxantrone; VCR, vincristine; DTIC, dacarbacine; HXM, hexamethylmelamine; Rz, razoxane.

Excluded: [a]one survivor (6%), [b]41 survivors (73%), [c]two survivors (19%), [d]10 survivors (19%), and [e]two survivors (4%).

Table 3.6 Effectiveness of ICRF-159 plus doxorubicin (DOX) against early mouse leukemia L1210[a]

Treatment (intraperitoneally every 3 h/24 h on day 1 only)	Optimal dose (mg/kg/injection)	ILS (%)	Survivors/total
ICRF-159 alone	50	60	0/8
DOX alone	2	60	0/8
ICRF-159 + DOX	25–2	>500	6/8

ILS, increase in life span over controls.

[a]Unpublished data of I Kline, R Woodman, and JM Venditti (1973, personal communication).

in accord with the preclinical findings of Wang et al. [7], who showed that the protection that Rz gave to animals receiving a lethal dose of daunorubicin was due to the Rz's protection of the mucosa of the small bowel. Cytoprotection by Rz is also seen in the reduction of the anthracycline-induced nephrotoxicity [8]. Therefore, DXRz not only protects a wide range of replicating and nonreplicating tissues from anthracycline toxicity, but also protects against the toxicities of other drugs, such as mitoxantrone [9], cisplatin [10], bleomycin [11], and VP-16 [12].

7. It has been suggested that the highly effective protection due to DXRz might extend to protection of tumor cells, and in vitro experiments conducted by Supino [13], Sehested et al. [14], and Hasinoff et al. [15] have supported this hypothesis. However, these investigators also showed that any protection by DXRz of the tumor cells is very conditional, antagonism varying with synergism depending on conditions and the anthracycline used. Extensive in vivo experiments by Verhoef et al. [16], however, showed conclusively that DXRz did not interfere with the antitumor activity of DOX in nine different preclinical tumor models (six mouse and three human) tested under a variety of different conditions.

Dexrazoxane is highly effective at preventing anthracycline-induced cardiotoxicity. However, most probably one has to give DXRz not after having done the damage, but before the first dose of doxorubicin.

Dexrazoxane significantly increases survival of breast cancer patients who respond to FAC, thus keeping responders responding. For nonresponders, the cardiac protection permits greater therapeutic flexibility by allowing treatment with second-line cardiotoxic drugs. Moreover, patients with pre-existing cardiac risk factors, including age more than 65 years, can tolerate full doses of DOX, as well as those patients without these risk factors.

If cardioprotection were the sole reason for giving DXRz, many oncologists would be justified in thinking that they have learned over many years how to avoid cardiotoxicity of the anthracyclines without any cardioprotectors. The increase in survival of breast cancer patients, which is in line with what has been seen with DXRz and Rz in other malignancies, should, however, if nothing else, call for a wider appraisal of DXRz.

References

1. Speyer JL, Green MD, Zeleniuch-Jacquotte A et al (1997) ICRF-187 permits longer treatment with doxorubicin in women with breast cancer. J Clin Oncol 10:117–27

2. Swain SM, Whaley FS, Gerber MC et al (1997) Cardioprotection with dexrazoxane for doxorubicin-containing therapy in advanced breast cancer. J Clin Oncol 15:1318–32

3. Swain SM, Whaley FS, Gerber MC et al (1997) Delayed administration of dexrazoxane provides cardioprotection for patients with advanced breast cancer treated with doxorubicin-containing therapy. J Clin Oncol 15:1333–40

4. Lipshultz SE, Colan SD, Gelber RD et al (1991) Late cardiac effects of doxorubicin therapy for acute lymphoblastic leukemia in childhood. N Engl J Med 324:808–15

5. Swain SM, Whaley FS, Gams RA (1997) Reply: delayed administration of dexrazoxane provides cardioprotection against anthracyclines in breast cancer or acute myeloid leukemia. J Clin Oncol 15:3292 (letter)

6. Wampler G (1975) NCI New Drug Liaison Meeting, Bethesda, 25 Feb 1975

7. Wang G, Finch MD, Trevan D et al (1981) Reduction of daunomycin toxicity by razoxane. Br J Cancer 43:871–7

8. Herman EH, El-Hage A, Ferrans VJ (1988) Protective effect of ICRF 187 on doxorubicin-induced cardiac and renal toxicity in spontaneously hypertensive (SHR) and normotensive (WKY) rats. Toxicol Appl Pharmacol 92:42–53

9. Hellmann K, Hutchinson GE, Henry K (1987) Mitoxantrone therapeutic index improvement by gangliosides. Cancer Treat Rev 14: 373–8

10. Woodman RJ (1974) Enhancement of antitumor effectiveness of ICRF 159 against early L1210 by combination with cis-diamminedichloroplatinum (NSC-82151). Cancer Chemother Rep 4:45–52

11. Herman EH, Hasinoff BB, Zhang J et al (1995) Morphologic and morphometric evaluation of the effect of ICRF-187 on bleomycin-induced pulmonary toxicity. Toxicology 98:163–75

12. Holm B, Jensen PB, Sehested M (1996) ICRF-187 rescue in etoposide treatment in vivo. A model targeting high-dose topoisomerase II poisons to CNS tumors. Cancer Chemother Pharmacol 38:203–9

13. Supino R (1984) Influence of ICRF-159 or ICRF-186 on cytotoxicity of daunomycin and doxorubicin. Tumori 70:121–6

14. Sehested M, JensenPB, Sorensen BS et al (1993) Antagonistic effect of the cardioprotector (+)-1,2-bis(3,5-dioxopiperazinyl-1-yl) propane (ICRF-187) on DNA breaks and cytotoxicity induced by the topoisomerase II directed drugs daunorubicin and etoposide (VP-16). Biochem Pharmacol 46:389–93

15. Hasinoff BB, Yalowich JC, Ling Y et al (1996) The effect of dexrazoxane (ICRF-187) on doxorubicin and daunorubicin-mediated growth inhibition of Chinese hamster ovary (CHO) cells. Anticancer Drugs 7:558–67

16. Verhoef V, Bell V, Filppi J (1988) Effect of the cardioprotective agent ADR-529 (ICRF-187) on the antitumor activity of doxorubicin. Proc Am Assoc Cancer Res 29:273

3.6.3 Studies of Dexrazoxane Against the Cardiotoxicity of Anthracyclines in Adult and Paediatric Patients – An Update

Robin L. Jones

From the results of numerous randomised trials described previously [1–4], the cardioprotective efficacy of dexrazoxane in breast cancer treated with anthracyclines is well established. Furthermore, two meta-analyses, of all the randomised trials, have confirmed that patients treated with dexrazoxane (compared to those not treated with the drug) are significantly less likely to develop cardiotoxicity secondary to anthracyclines [5, 6]. Crucially, these meta-analyses confirmed no significant difference in response rate or survival between those treated and not treated with dexrazoxane. Further supporting the utility of dexrazoxane are the results of three pharmacoeconomic analyses demonstrating that dexrazoxane is a cost effective cardioprotectant [7–9].

Randomised Trials of Dexrazoxane in Adult Malignancies

The results of randomised trials of dexrazoxane are summarised in Table 3.7. The most recently published trial, conducted by Marty and colleagues, randomised 164 relapsed breast cancer patients to receive doxorubicin/epirubicin either alone or in combination with dexrazoxane [10]. The dose ratio of dexrazoxane to doxorubicin was 20:1 and for epirubicin the ratio was 10:1. Seventy nine patients were randomised to the anthracycline alone arm and 85 to the dexrazoxane arm. There was a significant difference in the number of cardiac events, 10 (13%, 95%CI 6–21%) in the dexrazoxane arm and 29 (39%, 95%CI 28–51%) in the anthracycline alone arm, $p < 0.001$. There was also a significant difference in the incidence of congestive cardiac failure (CHF), 1 (1%, 95%CI 0.03–7%) in the dexrazoxane arm and 8 (11%, 95%CI 5–20%) in the anthracycline alone arm, $p < 0.015$. No significant difference in non-cardiac toxicity was observed between both arms. This trial also demonstrated no significant difference in response rate, progression-free and overall survival between those treated with anthracycline alone and those treated with anthracycline in combination with dexrazoxane.

Most of the randomised trials have been performed in breast cancer, but Lopez and colleagues randomised 95 patients with advanced breast cancer and 34 with soft tissue sarcoma to receive high dose epirubicin (160 mg/m^2 every 3 weeks) with or without dexrazoxane [11]. None of the patients treated with dexrazoxane developed congestive cardiac failure, whereas, 4 did in the control arm. Furthermore, the decrease in left ventricular ejection fraction (LVEF) form baseline was significantly greater in the control compared to dexrazoxane arm. In both disease subgroups no significant difference in response rate, time to progression and overall survival was observed between those treated with epirubicin alone or in combination with dexrazoxane.

There are a number of ongoing clinical trials evaluating dexrazoxane in sarcoma patients (http://clinicaltrials.gov/ct2/results?term=dexrazoxane).

Feldmann et al. randomised 155 patients with small cell lung cancer to receive cyclophosphamide, doxorubicin, vincristine (CAV) with or without dexrazoxane [12]. The doxorubicin dose in this study was 50 mg/m^2 and dexrazoxane was administered with anthracycline at a ratio of 10:1. Cardiac events were recorded in 12% of patients in the dexrazoxane arm and 29% in the placebo arm ($p < 0.05$). There was no significant difference in the incidence of CHF between the dexrazoxane (4%) and placebo (10%) arm. No significant difference in response rate, progression-free and overall survival was observed between the dexrazoxane and placebo arm of this trial.

Table 3.7 Randomised trials of dexrazoxane

Trial	Patient number	Regimen (anthracycline dose in mg/m^2)	CHF (%)	Cardiac events	Response rate
Trials in advanced breast cancer					
Comparison with placebo					
Swain et al. [3]	168	CDF (50) + DEX (10:1)	0	15	47
	181	CDF (50) + Placebo	8 $p < 0.001$	32 $p < 0.001$	61 $p = 0.019$
Swain et al. [2]	81	CDF (50) + DEX (10:1)	3	14	54
	104	CDF (50) + Placebo	7	31 $p < 0.01$	49
Comparison with chemotherapy alone					
Marty et al. [10]	85	Dox (50) + DEX (20:1) or Epi (90) + DEX (10:1)	1	13	35
	79	Dox (50) or Epi (90)	11 $p < 0.05$	39 $p < 0.001$	35
Speyer et al. [1]	76	CDF (50) + DEX (20:1)	3	8	37
	74	CDF (50)	27 $p < 0.001$	50 $p < 0.001$	41
Venturini et al. [4]	82	CEF (60) + DEX (10:1) or Epi (120) + DEX (10:1)	2	7	48
	78	CEF (60) or Epi (120)	5	23 $p < 0.01$	46
Trials in small cell lung cancer and soft tissue sarcoma					
Feldmann et al. [12] (small cell lung cancer)	73	CAV + DEX (10:1)	4	12	45
	82	CAV (50) + Placebo	10	29 $p < 0.05$	59
Lopez et al. [16] (soft tiss. sarc.)	18	Epi (160) + DEX (6.25:1)	7	9	11
	16	Epi (160)	24	29 $p < 0.01$	38

CHF, congestive heart failure; CDF, cyclophosphamide, doxorubicin, 5-FU; DEX, dexrazoxane; Dox, doxorubicin; Epi, epirubicin; CEF, cyclophosphamide, epirubicin, 5-FU; CAV, cyclophosphamide, doxorubicin, vincristine.

None of these trials documented a significant difference in progression-free or overall survival between patients who received or did not receive dexrazoxane.

Galetta et al. have reported the results of a trial randomising 20 non-Hodgkin lymphoma patients to receive epirubicin-based therapy with or without dexrazoxane (400 mg/m^2) [13]. Dexrazoxane was found to have a beneficial effect on QT interval dispersion.

Randomised Trials of Dexrazoxane in Paediatric Patients

A trial performed by Wexler and colleagues randomised 38 paediatric sarcoma patients to receive doxorubicin with or without dexrazoxane [14]. Fifteen patients were randomised to the control arm and 18 to the dexrazoxane arm. There was a significantly lower incidence of subclinical cardiotoxicity in the dexrazoxane arm (22%) compared to the placebo arm (67%, $p < 0.01$). Of note, patients in the dexrazoxane arm (1.0) had a significantly smaller decrease in LVEF per 100 mg/m^2 of doxorubicin compared to those in the control arm (2.7, $p = 0.02$). Furthermore, patients in the dexrazoxane arm received a significantly higher median doxorubicin cumulative dose (410 mg/m^2) compared to the control arm (310 mg/m^2, $p < 0.05$). No significant difference in response rate, median event-free and median overall survival was observed in this small trial.

A trial of 206 paediatric patients with acute lymphoblastic leukaemia, randomised between doxorubicin 30 mg/m^2 every 3 weeks for ten doses or dexrazoxane (300 mg/m^2) followed by doxorubicin [15]. One hundred and five were randomised to the dexrazoxane arm and 101 to the control arm. Serial cardiac troponin T measurements were obtained in 82 of the dexrazoxane arm patients and 76 of the control arm. Patients treated in the control arm (50%) were significantly more likely to have elevated cardiac troponin T levels compared to those in the dexrazoxane arm (21%, $p < 0.001$). In addition, patients in the control arm were significantly more likely to have extremely elevated cardiac troponin T levels (32%) compared to those in the dexrazoxane arm (10%, $p < 0.001$). No significant difference in mean left ventricular dimension, fractional shortening or contractility before, during or following therapy was documented between the two arms of this trial. No significant difference in event-free survival between the control and dexrazoxane arms was recorded (83% in both arms, $p = 0.87$).

Another trial (95-01) randomising 205 high-risk ALL patients to receive doxorubicin with or without dexrazoxane recorded no episodes of CHF in either arm [16]. In concordance with other randomised trials no significant difference in 5-year disease-free survival between the two arms was observed. Recently published, updated results of the Dana-Farber Cancer Institute ALL Consortium, have confirmed the initial findings of the 95-01 trial demonstrating no significant difference in event-free and overall survival between patients treated within the control and dexrazoxane arms and no increase in second malignant neoplasia in the dexrazoxane arm [17]. Furthermore, gender differences were seen in long-term dexrazoxane cardioprotection. While its impact was seen in all groups, females primarily benefited from the long-term cardioprotective effect and exhibited more normal left ventricular (LV) dimensions and more appropriate wall thickness for LV dimension, both of which are consistent with less LV remodelling. Dexrazoxane/doxorubicin treated females also had more normal LV function than females who received doxorubicin alone or males of either group [15].

Non-randomised Studies

A number of other studies have confirmed the favourable toxicity profile and cardioprotective value of dexrazoxane in paediatric/ adolescent [18–20] and adult patients [21, 22].

Other Methods of Cardioprotection

A number of other potential cardioprotective agents have been suggested [6]. However, dexrazoxane is the only cardioprotectant that has been shown to significantly decrease the cardiotoxicity of anthracyclines in randomised clinical trials [23]. Other techniques have been adopted to try and minimise the risk of anthracycline-induced cardiotoxicity including the limitation of cumulative dose, the development and use of anthracycline analogues and administration as a slow infusion rather than bolus [23].

It has been proposed that the cardiotoxicity of doxorubicin is related to peak plasma concentrations, but that the anti cancer action is dependent on the tissue concentration over time [24]. Studies investigating the effect of decreasing peak plasma levels by prolonging infusion time, in metastatic breast cancer [25] and adult soft tissue sarcoma [26] found similar anti-tumour activity but reduced cardiotoxicity. However, more recent studies have found no significant difference in cardiotoxicity by varying the duration of anthracycline administration [17, 27, 28].

In attempts to reduce the cardiotoxicity of conventional anthracyclines, the development of liposome encapsulation of these drugs has been undertaken [29]. Randomised trials have demonstrated that pegylated and non-pegylated liposomal doxorubicin have similar efficacy to conventional doxorubicin but with significantly lower incidence of cardiotoxicity, but also with the significant risk of palmar plantar erythema. It should, however, be noted that there have been no randomised trials comparing the activity of dexrazoxane and any liposomal anthracycline preparation, but such investigation is warranted.

Conclusion

There is an undoubted need for effective cardioprotection with dexrazoxane due to the increasing number of long-term cancer survivors treated with doxorubicin. The role of dexrazoxane in breast cancer patients treated with adjuvant chemotherapy also requires investigation. In addition, dexrazoxane has a valuable role to play in the elderly cancer population, who may have considerable cardiac co-morbidity and poor baseline cardiac function. Dexrazoxane may have an impact in conditions that anthracycline-based regimens form the backbone of management such as soft tissue sarcomas.

References

1. Speyer JL, Green MD, Zeleniuch-Jacquotte A, Wernz JC, Rey M, Sanger J, Kramer E, Ferrans V, Hochster H, Meyers M et al (1992) ICRF-187 permits longer treatment with doxorubicin in women with breast cancer. J Clin Oncol 10:117–27
2. Swain SM, Whaley FS, Gerber MC, Ewer MS, Bianchine JR, Gams RA (1997) Delayed administration of dexrazoxane provides cardioprotection for patients with advanced breast cancer treated with doxorubicin-containing therapy. J Clin Oncol 15: 1333–40
3. Swain SM, Whaley FS, Gerber MC, Weisberg S, York M, Spicer D, Jones SE, Wadler S, Desai A, Vogel C, Speyer J, Mittelman A, Reddy S, Pendergrass K, Velez-Garcia E, Ewer MS, Bianchine JR, Gams RA (1997) Cardioprotection with dexrazoxane for doxorubicin-containing therapy in advanced breast cancer. J Clin Oncol 15:1318–32
4. Venturini M, Michelotti A, Del Mastro L, Gallo L, Carnino F, Garrone O, Tibaldi C, Molea N, Bellina RC, Pronzato P, Cyrus P, Vinke J, Testore F, Guelfi M, Lionetto R, Bruzzi P, Conte PF, Rosso R (1996) Multicenter randomized controlled clinical trial to evaluate cardioprotection of dexrazoxane versus no cardioprotection in women receiving epirubicin chemotherapy for advanced breast cancer. J Clin Oncol 14:3112–20
5. Seymour L, Bramwell V, Moran LA (1999) Use of dexrazoxane as a cardioprotectant in patients receiving doxorubicin or epirubicin chemotherapy for the treatment of cancer. The Provincial Systemic Treatment Disease Site Group. Cancer Prev Control 3: 145–59
6. van Dalen EC, Caron HN, Dickinson HO, Kremer LC (2008) Cardioprotective interventions for cancer patients receiving anthracyclines. Cochrane Database Syst Rev (2):Article No. CD003917
7. Bates M, Lieu D, Zagari M, Spiers A, Williamson T (1997) A pharmacoeconomic evaluation of the use of dexrazoxane in preventing anthracycline-induced cardiotoxicity in patients with stage IIIB or IV metastatic breast cancer. Clin Ther 19:167–84
8. Limat S, Demesmay K, Fagnoni P, Voillat L, Bernard Y, Deconinck E, Brion A, Arveux P, Cahn JY, Woronoff-Lemsi MC (2005) Cost Effectiveness of cardioprotective strategies in patients with aggressive non-Hodgkin's lymphoma. Clin Drug Investig 25: 719–29
9. Tonkin K, Bates M, Lieu D, Arundell E, Williamson T, Zagari M (1996) Dexrazoxane cardioprotection for patients receiving FAC chemotherapy: a pharmacoeconomic evaluation. Can J Oncol 6:458–73
10. Marty M, Espie M, Llombart A, Monnier A, Rapoport BL, Stahalova V (2006) Multicenter randomized phase III study of the cardioprotective effect of dexrazoxane (Cardioxane) in advanced/metastatic breast cancer patients treated with anthracycline-based chemotherapy. Ann Oncol 17:614–22
11. Lopez M, Vici P, Di Lauro K, Conti F, Paoletti G, Ferraironi A, Sciuto R, Giannarelli D, Maini CL (1998) Randomized prospective clinical trial of high-dose epirubicin and dexrazoxane in patients with advanced breast cancer and soft tissue sarcomas. J Clin Oncol 16: 86–92
12. Feldman JE, Jones SE, Weisberg SR et al (1992) Advanced small cell lung cancer treated with CAV (cyclophosphamide, adriamycin, vincristine) chemotherapy and the cardioprotective agent dexrazoxane. In: Proceedings of the American Society of Clinical Oncology Annual Meeting, vol 11. p 296, Abstract 993
13. Galetta F, Franzoni F, Cervetti G, Cecconi N, Carpi A, Petrini M, Santoro G (2005) Effect of epirubicin-based chemotherapy and dexrazoxane supplementation on QT dispersion in non-Hodgkin lymphoma patients. Biomed Pharmacother 59:541–4
14. Wexler LH, Andrich MP, Venzon D, Berg SL, Weaver-McClure L, Chen CC, Dilsizian V, Avila N, Jarosinski P, Balis FM, Poplack DG, Horowitz ME (1996) Randomized trial of the

cardioprotective agent ICRF-187 in pediatric sarcoma patients treated with doxorubicin. J Clin Oncol 14:362–72

15. Lipshultz SE, Scully RE, Lipsitz SR, Sallan SE, Silverman LB, Miller TL, Orav EJ, Colan SD, Dana-Farber Cancer Institute Acute Lymphoblastic Leukemia Consortium (2009) Gender differences in long-term dexrazoxane cardioprotection in doxorubicin-treated children with acute lymphoblastic leukemia. J Clin Oncol 27:Abstract 10005

16. Moghrabi A, Levy DE, Asselin B, Barr R, Clavell L, Hurwitz C, Samson Y, Schorin M, Dalton VK, Lipshultz SE, Neuberg DS, Gelber RD, Cohen HJ, Sallan SE, Silverman LB (2007) Results of the Dana-Farber Cancer Institute ALL Consortium Protocol 95-01 for children with acute lymphoblastic leukemia. Blood 109:896–904

17. Silverman LB, Stevenson KE, O'Brien JE, Asselin BL, Barr RD, Clavell L, Cole PD, Kelly KM, Laverdiere C, Michon B, Schorin MA, Schwartz CL, O'Holleran EW, Neuberg DS, Cohen HJ, Sallan SE (2009) Long-term results of Dana-Farber Cancer Institute ALL Consortium protocols for children with newly diagnosed acute lymphoblastic leukemia (1985–2000). Leukemia 24(2):320–34

18. Bu'Lock FA, Gabriel HM, Oakhill A, Mott MG, Martin RP (1993) Cardioprotection by ICRF187 against high dose anthracycline toxicity in children with malignant disease. Br Heart J 70:185–8

19. de Matos Neto RP, Petrilli AS, Silva CM, Campos Filho O, Oporto VM, Gomes Lde F, Paiva MG, Carvalho AC, Moises VA (2006) Left ventricular systolic function assessed by echocardiography in children and adolescents with osteosarcoma treated with doxorubicin alone or in combination with dexrazoxane. Arq Bras Cardiol 87:763–71

20. Elbl L, Hrstkova H, Tomaskova I, Blazek B, Michalek J (2005) Long-term serial echocardiographic examination of late anthracycline cardiotoxicity and its prevention by dexrazoxane in paediatric patients. Eur J Pediatr 164: 678–84

21. Kolaric K, Bradamante V, Cervek J, Cieslinska A, Cisarz-Filipcak E, Denisov LE, Donat D, Drosik K, Gershanovic M, Hudziec P et al (1995) A phase II trial of cardioprotection with Cardioxane (ICRF-187) in patients with advanced breast cancer receiving 5-fluorouracil, doxorubicin and cyclophosphamide. Oncology 52:251–5

22. Testore F, Milanese S, Ceste M, de Conciliis E, Parello G, Lanfranco C, Manfredi R, Ferrero G, Simoni C, Miglietta L, Ferro S, Giaretto L, Bosso G (2008) Cardioprotective effect of dexrazoxane in patients with breast cancer treated with anthracyclines in adjuvant setting: a 10-year single institution experience. Am J Cardiovasc Drugs 8:257–63

23. Jones RL (2008) Utility of dexrazoxane for the reduction of anthracycline-induced cardiotoxicity. Expert Rev Cardiovasc Ther 6:1311–7

24. Legha SS, Benjamin RS, Mackay B, Ewer M, Wallace S, Valdivieso M, Rasmussen SL, Blumenschein GR, Freireich EJ (1982) Reduction of doxorubicin cardiotoxicity by prolonged continuous intravenous infusion. Ann Intern Med 96:133–9

25. Hortobagyi NG, Frye D, Buzdar AU et al (1989) Decreased cardiotoxicity of doxorubicin administered by continuous intravenous infusion in combination chemotherapy for metastatic breast carcinoma. Cancer 63(1):37–45

26. Zalupski M, Metch B, Balcerzak S, et al (1991) Phase III comparison of doxorubicin and dacarbacine given by bolus versus infusion in patients with soft-tissue sarcomas: a Southwest Oncology Group study. J Natl Cancer Inst 83(13):926–32

27. Levitt GA, Dorup I, Sorensen K, Sullivan I (2004) Does anthracycline administration by infusion in children affect late cardiotoxicity? Br J Haematol 124:463–8

28. Lipshultz SE, Giantris AL, Lipsitz SR, Kimball Dalton V, Asselin BL, Barr RD, Clavell LA, Hurwitz CA, Moghrabi A, Samson Y, Schorin MA, Gelber RD, Sallan SE, Colan SD (2002) Doxorubicin administration by continuous infusion is not cardioprotective: the Dana-Farber 91-01 acute lymphoblastic leukemia protocol. J Clin Oncol 20:1677–82

29. Jones RL, Swanton C, Ewer MS (2006) Anthracycline cardiotoxicity. Expert Opin Drug Saf 5:791–809

3.7 Non-cardioprotective Efficacy

Seppo W. Langer

3.7.1 Anthracycline Extravasation

Summary

For more than 40 years accidental extravasation of anthracyclines has been a feared complication of anticancer therapy. Much effort has been made to prevent and treat the devastating effects of extravasated anthracyclines. Dexrazoxane has proven to be an effective antidote in anthracycline extravasation and is now widely established as the treatment of choice (Fig. 3.4).

Anthracycline Extravasation

Extravasation is the unintentional leakage into the subcutaneous spaces of a drug during administration. Extravasation of anthracyclines (i.e. doxorubicin, daunorubicin, epirubicin, and idarubicin) may lead to severe, long-lasting ulceration and necrosis. Needless to say, extravasation of anthracyclines is a devastating and feared complication. Fortunately, the incidence is low. The use of central venous

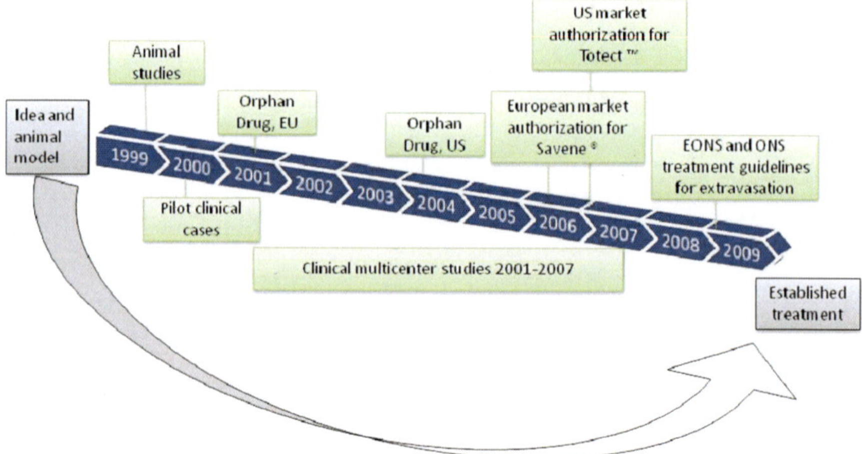

Fig. 3.4 From bench to bed. Starting in 1999 with the idea and the developement of a reproducible murine test model, through pilot patients and clinical trials, and ending with implementation into treatment guidelines. Orphan drug designation was granted for dexrazoxane by the European Medicines Agency (EMEA) in 2001 for the treatment of anthracycline extravasation. In 2004, The US Food and Drug Administration (FDA) also granted dexrazoxane orphan drug designation. In 2006, European marketing approval was obtained by TopoTarget A/S for Savene® in Europe. FDA approved Totect™ in US in 2007. Abbreviations: EU, Europe; US, United States of America; EONS, European Oncology Nurses Society; ONS, Oncology Nurses Society

access devices for infusion of chemotherapy may reduce but do not eliminate the risk of extravasation [1].

The clinical course of anthracycline extravasation is characterized by swelling and redness, often pain, and it may progress into blistering. In days to weeks it may cause progressive ulceration and necrosis [2–4]. Anthracyclines has the ability to persist in tissues for weeks or even months following extravasation.

Potential long-term sequelae in patients experiencing anthracycline extravasation include pain and serious joint and nerve damage, permanent disfigurement, and other cosmetic changes. Interruption or discontinuation of further scheduled cancer chemotherapy is also an important effect of the complication. In potentially severe cases of anthracycline extravasation, surgical debridement may be performed, which often requires subsequent skin grafting [4, 5].

Treatment options for anthracycline extravasation were limited before dexrazoxane was found to be a potent antidote against anthracycline extravasation injuries. The discovery and preclinical experiments were done by Buhl Jensen, Maxwell Sehested, and myself at Rigshospitalet in Copenhagen.

Preclinical Experiments

We published the first report on the experimental amelioration of subcutaneous injuries caused by anthracyclines with dexrazoxane in 2000 [6]. In a large series of mouse experiments, we induced ulcers with doxorubicin, daunorubicin, and idarubicin in female B6D2F1 hybrid mice by injections of different doses and types of anthracyclines. To mimic the clinical incident of an accidental extravasation, the anthracyclines were injected subcutaneously beneath the panniculus carnosus (the skin muscle layer of rodents) rather than intradermally. A wide range of treatment schedules with dexrazoxane were tested (Fig. 3.5). It turned out, that systemic treatment was highly efficacious in protecting against ulcers, and the protection obtained by triple-dose dexrazoxane was clearly superior to the protection obtained by a single-dose. The effect was highly significant regarding both the frequency of wounds and the sizes and healing period of occurring wounds. Moreover, early treatment was clearly more protective than late treatment. Hence, we could demonstrate that the protection depended on the dose of dexrazoxane as well as on the time and frequency of administration.

For the purpose of investigating a possible mechanism of action, dexrazoxane was tested against ulcers induced by hydrogen peroxide (i.e. ulcers induced by formation of toxic hydroxyl radicals), however with no signs of protection. Later, additional experiments were carried out to further explore the possible mechanism of action [7]. It turned out, that neither N-acetylcysteine, alfa-tocopherol, amifostine, merbarone, aclarubicin, ADR-925, nor EDTA had any inhibitory effect on anthracycline-induced ulcers in mice (Fig. 3.6). As a result, the experiments did not provide the fine clue to explain the exact mechanism of action behind the protective effect.

An often-used clinical treatment, topical cooling, did not show signs of reduction of the frequency, sizes, or duration of experimentally induced ulcers. Other

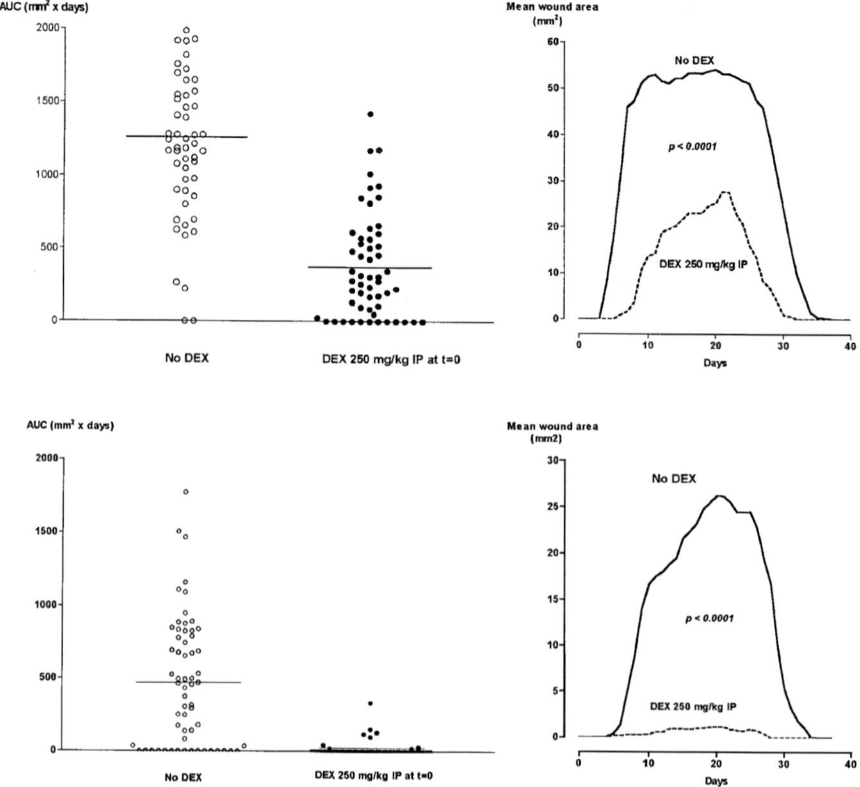

Fig. 3.5 A single systemic injection of DEX significantly reduces the wounds induced by SC daunorubicin and doxorubicin. *Top panel*: *Left*, scatter plot showing the distribution of the AUCs of individual mice after 3 mg/kg daunorubicin SC followed by saline IP (o; $n = 56$) or 250 mg/kg DEX IP at $t = 0$ (•; $n = 55$). *Right*, mean wound area vs. time of the same data as in the left graph. The difference in AUCs is highly significant, and the duration of wounds significantly shorter. *Lower panel*: Scatter plot and mean wound area vs. time curve showing the distribution of the AUCs of individual mice after 2 or 3 mg/kg doxorubicin SC followed by saline IP (o; $n = 56$) or 250 mg/kg DEX IP at $t = 0$ (•; $n = 55$). From [6]

treatments, that had some clinical impact was topical treatment with DMSO and topical or intralesional corticosteroid. Hence, these agents were investigated experimentally as well. In an additional series of mouse experiments, where IP dexrazoxane now served as the golden standard, it turned out, that DMSO and hydrocortisone showed no signs of ameliorating anthracycline induced ulcers [8] (Fig. 3.7).

In conclusion, the experiments in mice showed a hitherto unseen prevention of ulcers with systemic dexrazoxane against experimental extravasation injuries with several types of anthracyclines. The mechanism was not fully elucidated.

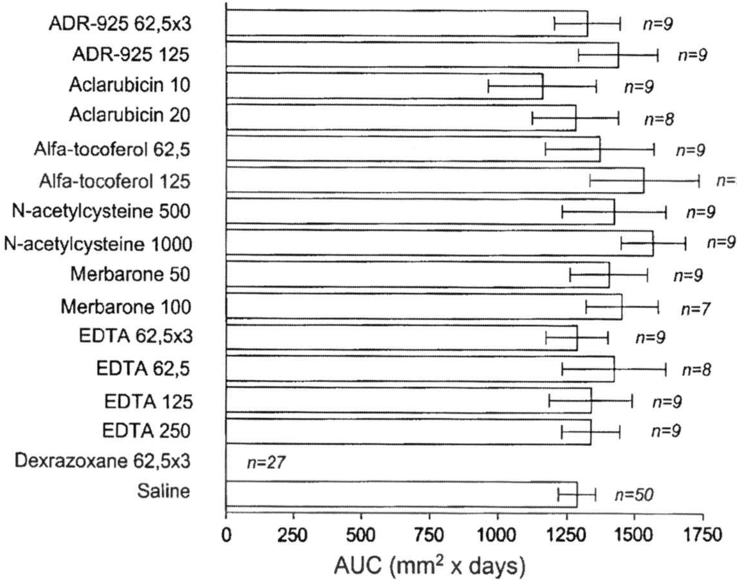

Fig. 3.6 Comparison of the mean area of ulceration over time (mean AUC in mm^2 × days) in mice receiving 3 mg/kg daunorubicin SC in 50 ml and IP adjuvants. Triple-dosage of IP dexrazoxane completely prevented wounds. Numbers after the adjuvants (*left*) correspond to the doses in mg. Bars represent SEM. From [7]

Clinical Studies

Based on the convincing murine data, a number of patients were treated as pilot cases. The publication of treatment success in the first three patients [9] was soon followed by several other cases that supported systemic dexrazoxane as an acute antidote [10–12].

Soon hereafter, three prospective multicenter clinical trials were initiated: TT01, TT02, and TT04. In order for the patients to be evaluable for measuring clinical efficacy, the presence of anthracycline in the extravasation area had to be verified by the positive fluorescence microscopy in punch biopsies taken from the injury site. Patients were entered from 24 different European oncology centres.

The primary objective of the first study, TT01, was to determine the efficacy of dexrazoxane in avoiding surgical intervention following anthracycline extravasation, thus preventing the development of late effects of the extravasation incident [13]. The secondary objectives were to avoid postponement of further planned cancer chemotherapy, to describe and evaluate subjective and objective symptoms/signs following dexrazoxane treatment, as well as evaluating the tolerability and/or toxicity of dexrazoxane. Twenty-three patients with suspected anthracycline extravasation entered this study. Dexrazoxane was administered as an IV infusion over 1–2 h: 1,000 mg/m^2 was given within 6 h after the extravasation injury; 1,000 mg/m^2 24 h later, and 500 mg/m^2 another 24 h later. Eighteen patients

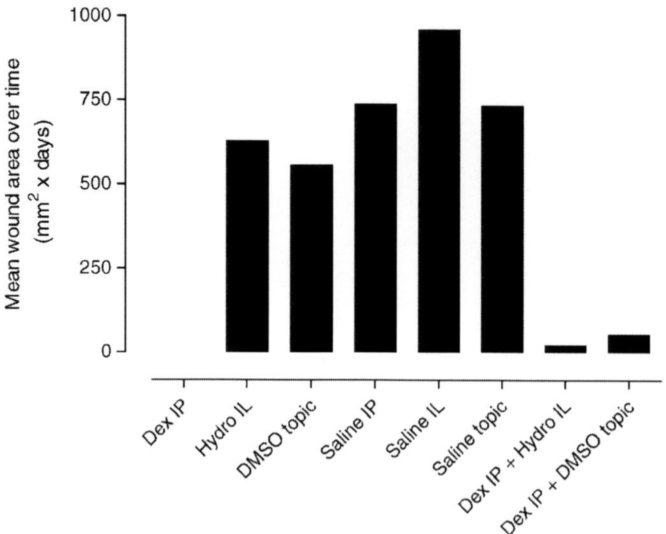

Fig. 3.7 Barplot of mean wound area over time (Y-axis in mm² × days) after subpannicular injection of daunorubin in mice and acute treatment with various potential antidotes. Dexrazoxane is the only antidote with statistically significant antidotal effect – and actually completely prevents ulceration (*far left*). IP dexrazoxane completely prevents ulcers. Intralesional saline increases ulceration. Adding of hydrocortisone or DMSO tends to decrease the efficacy of dexrazoxane. Abbreviations: Dex, dexrazoxane; Hydro, hydrocortisone; DMSO, dimethylsulfoxide; IP, intraperitoneal; IL, intralesional; topic, topical. From [8]

were evaluable for efficacy, and none of these patients underwent surgery due to the extravasation. Furthermore, no patients developed blistering after dexrazoxane treatment. Two-thirds of the patients continued their planned chemotherapy without delay. One third of the patients experienced a delay in planned chemotherapy of median 6 days. Two patients experienced mild, long-term sequelae as a result of the extravasation. No patients developed limitations of limb movement.

In the second study, TT02, 57 cancer patients with anthracycline extravasation as evidenced by pain, swelling and/or redness at the suspected leakage site and confirmed by fluorescence microscopy of two punch biopsies entered [13]. The endpoints were very similar to the first clinical study, and dexrazoxane was administered according to the same schedule. The efficacy parameters were nearly as encouraging as in the first study:

Thirty-six dexrazoxane treated patients were evaluable for efficacy. One of these patients underwent surgery as a consequence of a large extravasation, despite dexrazoxane therapy. Thirty-five out of 36 patients treated with dexrazoxane did not show signs of tissue destruction and thus did not require surgery.

Less than 30% of the patients experienced a delay in scheduled further chemotherapy; the median delay was of 9 days. Approximately one-third of the efficacy population had one or more sequelae, most commonly mild pain. The one patient who failed treatment and underwent surgical treatment experienced disfigurement, sensory disturbances, limitation of movements, and pain.

The third trial, TT04 [14], was a small pharmacokinetic study in which the patients with biopsy verified anthracycline extravasation had the same treatment as in TT01 and TT02, and furthermore had blood test taken for pharmacokinetic testing. The question was if the 3-day dexrazoxane schedule would lead to drug accumulation and possible increased side-effects. Six biopsy positive patients entered the study and received same treatment as in TT01 and TT02. No patients required surgical intervention. The pharmacokinetic data showed that there was no accumulation of dexrazoxane during the 3-day schedule. In all three clinical studies, the side effects were mild and manageable. The first treatments in study TT01 were associated with a high proportion of patients with local pain upon injection of dexrazoxane. However, after the diluent was changed, the reported incidence of injection site reactions decrease from approximately 60 to 14%. In all three trials, a slight, reversible increase in liver transaminases was noted.

Subsequent case stories have supported the study findings, and the treatment has also shown value in extravasations from central venous catheters [15–17]. Recently, treatment with Savene® became the recommended treatment of anthracycline extravasation by the European Nurses Society, and treatment with Totect™ by the Oncology Nurses Society [18, 19].

3.7.2 Other Possible Indications

Summary

A few studies have investigated dexrazoxane as an antineoplastic drug. They were however negative. A couple of other possible indications for the use of dexrazoxane as a protective drug have been pursued, but have not translated into clinical practice.

Experimental Studies

In experiments with doxorubicin-treated female Fischer rats, the addition of dexrazoxane resulted in reduced normal tissue toxicity, i.e. weight loss, renal toxicity, cardiomyopathy, and neuropathy [20]. Specific pulmonary protection was investigated in several studies: Dexrazoxane failed to show protection against pulmonary damage induced by bleomycin or a combination of bleomycin and hyperoxia in a hamster model [21]. On the other hand, in C57/BL6 mice dexrazoxane was shown to significantly reduce bleomycin-induced pulmonary fibrosis [22]. Dexrazoxane-treated hypoxic rats additionally showed decreased acute pulmonary changes compared to controls [23]. Liver protection has also been addressed. Hence, dexrazoxane administered before acetaminophen attenuated the increases in enzyme activities and reduced the incidence and severity of hepatocellular injury in male Syrian golden hamsters that received an acetaminophen overdose [24]. None of these experimental findings have undergone systematic clinical evaluation.

Clinical Studies

The antineoplastic effect of dexrazoxane was studied in a phase II study in 25 patients with squamous cell cancer of the head and neck region [25]. There were 2 partial responses. In a phase II study of 29 patients with advanced NSCLC, of whom 18 were chemonaive, no responders were seen after dexrazoxane [26]. A response rate of 0 was also the outcome of a phase II study of dexrazoxane administered to 40 patients with renal adenocarcinomas [27]. In a large paediatric phase I study with 46 patients, there was only one short-lived partial response in the bone marrow of a child with acute lymphocytic leukemia [28].

The results of human studies in non-proliferating conditions other than prevention of cardiomyopathy and extravasation injuries are sparse. In a phase I crossover study of epirubicin $\pm$ dexrazoxane, a small but significant reduction in severe vomiting and stomatitis was noted in patients receiving dexrazoxane [29].

By measuring the flow-mediated dilation in the brachial artery in a randomized, double-blind, crossover study in ten healthy volunteers, it was furthermore indicated that dexrazoxane abrogates homocysteine-induced endothelial dysfunction [30]. Further investigation of dexrazoxane as a protector against oxidative stress in certain cardiovascular diseases is therefore probably justified.

References

1. Langstein HN, Duman H, Seelig D, Butler CE, Evans GR (2002) Retrospective study of the management of chemotherapeutic extravasation injury. Ann Plast Surg 49: 369–74
2. Langer SW, Sehested M, Jensen PB (2009) Anthracycline extravasation: a comprehensive review of experimental and clinical treatments. Tumori 95(3):273–82
3. Sonneveld P, Wassenaar HA, Nooter K (1984) Long persistence of doxorubicin in human skin after extravasation. Cancer Treat Rep 68(6):895–6
4. Reilly JJ, Neifeld JP, Rosenberg SA (1977) Clinical course and management of accidental adriamycin extravasation. Cancer 40(5):2053–6
5. Preuss P, Partoft S (1987) Cytostatic extravasations. Ann Plast Surg 19(4): 323–9
6. Langer SW, Sehested M, Jensen PB. (2000) Treatment of anthracycline extravasation with dexrazoxane. Clin Cancer Res 6(9):3680–6
7. Langer SW, Sehested M, Jensen PB (2001) Dexrazoxane is a potent and specific inhibitor of anthracycline induced subcutaneous lesions in mice. Ann Oncol 12(3):405–10
8. Langer SW, Thougaard AV, Sehested M, Jensen PB. (2006) Treatment of anthracycline extravasation in mice with dexrazoxane with or without DMSO and hydrocortisone. Cancer Chemother Pharmacol 57(1):125–8
9. Langer SW, Sehested M, Jensen PB, Buter J, Giaccone G (2000) Dexrazoxane in anthracycline extravasation. J Clin Oncol Aug;18(16):3064
10. Jensen JN, Lock-Andersen J, Langer SW, Mejer J (2003) Dexrazoxane-a promising antidote in the treatment of accidental extravasation of anthracyclines. Scand J Plast Reconstr Surg Hand Surg 37(3):174–5
11. El-Saghir N, Otrock Z, Mufarrij A, Abou-Mourad Y, Salem Z, Shamseddine A, Abbas J (2004) Dexrazoxane for anthracycline extravasation and GM-CSF for skin ulceration and wound healing. Lancet Oncol May;5(5):320–1.

12. Bos AM, van der Graaf WT, Willemse PH (2001) A new conservative approach to extravasation of anthracyclines with dimethylsulfoxide and dexrazoxane. Acta Oncol 40(4): 541–2

13. Mouridsen HT, Langer SW, Buter J, Eidtmann H, Rosti G, de Wit M, Knoblauch P, Rasmussen A, Dahlstrøm K, Jensen PB, Giaccone G (2007) Treatment of anthracycline extravasation with Savene (dexrazoxane): results from two prospective clinical multicentre studies. Ann Oncol Mar;18(3): 546–50

14. Mouridsen HT, Langer SW, Tjoernelund J, Grundtvig P, Buter J, Knoblauch P, Jensen PB (2008) Anthracycline extravasation in breast cancer patients. Effective treatment with dexrazoxane in three multicenter trials. Eur J Cancer Suppl 6(7):197

15. Jordan K, Behlendorf T, Mueller F, Schmoll HJ (2009) Anthracycline extravasation injuries: management with dexrazoxane. Ther Clin Risk Manag Apr;5(2):361–6

16. Hünerlitürkoglou AN, Tapprich C, Langer SW, Sehested M, Jensen PB, Heintges T (2009) Successful use of dexrazoxane in two cases of anthracycline containing polychemotherapy. Eur J Clin Med Oncol Oct;1(1):13–5

17. Langer SW (2008) Treatment of anthracycline extravasation from centrally inserted venous catheters. Oncol Rev 2:115–7

18. Wengström Y, Margulies A (2008) European oncology nursing society task force. European Oncology Nursing Society extravasation guidelines. Eur J Oncol Nurs Sep;12(4): 357–61

19. Polovich M, Whitford JM, Olsen M (2009) Chemotherapy and biotherapy guidelines and recommendations for practice, 3rd edn. Oncology Nursing Society, Pittsburg

20. Baba H, Stephens LC, Strebel FR, Siddik ZH, Newman RA, Ohno S, Bull JM (1991) Protective effect of ICRF-187 against normal tissue injury induced by adriamycin in combination with whole body hyperthermia. Cancer Res 51(13):3568–77

21. Tryka AF (1989) ICRF 187 and polyhydroxyphenyl derivatives fail to protect against bleomycin induced lung injury. Toxicology Dec 1;59(2):127–38

22. Herman EH, Hasinoff BB, Zhang J, Raley LG, Zhang TM, Fukuda Y, Ferrans VJ (1995) Morphologic and morphometric evaluation of the effect of ICRF-187 on bleomycin-induced pulmonary toxicity. Toxicology Apr 12;98(1–3):163–75

23. Fukuda Y, Herman EH, Ferrans VJ (1992) Effect of ICRF-187 on the pulmonary damage induced by hyperoxia in the rat. Toxicology Sep;74(2–3):185–202

24. El-Hage AN, Herman EH, Ferrans VJ (1983) Examination of the protective effect of ICRF-187 and dimethyl sulfoxide against acetaminophen-induced hepatotoxicity in Syrian golden hamsters. Toxicology Nov;28(4):295–303

25. Wheeler RH, Bricker LJ, Natale RB, Baker SR (1984) Phase II trial of ICRF-187 in squamous cell carcinoma of the head and neck. Cancer Treat Rep 68:427–8

26. Natale RB, Wheeler RH, Liepman MK, Sauder A, Bricker L (1983) Phase II trial of ICRF-187 in non-small cell lung cancer. Cancer Treat Rep Mar;67(3):311–3

27. Brubaker LH, Vogel CL, Einhorn LH, Birch R (1986) Treatment of advanced adenocarcinoma of the kidney with ICRF-187: a Southeastern Cancer Study Group trial. Cancer Treat Rep 70(7):915–6

28. Holcenberg JS, Tutsch KD, Earhart RH, Ungerleider RS, Kamen BA, Pratt CB, Gribble TJ, Glaubiger DL (1986) Phase I study of ICRF-187 in pediatric cancer patients and comparison of its pharmacokinetics in children and adults. Cancer Treat Rep Jun;70(6): 703–9

29. Basser RL, Sobol MM, Duggan G, Cebon J, Rosenthal MA, Mihaly G, Green MD (1994) Comparative study of the pharmacokinetics and toxicity of high-dose epirubicin with or without dexrazoxane in patients with advanced malignancy. J Clin Oncol 12(8): 1659–66

30. Zheng H, Dimayuga C,Hudaihed A, Katz SD (2002) Effect of dexrazoxane on homocysteine-induced endothelial dysfunction in normal subjects. Arterioscler Thromb Vasc Biol 22: e15–8

3.8 Neurodegenerative Diseases, a Future Avenue for Razoxane and Dexrazoxane Therapeutic Use?

Nigel H. Greig, Robert E. Becker, Kurt Hellmann

Neurodegeneration is a complex and multifaceted process that leads to numerous different but increasingly common pathological conditions. Although a variety of factors, both genetic and environmental, are involved in neurodegeneration and define the vulnerable cell population affected and, thereby, the resulting specific disorder, they likely share similar critical metabolic processes. Key among those that induce neuronal dysfunction and death are protein aggregation and oxidative stress, both of which are associated with the involvement of metal ions. This provides rationale for the use of chelation therapy as a treatment strategy for specific neurodegenerative conditions, such as Alzheimer's disease, and the use of clinically effective multi-functional drugs that combine this with other potentially valuable pharmacological actions, as found with the cancer drugs razoxane and dexrazoxane.

3.8.1 Background

Improvements in preventative, diagnostic and therapeutic measures for numerous forms of cancer as well as of cardiovascular disease over past decades, have led to a gradual rise in average lifespan in North American and European populations. Undesirably, accompanying this increase there has been a progressive rise in the number of individuals afflicted with age-related neurodegenerative disorders, epitomized by Alzheimer's disease (AD) and Parkinson's disease (PD). Although different cell types and brain areas are vulnerable between these, each disorder likely develops from activation of a common final cascade of biochemical and cellular events that ultimately lead to neuronal dysfunction and death. Different triggers, including oxidative damage to DNA, the over-activation of glutamate receptors and disruption of cellular calcium homeostasis, although initiated by different genetic and/or environmental factors, can instigate a shared cascade of intracellular events causing either acute or chronic perturbation of physiological function that leads to neuronal cell death and clinical manifestations. Invariably, inflammatory reaction accompanies the pathological processes that underpin neurodegeneration, and can be initiated in numerous ways [1–3]. For example, glial activation occurs as part of a defense mechanism to remove unwanted cell debris in order to facilitate tissue repair. Regrettably, such beneficial actions can go amiss. Activated microglial cells can generate and secrete pro-inflammatory cytokines in addition to cytotoxic elements, like reactive oxygen species (ROS), nitric oxide (NO) and excitatory amino acids (e.g., glutamate), to thereby instigate a self-propagating cycle of events that drives disease progression. Intervening in pathways early during this process has the potential to slow disease course. Recent studies suggest that multi-functional agents that effectively combine metal-chelating with antioxidant and other actions may intervene in the processes that lead to AD and PD and, thereby, represent a useful treatment [4]. The clinically relevant bisdioxopiperazines, razoxane

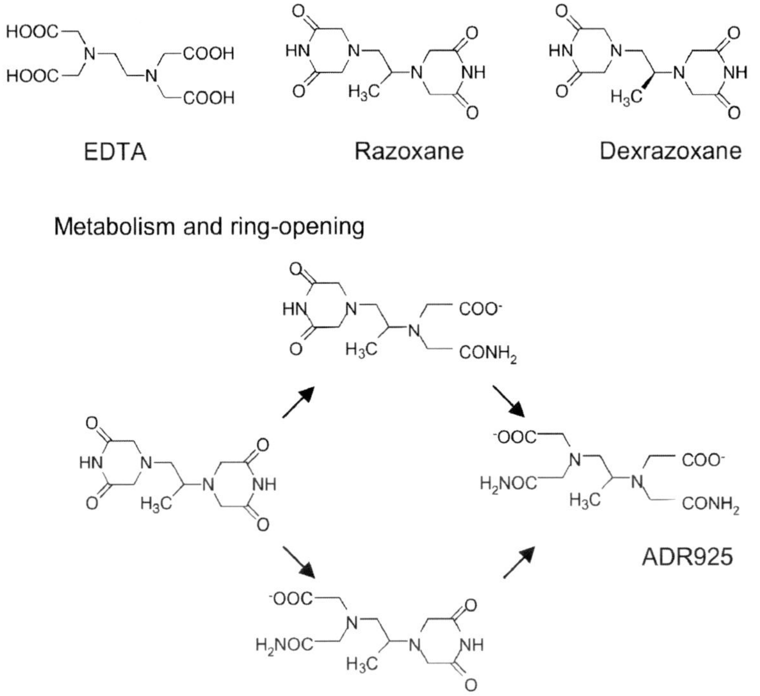

Fig. 3.8 *Top*: Chemical structures of razoxane, dexrazoxane and the classical chelating agent, EDTA. *Bottom*: Metabolism and ring opening of razoxane/dexrazoxane [61–63]. Initial hydrolysis converts razoxane/dexrazoxane to a single ring-opened analogue by zinc hydrolase DHPase (EC, Section 3.5.2.2). Subsequent hydrolysis by zinc hydrolase DHOase (EC, Section 3.5.2.3) generates the fully ring-opened metabolite, ADR925 (for dexrazoxane or ICRF 198 in the case of razoxane). The ring-opened species are purported to be the active chelating entities, with razoxane and dexrazoxane acting as prodrugs to generate favorable pharmacokinetic and pharmacodynamic profiles for the active species [61–65]. Other actions of these drugs (e.g., topoisomerase II action) likely are mediated via the primary compounds themselves

and dexrazoxane (Fig. 3.8), share these features and may hence have utility in the treatment of neurodegenerative conditions. AD and PD provide examples of interventions currently under study in neurodegeneration.

3.8.2 Alzheimer's Disease

AD, a progressive, degenerative disorder of the brain, is the most widespread cause of dementia amongst the elderly. Typified by an increasing impairment of memory that is accompanied by psychiatric disturbances, behavioral anomalies arise from dysfunction and death of neurons in brain regions implicated in cognition and mood, exemplified by the hippocampus, entorhinal cortex, basal forebrain, and frontal and parietal lobes [5]. Neuropathologically, AD is distinguished by the presence of amyloid deposits, neurofibrillary tangles (NFTs), synapse loss and a deficit of presynaptic markers of the cholinergic system in the described brain regions [5–8]. AD has a heterogeneous etiology: a large percent of AD occurs with unknown causes, sporadically and with increasing aging after 65; a smaller fraction, early onset familial AD (FAD), arises before age 65 in association with mutations in one of several genes linked with β-amyloid precursor protein (APP) and presenilins (PS1, PS2) [5–9]. A shared feature of these numerous mutations is that, albeit via different mechanisms, they promote an increase in Aβ synthesis; particularly of the longer more hydrophobic $A\beta_{42}$ form [5–9]. Data indicates that soluble aggregates of Aβ, exemplified by dimers [10], the dodecameric (56 kDa) form [11] and Aβ-derived diffusible ligands (ADDLs) [12], target synapses and impair memory. These, collectively with the secretases (β and γ) involved in the cleavage of APP to generate Aβ are current AD drug targets, as are strategies focused to overcome the consequences of tau hyperphosphorylation, apolipoprotein E, and inflammation that have been reported to also drive the progression of AD [7, 13–16].

Although advances have increased our understanding of the genetic, cellular and molecular events that lead to AD, their translation to clinical management has been slow, and cholinesterase inhibitors (ChEIs) and the NMDA antagonist memantine remain the principal treatment of mild to moderate AD subjects [7, 17–19]. Such treatment is chiefly symptomatic, elevating neurotransmitter levels with ChEIs identified as affected in AD brain, and minimally influences disease course. On the contrary, the characterization of targets, such as Aβ, implicated in disease pathology provides the potential to slow disease progression [7, 20, 21].

Determining the efficacy of AD drug candidates is both arduous and blighted by numerous potential pitfalls that may negate efficacy [22–24]. Assessing a statistical change in the course of a chronic degenerative disease, whose progression is known to be variable between individuals, requires a long duration trial (18 months), compared to a symptomatic treatment (6 months), and numerous clinical trials have ended without the ability to differentiate drug efficacy from clinical trial design failure [23, 24]. This, together with the modest effect of currently available symptomatic AD drugs, has fueled two avenues of research: one to develop a new generation of ChEIs with activity beyond the modest symptomatic levels of initial agents, and another to optimize agents with a known history of effective clinical use in a different indication that are multi-functional to mechanisms relevant in AD. It has become increasingly clear that neuropathological events that underlie the development of AD and once considered separate, are complexly linked. Biochemical cascades that induce the awry production of Aβ may alter tau phosphorylation or mechanisms regulating neurotransmitter synthesis [5, 7, 8, 15]. In a reciprocal manner, alterations in cholinergic function may feed back on APP processing and subsequent Aβ and both N- and C-terminal fragment (CTF) generation, each of which has been reported toxic [25, 26].

3.8.3 Parkinson's Disease

PD, a major progressive neurological disorder that afflicts approximately 1% of the population over the age of 50 years, is the second most common neurodegenerative condition after AD. Unlike AD, which impacts memory and behavior, PD is distinguished by a progressive loss of control over voluntary movement. Its symptoms comprise tremors at rest, slow movement (termed bradykinesia) that can develop into a lack of movement/freezing (akinesia), muscle stiffness/rigidity, a loss of balance and gait/posture changes, speech/swallowing disturbances, muscle pains (myalgia) and abnormal contractions (dystonia), and fatigue [27]. In addition, neuropsychiatric disturbance can occur and impact cognition, mood and behavior [28]. In large part, the neuropathogenesis of PD is characterized by a preferential loss of dopaminergic neurons (dopamine producing and releasing neurons) within the substantia nigra pars compacta (SNpc), situated in the midbrain [27]. However, cell loss may additionally occur within the locus coeruleus, dorsal nuclei of the vagus, raphe nuclei, and nucleus basalis of Meynert (a prime area of neuronal loss in AD [5]), as well as in other catecholaminergic brain stem structures [29]. This is accompanied by the presence of proteinaceous inclusions, the Lewy body (LB), pale body, and the Lewy neurite that differ slightly in their morphology, with the latter as potential precursors of LBs. LBs are primarily composed of α-synuclein, generally an abnormal, post-translationally modified, and aggregated form of this presynaptic protein, and additionally contain ubiquitinated proteins [30]. Debate is ongoing as to whether LBs either directly cause neuronal death or sequester neurotoxic protein aggregates to, thereby, sustain neuronal viability, particularly as LBs are not constant features and may both appear and disappear in cells [29]. Interestingly, however, mutations in several genes linked to inherited rare forms of PD [29], such as with α-synuclein, *LRRK*-2, and *GBA* mutations [31], supports the notion that these mutations promote inclusion formation, ubiquitin-proteosome system dysfunction and nigral cell death, albeit via yet to be determined mechanisms.

Existing PD treatments provide only partial symptomatic relief of the disease to improve quality of life without affecting its progression. The drug levodopa (L-dopa), in combination with a peripheral decarboxylase inhibitor (benserazide or carbidopa), is the current most effective therapy and initial treatment option. This provides the brain a dopamine precursor that can cross the blood-brain barrier, which can be converted into dopamine by surviving dopaminergic neurons [27]. Dopamine loss within the brain can be potentially slowed by type B monoamine oxidase inhibitors, as this enzyme degrades dopamine into dihydroxyphenyl acetic acid (DOPAC) and hydrogen peroxide, the latter of which can induce oxidative stress. A further approach is the use of dopamine receptor agonists, exemplified by apomorphine, bromocriptine, pramipexole and ropinirole (Requip), separately or together with anticholinergic drugs [32]. Whereas all have demonstrated clinical value, none impact mechanisms likely involved in dopaminergic cell dysfunction and loss during PD.

3.8.4 Metals in AD and PD Brain

Increasing evidence indicates that metals, and particularly iron, accumulation in the brain can cause a wide range of neurological disorders. Iron, for example, progressively accumulates within brain during age, to potentially impact a number of finely balanced biological systems [33]. Key among these is iron-induced oxidative stress (OS) that can induce neurodegeneration [34]. It is well recognized that metal ions can accept and donate electrons that can lead to radical formation, reactive nitrogen and oxygen species (ROS) and oxidative attack of tissue components to induce and/or contribute to disease and aging processes. Specifically, free iron as well as copper can induce OS via its interaction with hydrogen peroxide (the Fenton reaction), which generates an increased formation of hydroxyl free radicals that, ultimately, cause molecular damage, cellular dysfunction and, eventually, cell death [34].

Steady-state levels of iron, copper and zinc appear to be elevated in AD brain, particularly in vicinity of amyloid plaques [35], and iron, likewise, has been reported elevated in AD transgenic mice [36]. Not only can these metals generate ROS, but they can additionally interact with Aβ to potentially mediate OS mechanisms associated with its toxicity. For example, in neuronal cultures, Cu^{2+} and Fe^{3+} ions elevate Aβ toxicity whereas Zn^{2+} ions attenuates it [37], in addition, these metals have been reported to impact the oligomerization rate of Aβ, whereby, more toxic species of this peptide are generated [38]. Multiple studies have now confirmed that iron promotes the aggregation of α-synuclein, and removal of free iron with a chelating agents blocks aggregation [39]. Interestingly, elevated levels of iron have also been reported in PD brain, particularly within the SNpc [37]. Indeed, this area, together with the basal ganglia, globus pallidus, caudate nucleus and putamen appear to have the highest brain iron content, and several of these same areas are associated with the dopaminergic system.

Metals, exemplified by iron, appear to be essential for fundamental brain processes like neurotransmission and myelination. As an example, iron is a cofactor for tyrosine hydroxylase that catalyses the hydroxylation of tyrosine to generate 3,4-dihydroxyphenylalanine (DOPA), the catecholamine precursor for dopamine, adrenaline and noradrenaline synthesis. Likewise, tryptophan hydroxylase contains iron to catalyse the initial step for generation of serotonin from tryptophan. Iron, additionally, is involved in neurotransmitter metabolism, as a cofactor for monoamine oxidase (MAO) in the oxidative deamination of serotonin, adrenaline and noradrenaline [39].

Iron Homeostasis and Interactions with APP and α-Synuclein

Cellular iron homeostasis is orchestrated by the coordinated expression of a series of proteins involved in its uptake, storage, utilization and export. Whereas, these proteins are regulated at multiple levels, a fundamental central one occurs at a post-transcriptional level via an iron-responsive element (IRE)/iron-regulatory protein (IRP) regulatory network [40]. IRP1 and –2 interact with cis-regulatory mRNA motifs, highly conserved RNA stem loops that are IREs, which are localized within the 3′ or 5′ untranslated regions (UTRs) of specific mRNAs that encode proteins involved iron uptake/transport (transferrin receptor 1 and divalent metal transporter1) iron storage (ferritin H and L), iron utilization (erythroid 5-aminolevulinic acid synthase, mitochondrial aconitase, and hypoxia-inducible factor2) and iron excretion (ferroportin1) [39, 40]. IRP activity is regulated post-translationally by cellular iron availability, and so when iron abundance is depleted, IRP binding to the 3′ UTR of ferritin H or ferroportin1 can inhibit mRNA translation. By contrast, IRP binding to the 3′ UTR of transferrin receptor 1 regulates its mRNA stability, and in a careful coordinated manner iron levels can be raised by modulating each of the numerous proteins involved in its uptake, export, utilization and storage. Studies by Pinero et al. [41], in AD brain, suggest that the normally well orchestrated iron regulatory system can become dysfunctional, leading to a change in RNA binding affinity, enhanced iron transport into cells in the presence of lowered ferritin levels, and thus a diminished cellular iron storage capacity lead to potential cellular dysfunction and death (for review see [42, 43]).

Interestingly, like ferritin, APP is a metalloprotein, and intracellular levels of APP are reported to be regulated by a mechanism that resembles the mechanism via which iron modulates the translation of ferritin H and L mRNAs through IRE stem loops in their 5′ UTR [44]. Alignment studies between APP mRNA and ferritin showed high homology (greater than 70% sequence similarity) in key areas within their 5′ UTRs, with that in APP being a non-canonical 'type-II' IRE structure characterized by an 11 base stem loop (differing from a 6 base one in all other thus far identified IRE-containing mRNAs) [45]. This IRE within APP mRNA appears to be fully functional, binds IRP-1, and not only appears to up and down regulate APP protein expression in high and low iron conditions in cell cultures, but has additionally provided a target to support drug screening [46, 47]. Additional studies of this same 5′ UTR region of APP mRNA have localized an IL-1 responsive element, through which inflammation appears to up regulate APP holoprotein levels, as well as a 'CAGA' box that appears to be amyloid species specific [48] and has the same sequence responsible for binding of IRP-1 to the core IRE sequence in the APP transcript [45–47]. Of note, these features are not shared by APP-like proteins (APLP-1 and −2) [48], and thus provide a handle to pharmacologically down-regulate APP protein levels, and peptides, like Aβ, that derive from APP cleavage. In keeping with this, cellular drug screens targeted to the APP 5/ UTR have consistently identified chelating agents as a drug class that lowers APP and Aβ [46, 47].

More recent parallel studies by Friedlich et al. [49] have identified sequences in the 5′ UTR of α-synuclein mRNA that are also homologous to the IRE stem-loops encoded in the 5′ UTR of ferritin L and H, and open the opportunity of similar pharmacological manipulation of α-synuclein protein levels.

Tau Hyperphosphorylation and APP, a Metalloprotein Whose Cleavage Yields Aβ

Major histopathologic abnormalities in AD brain are the neurofibrillary pathology comprising of NFTs of neurofibrillary tangles (NFTs), dystrophic neurons associated with senile plaques comprised of Aβ, and neuropil threads [5–8]. The former contain filamentous structures, termed paired helical filaments (PHFs), and the allied straight filaments. Both of these are formed from hyperphosphorylated tau (PHFτ). Tau protein (τ) is fundamental in maintaining the microtubular integrity of neurons. The hyperphosphorylation of τ impedes its ability to bind with microtubules, and the resulting aggregation of PHFτ results in its abnormal accumulation as neurofibrillary pathology. Metals, and in particular ionic aluminium and iron, have been shown to accumulate in association with NFTs, and specific binding sites (Ser 202, Ser 396, Ser 404 and Ser 422) have been defined [50–52]. This binding appears to induce aggregation of PHFτ and generation of NFTs, and, similarly, is involved in Aβ aggregation [53].

Aβ, together with N-terminal APP and intracellular C-terminal products that have, additionally, been reported to possess neurotoxicity [5, 25, 26], are proteolytic cleavage products of APP. APP is an abundant and ubiquitously expressed metalloprotein that spans membranes in the endoplasmic reticulum, trans-Golgi and cell membrane of most cells [5, 43]. It has been shown to be a redox active copper-zinc binding protein [54] and may additionally bind iron [43]. During health, APP is predominantly cleaved by a group of metalloproteases, the α-secretases, to generate a secreted ectodomain that is non-amyloidogenic [5]. By contrast, during the course of AD and to a small extent during health, APP is proteolysed via a different pathway to generate the 40–42 amino acid Aβ peptide [5–8]. Key in this cleavage is γ-secretase, a multimer that comprises Pen2, presenilin, APH1 and nicastrin [5–9], and β-secretase [5–9]. Chaperones, such as ApoE and alpha-1 antichymotrypsin, together with metals, clearly add to this environment to augment dimerization and aggregation [37, 38].

3.8.5 Chelating Agents in AD and PD

Metal ion chelators have long been utilized in the treatment of diseases involving metal ion imbalance and, for neurodegenerative disease, could potentially work on multiple levels, such as to lower ROS and reduce interactions with Aβ and PHFτ known to augment aggregation, as well as to limit interactions with elements in APP and α-synuclein mRNA involved in translational regulation. Chelation therapy has been assessed in a number of neurodegenerative conditions, including AD and PD and is the subject of several recent reviews [4, 43, 55]. Consequent to their availability, several AD clinical trials have focused on desferrioxamine and clioquinol.

Desferrioxamine is a water-soluble chelator with an affinity for copper, zinc, iron and aluminium, which has a history of use in iron overload disorders. It is a bacterial siderophobe with a relatively high molecular weight (MW 580.68) that is generally administered subcutaneously, but has a relatively poor brain entry. Administered to subjects with AD by intramuscular route twice daily, 5 days a week over 2 years, it appeared to slow disease course as assessed by measures of activities of daily living recorded by video [56]. Albeit, that the study utilized a relatively blunt instrument to define differences in disease progression, more recent preclinical studies have shown desferrioxamine-induced reductions in APP levels in neuronal cell cultures [46, 47]. By contrast, clioquinol (MW 305.5) is an antifungal/antiprotozoal agent that has zinc and copper chelating specificity, together with a reasonable lipophilicity to allow brain entry but evidence of neurotoxicity following high dose

[57]. Administered in a double-blind AD pilot clinical trial of 36 weeks, the compound lowered plasma Aβ levels, relative to controls, and improved cognition in severe subjects [58]. In accord with this, in preclinical animal and cell culture studies, clioquinol, together with a second-generation compound, PBT2, has lowered Aβ levels [59].

The earliest reported effective use of chelation therapy was in 1946, utilizing ethylene diamine tetracetic acid (EDTA) to remove plaque producing calcium products [60], and this highly water-soluble agent (MW 292.2) has been extensively utilized since in industry and medicine. EDTA is highly charged and does not have appropriate physicochemical characteristics for use as a long-term drug. By contrast, the clinically approved drugs, razoxane and dexrazoxane that are highly effective in reducing anthracycline-induced cardiotoxicity and extravasation injury and additionally possess anticancer activity, can be considered to be prodrugs that on hydrolysis/ring-opening generate iron chelating EDTA-type structures [61] (Fig. 3.8). The ring-opened metabolite, ADR-925 (of dexrazoxane, that is known as ring-opened metabolite ICRF 198 for razoxane), binds Fe^{2+} and Fe^{3+} with potent formation constants of 10^{10} and $10^{18.2}$ M^{-1}, respectively, as compared to EDTA with formation constants of $10^{14.3}$ and $10^{25.1}$ M^{-1} [62, 63]. However, razoxane and dexrazoxane are considerably more permeable to cells and can thus be considered to function as chelating prodrugs [64].

Razoxane and dexrazoxane hence exert their numerous pharmacological actions via two or more mechanisms. First, by virtue of their EDTA-like structures, they are potent chelators of iron and other metal ions. It is this action that is predicted to impart their cardioprotective effects by either binding free or loosely bound iron, or iron complexed to an anthracycline (e.g., doxorubicin) that, thereby prevents or reduces site-specific oxygen radical generation that then damages cellular components. Second, razoxane and dexrazoxane are potent inhibitors of topoisomerase II, a mechanism that likely underpins their anticancer actions [62, 65]. To be expected, additional as yet unknown mechanisms likely underpin the actions of razoxane on the normalization of tumor blood vessels [65, 66], and hence it and its clinically effective analogues can truly be considered as multi-functional agents.

Razoxane/Dexrazoxane in AD Models

Following a methodological design previously utilized to characterize the activity of experimental AD drugs to lower APP and Aβ levels in cell culture and small animal models [67], the actions of razoxane were assessed in SH-SY5Y human neuroblastoma cells and in AD double transgenic (APP Swedish mutation + PS1 mutation [68]) mice. Razoxane, in addition to dexrazoxane (not shown), induced a dose-dependent decline in secreted APP levels without loss of cell viability (assessed by MTS assay, not shown). As illustrated in Fig. 3.9, razoxane proved to be more effective in lower secreted APP levels than equimolar clioquinol, that latter inducing a loss of cell viability in concentrations above 10 μM. To assess whether or not actions of razoxane translated from cell cultures to animals, AD transgenic mice that generate human Aβ were administered once daily doses of 20 and 30 mg/kg razoxane for 21 consecutive days. As illustrated in Fig. 3.10, the latter dose induced a significant

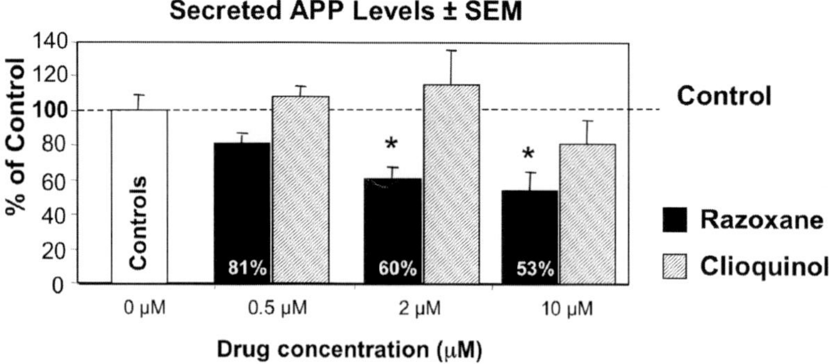

Fig. 3.9 Razoxane dose-dependently lowers APP levels in an AD cellular model. SH-SY5Y human neuroblastoma cells (American Type Culture Collection, Manassa, VA), were grown in a 1:1 mixture of Eagle's Minimum Essential Medium and Ham's F12 Medium supplemented with 10% heat-inactivated fetal bovine serum (FBS) and 100 U/mL penicillin/streptomycin (Invitrogen, Carlsbad, CA) at 37°C in a humidified incubator with 5% CO_2 and 95% air. On reaching 80% confluence, cells were cultured in low FBS (1%) for 16 h alone or in the presence of either razoxane or clioquinol. For protein analysis, an equal volume of conditioned media (CM) was mixed with Laemmli protein sample loading buffer and denatured at 95°C. Fifteen microliter of the denatured CM was then loaded on to a polyacrylamide-SDS criterion gel (10%) (Bio-Rad, CA) and electrophoresis was run for 1.2 h. Proteins from the gel were then transferred onto a PVDF membrane. This was blocked with 5% non-fat milk, and the membrane was then probed with a primary antibody (22C11, Chemicon, MA, USA) against the N-terminus 66–81 residues of APP. Chemoluminescence signals were obtained, and densitometric quantification of the protein bands was performed by using a PC version of NIH IMAGE (ImageJ software). For complete methodology see [67]

decline in the more hydrophobic and toxic $A\beta_{1-42}$ form, without evident side effects. Given earlier research with similar effects from phenserine and posiphen operating through the 5′ UTR [69–71], but with some dissociation of effects on APP and AB_{42}, razoxane may also be expected to affect α-synuclein synthesis directly by what the APP-AB_{42} dissociation suggest are not yet fully characterized mechanisms of action.

Although razoxane, unlike dexrazoxane, is only sparingly water-soluble, neither are expected to dramatically enter the brain [72], with a reported CSF/plasma of 0.1 [73] and a computed log p value of –2.7 (log octanol/water partition coefficient value: an indication of brain uptake). Interestingly, dexrazoxane has been found to exert neuroprotective activity in mice subjected to global cerebral ischemia inducing both a reduction in mortality and neurological deficits [73]. This activity, additionally, proved to be superior to desferrioxamine as well as the N-methyl-d-aspartate (NMDA) receptor antagonist, dizocilpine, and the calcium channel blocker, nimodipine [73], and suggests that centrally effective concentrations of razoxane can be achieved in vivo. In addition, the doses chosen for assessment in the AD mouse model illustrated in Fig. 3.10 are far lower than maximum tolerated ones that are in excess of 200 mg/kg. Together, these data suggest that albeit razoxane and dexrazoxane were not developed as centrally mediated drugs, they may prove useful in specific neurological disorders.

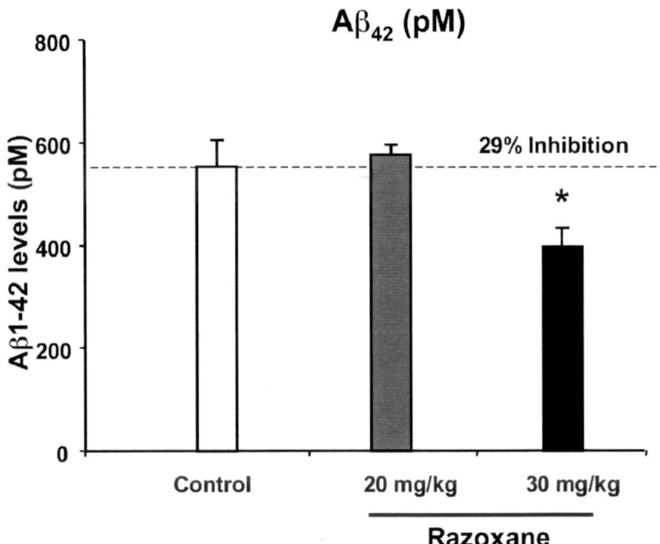

Fig. 3.10 Razoxane lowers brain levels of Aβ in AD transgenic mice. Razoxane was administered to male double transgenic mice (APP Swedish mutation + PS1 mutation), 5–6 months of age and approximately 22–28 g weight, by the intraperitoneal (IP) route once daily for 21 consecutive days. Control animals received vehicle. Within 3 h of the final razoxane/vehicle administration, animals were killed, and a 50 mg sample of cerebral cortex was collected and frozen to -80°C. Thereafter, samples were probed for human Aβ by a specific sandwich ELISA for Aβ$_{1-42}$, after formic acid extraction (for methodology see [67])

Synopsis

A consistent factor in AD and PD is the dysregulation and brain accumulation of metal ions, and in particular iron, that appears to be central to the progression of these conditions. Due to its redox nature, iron and other metals can react with endogenous hydrogen peroxide to produce hydroxyl radicals, which are both labile and harmful to neurons, inducing cellular dysfunction, apoptosis and neuroinflammation. In addition, free iron in brain can impact the expression of key proteins, such as APP and α-synuclein, that are central to the biological cascades leading to AD and PD, and together with other metal ions can both drive the dimerization and amplify the toxicity of Aβ. An additional commonality that links AD and PD is the presence of neuroinflammation that invariably drives the disease process at numerous levels [1–3]. One, amongst many, is the induction of inflammatory cytokines by Aβ [1–3, 5], which can then feed back to elevate APP synthesis [44, 45] and Aβ levels, thereby creating a self-propagating cycle. Interestingly, razoxane appears to possess clinically relevant antiinflammatory activity [74], that could provide additional value in AD or PD treatment. Clearly, strategies to safely reduce APP and Aβ in AD, α-synuclein in PD and other pathological metabolic processes in neurodegeneration, whether via potential chelating activity or through yet to be characterized mechanisms, are of major clinical interest. The induction of such activity at well tolerated doses by a drug developed for an alternate indication that has a long history of successful clinical use and a well characterized toxicological profile provides an appealing rationale to assess the therapeutic potential of razoxane's and analogues in AD and PD.

Acknowledgments This work was supported in part by the Intramural Research Program of the National Institute on Aging, National Institutes of Health. Conflicts of interest for authors: none.

References

1. Tweedie D, Sambamurti K, Greig NH (2007) TNF-alpha inhibition as a treatment strategy for neurodegenerative disorders: new drug candidates and targets. Curr Alzheimer Res 4:378–85
2. Maccioni RB, Rojo LE, Fernández JA, Kuljis RO (2009) The role of neuroimmunomodulation in Alzheimer's disease. Ann N Y Acad Sci 1153: 240–6
3. Frank-Cannon TC, Alto LT, McAlpine FE, Tansey MG (2009) Does neuroinflammation fan the flame in neurodegenerative diseases? Mol Neurodegener 16:4:47
4. Avramovich-Tirosh Y, Amit T, Bar-Am O, Weinreb O, Youdim MB (2008) Physiological and pathological aspects of Abeta in iron homeostasis via 5ʹ UTR in the APP mRNA and the therapeutic use of iron-chelators. BMC Neurosci 9(Suppl 2):S2
5. Sambamurti K, Greig NH, Lahiri DK (2002) Advances in the cellular and molecular biology of the beta-amyloid protein in Alzheimer's disease. Neuromolecular Med 1:1–31
6. Selkoe DJ (2005) Defining molecular targets to prevent Alzheimer disease. Arch Neurol 62:192–5
7. Lahiri DK, Farlow MR, Greig NH, Giacobini E, Schneider LS (2003) A critical analysis of new molecular targets and strategies for drug developments in Alzheimer's disease. Curr Drug Targets 4(2):97–112
8. Sambamurti K, Suram A, Venugopal C, Prakasam A, Zhou Y, Lahiri DK, Greig NH (2006) A partial failure of membrane protein turnover may cause Alzheimer's disease: a new hypothesis. Curr Alzheimer Res 3(1): 81–90
9. Sisodia SS, St George-Hyslop PH (2002) gamma-Secretase, Notch, Abeta and Alzheimer's disease: where do the presenilins fit in? Nat Rev Neurosci 3:281–90
10. Li S, Hong S, Shepardson NE, Walsh DM, Shankar GM, Selkoe D (2009) Soluble oligomers of amyloid Beta protein facilitate hippocampal long-term depression by disrupting neuronal glutamate uptake. Neuron 62: 788–801
11. Lesne S, Koh MT, Kotilinek L, Kayed R, Glabe CG, Yang A, Gallagher M, Ashe KH (2006) A specific amyloid-beta protein assembly in the brain impairs memory. Nature 440:352–7
12. Lacor PN et al (2004) Synaptic targeting by Alzheimer's-related amyloid beta oligomers. J Neurosci 24:10191–200
13. Hardy J (2006) Has the amyloid cascade hypothesis for Alzheimer's disease been proved? Curr Alzheimer Res 3:71–3
14. Lahiri DK, Chen DM, Lahiri P, Bondy S, Greig NH (2005) Amyloid, cholinesterase, melatonin, and metals and their roles in aging and neurodegenerative diseases. Ann N Y Acad Sci 1056:430–49
15. Sambamurti K et al (2004) Cholesterol and Alzheimer's disease: clinical and experimental models suggest interactions of different genetic, dietary and environmental risk factors. Curr Drug Targets 5:517–28
16. Lahiri DK, Greig NH (2004) Lethal weapon: amyloid beta-peptide, role in the oxidative stress and neurodegeneration of Alzheimer's disease. Neurobiol Aging 25:581–7
17. Cummings JL (2003) Use of cholinesterase inhibitors in clinical practice: evidence-based recommendations. Am J Geriatr Psychiatry 11:131–45

18. Farlow MR (2004) Utilizing combination therapy in the treatment of Alzheimer's disease. Expert Rev Neurother 4:799–808
19. Doraiswamy PM, Xiong GL (2006) Pharmacological strategies for the prevention of Alzheimer's disease. Expert Opin Pharmacother 7:1–10
20. Marlatt MW et al (2005) Therapeutic opportunities in Alzheimer disease: one for all or all for one? Curr Med Chem 12:1137–47
21. Courtney C et al, AD2000 Collaborative Group (2004) Long-term donepezil treatment in 565 patients with Alzheimer's disease (AD2000): randomised double-blind trial. Lancet 363(9427):2105–15
22. Becker RE, Greig NH (2008) Alzheimer's disease drug development in 2008 and beyond: problems and opportunities. Curr Alzheimer Res 5:346–57
23. Becker RE, Greig NH, Giacobini E (2008) Why do so many drugs for Alzheimer's disease fail in development? Time for new methods and new practices? J Alzheimers Dis 15:303–25
24. Becker RE, Greig NH (2008) Alzheimer's disease drug development: old problems require new priorities. CNS Neurol Disord Drug Targets 7:499–511
25. Nikolaev A, McLaughlin T, O'Leary DD, Tessier-Lavigne M (2009) APP binds DR6 to trigger axon pruning and neuron death via distinct caspases. Nature 457(7232):981–9
26. Galvan V, Gorostiza OF, Banwait S, Ataie M, Logvinova AV, Sitaraman S, Carlson E, Sagi SA, Chevallier N, Jin K, Greenberg DA, Bredesen DE (2006) Reversal of Alzheimer's-like pathology and behavior in human APP transgenic mice by mutation of Asp664. PNAS 103(18):7130–5
27. Westerlund M, Hoffer B, Olson L (2009) Parkinson's disease: exit toxins; enter genetics. Prog Neurobiol Nov 16;90(2):146–56
28. Maetzler W, Liepelt I, Berg D (2009) Progression of Parkinson's disease in the clinical phase: potential markers. Lancet Neurol 8(12):1158–71
29. Lees AJ, Hardy J, Revesz T (2009) Parkinson's disease. Lancet 373:2055–66
30. Greffard S, Verny M, Bonnet AM, Seilhean D, Hauw JJ, Duyckaerts C (2010) A stable proportion of Lewy body bearing neurons in the substantia nigra suggests a model in which the Lewy body causes neuronal death. Neurobiol Aging 31:99–103
31. Gilks WP, Abou-Sleiman PM, Gandhi S (2005) A common LRRK2 mutation in idiopathic Parkinson's disease. Lancet 365:415–6
32. Wu K, Politis M, Piccini P (2009) Parkinson disease and impulse control disorders: a review of clinical features, pathophysiology and management. Postgrad Med J 85(1009):590–6
33. Zecca L, Youdim MB, Riederer P, Connor JR, Crichton RR (2004) Iron, brain ageing and neurodegenerative disorders. Nat Rev Neurosci 5:863–73
34. Reddy VP, Zhu X, Perry G, Smith MA (2009) Oxidative stress in diabetes and Alzheimer's disease. J Alzheimers Dis 16:763–74
35. Lovell MA, Robertson JD, Teesdale WJ, Campbell JL, Markesbery WR (1998) Copper, iron and zinc in Alzheimer's disease senile plaques. J Neurol Sci 158:47–52
36. Falangola MF, Lee SP, Nixon RA, Duff K, Helpern JA (2005) Histological co-localization of iron in Abeta plaques of PS/APP transgenic mice. Neurochem Res 30:201–5
37. Adlard PA, Bush AI (2006) Metals and Alzheimer's disease. J Alzheimer's Dis 10:145–63
38. Barnham KJ, Bush AI (2008) Metals in Alzheimer's and Parkinson's diseases. Curr Opin Chem Biol 12:222–8
39. Altamura S, Muckenthaler MU (2009) Iron toxicity in diseases of aging: Alzheimer's disease, Parkinson's disease and atherosclerosis. J Alzheimers Dis 16:879–95
40. Muckenthaler MU, Galy B, Hentze MW (2008) Systemic iron homeostasis and the iron-responsive element/iron-regulatory protein (IRE/IRP) regulatory network. Annu Rev Nutr 28:197–213
41. Piñero DJ, Hu J, Connor JR (2000) Alterations in the interaction between iron regulatory proteins and their iron responsive element in normal and Alzheimer's diseased brains. Cell Mol Biol (Noisy-le-grand) 46(4):761–76

42. Rogers JT, Lahiri DK (2004) Metal and inflammatory targets for Alzheimer's disease. Curr Drug Targets 5:535–51
43. Cahill CM, Lahiri DK, Huang X, Rogers JT (2009) Amyloid precursor protein and alpha synuclein translation, implications for iron and inflammation in neurodegenerative diseases. Biochim Biophys Acta 1790:615–28
44. Rogers JT, Leiter LM, McPhee J, Cahill CM, Zhan SS, Potter H, Nilsson LN (1999) Translation of the Alzheimer amyloid precursor protein mRNA is up-regulated by interleukin-1 through 5′-untranslated region sequences. J Biol Chem 274:6421–31
45. Rogers JT, Randall JD, Cahill CM, Eder PS, Huang X, Gunshin H, Leiter L, McPhee J, Sarang SS et al (2002) An iron-responsive element type II in the 5′-untranslated region of the Alzheimer's amyloid precursor protein transcript. J Biol Chem 277:45518–28
46. Bandyopadhyay S, Ni J, Ruggiero A, Walshe K, Rogers MS, Chattopadhyay N, Glicksman MA, Rogers JT (2006) A high-throughput drug screen targeted to the 5′-untranslated region of Alzheimer amyloid precursor protein mRNA. J Biomol Screen 11:469–80
47. Bandyopadhyay S, Huang X, Cho H, Greig NH Youdim Y, Rogers JT (2006) Metal specificity of an iron-responsive element in Alzheimer's APP mRNA 5′-untranslated region, tolerance of SH-SY5Y and H4 neural cells to desferrioxamine, clioquinol, VK-28, and a piperazine chelator. J Neural Transm Suppl (71):237–47
48. Maloney B, Ge YW, Greig NH, Lahiri DK (2004) Presence of a "CAGA box" in the APP gene unique to amyloid plaque-forming species and absent in all APLP-1/2 genes: implications in Alzheimer's disease. FASEB J 18:1288–90
49. Friedlich AL, Tanzi RE, Rogers JT (2007) The 5′-untranslated region of Parkinson's disease alpha-synuclein messengerRNA contains a predicted iron responsive element. Mol Psychiatry 12:222–3
50. Murayama H, Shin RW, Higuchi J, Shibuya S, Muramoto T, Kitamoto T (1999) Interaction of aluminum with PHFτ in Alzheimer's disease neurofibrillary degeneration evidenced by desferrioxamine-assisted chelating autoclave method. Am J Pathol 155:877–85
51. Yamamoto A, Shin RW, Hasegawa K, Naiki H, Sato H, Yoshimasu F, Kitamoto T (2002) Iron (III) induces aggregation of hyperphosphorylated τ and its reduction to iron (II) reverses the aggregation: implications in the formation of neurofibrillary tangles of Alzheimer's disease. J Neurochem 82: 1137–47
52. Shin RW, Kruck TP, Murayama H, Kitamoto T (2003) A novel trivalent cation chelator Feralex dissociates binding of aluminum and iron associated with hyperphosphorylated tau of Alzheimer's disease. Brain Res 961:139–46
53. Pratico D, Uryu K, Sung S, Tang S, Trojanowski JQ, Lee VMY (2002) Aluminum modulates brain amyloidosis through oxidative stress in APP transgenic mice. FASEB J 16:1138–40
54. Multhaup G, Schlicksupp A, Hesse L, Beher D, Ruppert T, Masters C, Beyreuther K (1996) The amyloid precursor protein of Alzheimer's disease in the reduction of copper(II) to copper(I). Science 271:1406–9
55. Bolognin S, Drago D, Messori L, Zatta P (2009) Chelation therapy for neurodegenerative diseases. Med Res Rev 29:547–70
56. McLachlan DR, Smith WL, Kruck TP (1993) Desferrioxamine and Alzheimer's disease: video home behavior assessment of clinical course and measures of brain aluminum. Ther Drug Monit 15:602–7
57. Cahoon L (2009) The curious case of clioquinol. Nat Med 15:356–9
58. Ritchie C, Bush AI, Mackinnon A et al (2003) Metal-protein attenuation with iodochlorhydroxyquin (clioquinol) targeting Aß amyloid deposition and toxicity in Alzheimer's disease: a pilot phase 2 clinical trial. Arch Neurol 60:1685–91
59. Adlard PA, Cherny RA, Finkelstein DI, Gautier E, Robb E, Cortes M, Volitakis I, Liu X, Smith JP, Perez K, Laughton K, Li QX, Charman SA, Nicolazzo JA, Wilkins S, Deleva K, Lynch T, Kok G, Ritchie CW, Tanzi RE, Cappai R, Masters CL, Barnham KJ, Bush AI (2008) Rapid restoration of cognition in Alzheimer's transgenic mice with 8-hydroxy quinoline analogs is associated with decreased interstitial Abeta. Neuron 59:43–55

60. Trowbridge JP, Walker M (1991) The healing powers of chelation therapy. New Way of Life Inc., Stamford
61. Schroeder PE, Hasinoff BB (2005) Metabolism of the one-ring open metabolites of the cardioprotective drug dexrazoxane to its active metal-chelating form in the rat. Drug Metab Dispos 33:1367–72
62. Hasinoff BB, Hellmann K, Herman EH, Ferrans VJ (1998) Chemical, biological and clinical aspects of dexrazoxane and other bisdioxopiperazines. Curr Med Chem 5:1–28
63. Huang Z-X, May PM, Quinlan KM, Williams DR, Creighton AMM etal (1982) binding by pharmaceuticals. Part 2. Interactions of Ca(II), Cu(II), Fe(II), Mg(II), Mn(II) and Zn(II) with the intracellular hydrolysis products of the antitumor agent ICRF-159 and its inactive homologue ICRF-192. Agents Actions 12:536–42
64. Dawson KM (1975) Studies on the stability and cellular distribution of dioxopiperazines in cultured BHK-21S cells. Biochem Pharmacol 24:2249–53
65. Hasinoff BB (1998) Chemistry of dexrazoxane and analogues. Semin Oncol 25 (Suppl 10):3–9
66. Karakiulakis G, Missirlis E, Maragoudakis ME (1989) Mode of action of razoxane: inhibition of basement membrane collagen-degradation by a malignant tumor enzyme. Methods Find Exp Clin Pharmacol 11:255–61
67. Greig NH, Utsuki T, Ingram DK, Wang Y, Pepeu G, Scali C, Yu QS, Mamczarz J, Holloway HW, Giordano T, Chen D, Furukawa K, Sambamurti K, Brossi A, Lahiri DK (2005) Selective butyrylcholinesterase inhibition elevates brain acetylcholine, augments learning and lowers Alzheimer beta-amyloid peptide in rodent. PNAS 102(47):17213–8
68. Borchelt DR, Ratovitski T, van Lare J, Lee MK, Gonzales V, Jenkins NA, Copeland NG, Price DL, Sisodia SS (1997) Accelerated amyloid deposition in the brains of transgenic mice coexpressing mutant presenilin 1 and amyloid precursor proteins. Neuron 19:939–45
69. Shaw KT, Utsuki T, Rogers J, Yu QS, Sambamurti K, Brossi A, Ge YW, Lahiri DK, Greig NH (2001) Phenserine regulates translation of beta-amyloid precursor protein mRNA by a putative interleukin-1 responsive element, a target for drug development. PNAS 98:7605–10
70. Lahiri DK, Chen D, Maloney B, Holloway HW, Yu QS, Utsuki T, Giordano T, Sambamurti K, Greig NH (2007) The experimental Alzheimer's disease drug posiphen [(+)-phenserine] lowers amyloid-beta peptide levels in cell culture and mice. J Pharmacol Exp Ther 320:386–96
71. Marutle A, Ohmitsu M, Nilbratt M, Greig NH, Nordberg A, Sugaya K (2007) Modulation of human neural stem cell differentiation in Alzheimer (APP23) transgenic mice by phenserine. PNAS 104:12506–11
72. Greig NH, Fu X-C, Hellmann K (1982) Razoxane penetration into the cerebrospinal fluid of rats. Cancer Chemother Pharmacol 8:251–2
73. Rodríguez R, Santiago-Mejía J, Fuentes-Vargas M, Ramírez-San Juan E (2003) Outstanding neuroprotective efficacy of dexrazoxane in mice subjected to sequential common carotid artery sectioning. Drug Dev Res 60:294–302
74. Mom A, Aresca S, Fuente G, Pecorini V, Pellerano G, Perez V (1982) Razoxane in the treatment of psoriatic patients resistant to or intolerant of PUVA, methotrexate and etretinate. Acta Derm Venereol 62:357–8

Chapter 4
Summary and Outlook

Kurt Hellmann and Walter Rhomberg

1. Razoxane (Rz) and dexrazoxane (DXRz) have several interesting *modes of action*. They block the cell division at the level of G2/M in the cell cycle, they have shown antiinvasive activity in preclinical experiments, and they are able to normalize pathological tumour blood vessels. In addition, the drugs inhibit topoisomerase II α (the full implications have yet to emerge) and they are powerful chelating agents which not only chelate iron, but also a variety of other ions such as zinc, magnesium, lead, cadmium and copper.

2. Razoxane acts mainly on dividing cells and prevents daughter cells from separating after the division. The responding cells can, therefore, greatly enlarge with twice or even 4 times the normal complement of nuclei before they disappear. This impacts on tumour measurement after Rz or DXRz treatment, and thus, an initial increase of tumor size may not always be a sign of tumor progression. Razoxane leads to a major *enhancement of the radiation response* which was repeatedly seen in animal experiments and in the clinic. The biologic basis for this has probably several reasons of which the G2/M block of the cell cycle (the most radio-sensitive phase) is almost certainly the most important mechanism. In comparison to Rz, DXRz has as yet not been adequately tested as radiosensitizer.

3. Razoxane and dexrazoxane are both *antiangiogenic* – not in a simplistic general way but in preventing selectively the emergence of a pathologic vasculature. In 1969, razoxane was first shown to exhibit this unique ability to prevent pathological, especially tumour angiogenesis, thereby preventing tumour hemorrhagic events such as bleeds and tumour cell dissemination. This, in addition, allows more anticancer drugs to reach the tumour and also more oxygen.

4. The strong *antimetastatic activity* of razoxane is mainly to be seen in this context although other modes of action may also contribute to this phenomenon, e.g. the antiinvasive properties of the drug. The suppression of distant metastasis has been proven in numerous preclinical test models but as yet only in colorectal cancer and soft tissue sarcomas in man. If razoxane is combined with

K. Hellmann (✉)
Windleshaw House, Withyham, East Sussex, TN7 4DB, UK

K. Hellmann, W. Rhomberg (eds.), *Razoxane and Dexrazoxane – Two Multifunctional Agents*, DOI 10.1007/978-90-481-9168-0_4, © Springer Science+Business Media B.V. 2010

the tubulin affinic drug vindesine (a semisynthetic vinca alkaloid) an unrivalled enhancement of the antimetastatic activity is observed at least in soft tissue sarcomas. Further clinical development and proof of this area holds great promise for clinical success.

5. Both drugs have been shown to *protect normal tissues* from toxic, even carcinogenic substances such as doxorubicin, dimethyl hydrazine, DMBA, and etoposide. If doxorubicin accidently extravasates, dexrazoxane will, if given as rapidly as possible thereafter, protect the tissue from necrosis.

6. Very important is the *protective effect by dexrazoxane on the heart* after doxorubicin without affecting the antitumor effect of doxorubicin. This has been shown in numerous clinical trials and in 2 independent meta-analyses. For reasons unknown, the ASCO (American Society of Clinical Oncology) Panel on Cardioprotectants prefers to accept the result of one trial with aberrant results which seem to indicate that DXRz reduces the response of the one trial out of some 15 trials. They advise that DXRz be withheld until after 300 mg/m^2 of DXRz has been given. Since every dose of doxorubicin is known to produce an increment of cardiomyocyte damage, this recommendation is tantamount to advising that the protectant should be given *after* the irreversible damage has been done! Lipshultz (Heart 16: 47, 2008) speaks of an epoch making importance of giving this cardioprotector in children, the consequences of which were be seen in years to come.

   When DXRz is given only to breast cancer patients who have responded to doxorubicin, DXRz appears to keep the responders responding, but it does not convert non-responders to responders. Of special importance is the fact that DXRz also protects the heart in breast cancer patients who have co-morbidites such as age more than 65, hypertension, diabetes mellitus or a previous cardiac condition and previous irradiation of the mediastinum or left chest wall.

7. Razoxane is active on a variety of *non-malignant conditions* such as psoriasis, Crohn's disease, ulcerative colitis; perhaps also on diabetic retinopathy and age related macular degeneration. Furthermore, Rz and DXRz reduces the formation of amyloid-beta-peptide which is important in the development of Alzheimer's dementia. Clinical trials have yet to be done to see if either or both are able to prevent the ravages of neurodegenerative conditions.

Outlook

Kurt Hellmann

In view of the results described in the foregoing chapters it is remarkable that knowledge of razoxane and dexrazoxane appears to be confined largely to those who have worked with these agents for the last 2–3 decades. Even more extra-ordinary is the fact that those who have not had first hand experience of these 2 drugs but only second hand opinions have been quick to condemn them for reasons which will not stand a moments examination. We hope that this book will help those oncologists who despite the plethora of new and experimental substances which they are asked to examine, still retain their impartial judgement and recognize the value of dexrazoxane and razoxane for cancer patients. It is to be hoped that the example of the United States FDA will now be followed elsewhere and the proper place in the oncologists armamentarium will be awarded to these remarkable drugs. Although numerous substances were given accelerated approval by the FDA as oncology products, only 6 went on to receive regular approval one of which was dexrazoxane. Withholding dexrazoxane where it would be useful as outlined in the book is a matter for oncologists education which it is hoped will be provided in the near future.

Walter Rhomberg

Examining razoxane results in detail shows that this drug would seem to deserve more clinical attention for its radiosensitizing and antimetastatic activity than it has received so far. It is absolutely necessary not to judge razoxane solely as a cytotoxic drug for cancer. The future of this drug is in the further discovery of activities which are not easily visible at first glance, e.g., its ability to prevent distant metastases and the power to slow down tumor growth. Both of these activities are not adequately appreciated by oncologists. From initial clinical experiences, it seems that the antimetastatic activity of razoxane may be greatly enhanced by the addition of other antineoplastic agents such as some tubulin-affinic substances, and probably other drugs yet to be determined.

There are reasons for further studies of razoxane in special cancer types. Apart from the striking improvements which were seen when razoxane was combined with vindesine and radiotherapy in soft tissue sarcomas, promising candidate tumors for investigating the hidden anticancer activities mentioned above (a reduction of metastasis, retardation of tumor growth and prolongation of life) would be, according to our experience, colorectal carcinomas, melanoma, and bladder cancer.

One of the limitations, however, for a wider interest in and acceptance of razoxane is its limited availability. There is an urgent necessity for a pharmaceutical company to take care of the approval process for the drug and the educational process of health care professionals.

Presently, dexrazoxane is indispensable to treat accidental extravasation of anthracylines. It is also unrivalled in its use as protector against anthracyline cardiotoxicity, the only drug approved by the FDA for this purpose. Nevertheless, its early administration for cardioprotection is not generally accepted, although it is known that each dose of anthracycline causes an increment of irreversible cardiac damage. Giving DXRz after the administration of anthracyclines is giving the protectant against the damage *after* the damage has been done. The magnitude of damage done thereby in long-term survivors of cancer, especially in children, requires detailed discussions.

The future of dexrazoxane may be determined by an exciting range of clinical applications beginning with its use in chronic inflammatory diseases that are characterized by proliferation of pathologic blood vessels up to an exploration of the drug in conditions such as haemochromatosis or Morbus Wilson (chelating activity). Also the possibility to influence neuro-degenerative diseases by the use of either razoxane or dexrazoxane opens up a large and important field for clinical investigations.

Index

K. Hellmann, W. Rhomberg (eds.), *Razoxane and Dexrazoxane – Two Multifunctional*   241
Agents, DOI 10.1007/978-90-481-9168-0, © Springer Science+Business Media B.V. 2010

Printed by Books on Demand, Germany